CONNECTIONS IN STEEL STRUCTURES

Behaviour, Strength and Design

Proceedings of a state-of-the-art workshop on Connections and the Behaviour, Strength and Design of Steel Structures held at the Laboratoire de Mécanique et Technologie, Ecole Normale Supérieure, Cachan, France, 25–27 May 1987.

CONNECTIONS IN STEEL STRUCTURES

Behaviour, Strength and Design

Edited by

REIDAR BJORHOVDE

University of Pittsburgh, Pittsburgh, Pennsylvania, USA

JACQUES BROZZETTI

Centre Technique Industriel de la Construction Métallique, St. Rémy lès Chevreuse, France

and

ANDRÉ COLSON

Laboratoire de Mécanique et Technologie, Ecole Normale Supérieure de Cachan, Université Paris VI, Cachan, France

ELSEVIER APPLIED SCIENCE
LONDON and NEW YORK

ELSEVIER APPLIED SCIENCE PUBLISHERS LTD
Crown House, Linton Road, Barking, Essex IG11 8JU, England

Sole Distributor in the USA and Canada
ELSEVIER SCIENCE PUBLISHING CO., INC.
52 Vanderbilt Avenue, New York, NY 10017, USA

WITH 21 TABLES AND 274 ILLUSTRATIONS

British Library Cataloguing in Publication Data

Connections in steel structures: behaviour, strength and design: proceedings of a state-of-the-art workshop on connections and the behaviour, strength and design of steel structures held at Laboratoire de Mécanique et Technologie Ecole Normale Supérieure, Cachan, France, May 25–27, 1987 edited by Reidar Bjorhovde, Jacques Brozzetti, André Colson.
1. Building, Iron and steel—Joints
I. Bjorhovde, Reidar II. Brozzetti, Jacques III. Colson, André
624.1′821 TA684

Library of Congress Cataloging in Publication Data

State-of-the-Art Workshop on Connections and the Behaviour, Strength, and Design of Steel Structures (1987: Cachan, France)
Connections in steel structures.

"Proceedings of a State-of-the-Art Workshop on Connections and the Behaviour, Strength, and Design of Steel Structures held at Laboratoire de Mécanique et Technologie, Ecole Normale Supérieure, Cachan, France, May 25–27, 1987"—Half t.p.
1. Welded joints—Congresses. 2. Bolted joints—Congresses. 3. Building, Iron and steel—Congresses.
I. Bjorhovde, Reidar. II. Brozzetti, Jacques. III. Colson, André. IV. Title.
TA492.W4S66 1987 672.5′2042 87-36401

ISBN 1-85166-177-8

No responsibility is assumed by the publisher for any injury and/or damage to persons or property as a matter of products liability, negligence or otherwise, or from any use or operation of any methods, products, instructions or ideas contained in the material herein.

Printed in Great Britain by Page Bros (Norwich) Ltd

PREFACE

This book is the proceedings of an international workshop that was organized by the editors for the period 25–27 May 1987 at the Laboratoire de Mécanique et Technologie, Ecole Normale Supérieure de Cachan, France, in collaboration with the Centre Technique Industriel de la Construction Métallique, St Rémy-lès-Chevreuse, France.

Financial support was provided by a grant from the Ministère de la Recherche et de l'Enseignement Supérieur in the framework of the civil engineering research programme.

The editors are sincerely appreciative of the help of the many individuals who contributed to the success of the conference. In particular, thanks are extended to:

— M. le Recteur Marcel Bonvalet, Directeur de l'Ecole Normale Supérieure de Cachan;
— M. Lucien Wahl, Directeur Général du Centre Technique Industriel de la Construction Métallique; and
— M. Mircea Predeleanu, Directeur du Laboratoire de Mécanique et Technologie.

The support and contributions of the workshop participants are further acknowledged, as is the input and assistance of Elsevier Applied Science Publishers Ltd.

REIDAR BJORHOVDE
JACQUES BROZZETTI
ANDRÉ COLSON
Cachan, France

CONTENTS

Session 2A: Mathematical Models

Session 2B: Classification

Session 3: Frame Analysis

Session 4: Frame Stability and Simplified Methods

Session 5: Design Requirements

Session 6: Data Base Organisation

Research and Development Needs

LIST OF CONTRIBUTORS

M. H. Ackroyd
Techtrol Ltd, 3870 Cote Vertu, St Laurent, Québec H4R 1VR, Canada

H. Agerskov
Department of Structural Engineering, Building 118, Technical University of Denmark, DK-2800 Lyngby, Denmark

J. M. Aribert
INSA, 20 Avenue des Buttes de Coesmes, 35043 Rennes Cedex, France

A. Azizinamini
Construction Technology Laboratories, Inc., Skokie, Illinois 60077, USA

G. Ballio
Department of Structural Engineering, Politecnico di Milano, Piazza Leonardo da Vinci 32, I-20133 Milan, Italy

D. Beaulieu
Département de Génie Civil, Université Laval, Ste Foy, Québec G1K 7P4, Canada

F. S. K. Bijlaard
TNO Institute for Building Materials and Structures, PO Box 49, 2600 AA Delft, The Netherlands

R. Bjorhovde
Department of Civil Engineering, University of Pittsburgh, 949 Benedum Hall, Pittsburgh, Pennsylvania 15261, USA

J. Bródka
Metal Structures Research and Design Centre 'Mostostal', Ul. Krucza 16/22, 00-926 Warsaw, Poland

J. BROZZETTI
CTICM, Domaine St Paul, 78470 St Rémy-lès-Chevreuse, France

M. CELIKAG
Department of Civil and Structural Engineering, The University, Mappin Street, Sheffield S1 3JD, UK

W. F. CHEN
School of Civil Engineering, Purdue University, West Lafayette, Indiana 47907, USA

A. COLSON
Laboratoire de Mécanique et Technologie, ENS de Cachan, 61 Avenue du Président Wilson, 94230 Cachan, France

N. E. COOK JR
Department of Civil, Environmental and Architectural Engineering, University of Colorado, Boulder, Colorado 80309, USA

A. CZECHOWSKI
Metal Structures Research and Design Centre 'Mostostal', Ul. Krucza 16/22, 00-926 Warsaw, Poland

J. B. DAVISON
Department of Civil and Structural Engineering, The University, Mappin Street, Sheffield S1 3JD, UK

R. O. DISQUE
American Institute of Steel Construction, 400 North Michigan Avenue, Chicago, Illinois 60611, USA

F. FREY
IREM/DGC, Swiss Federal Institute of Technology, CH-1015 Lausanne, Switzerland

Y. GALEA
CTICM, Domaine St Paul, 78470 St Rémy-lès-Chevreuse, France

K. H. GERSTLE
Department of Civil, Environmental and Architectural Engineering, University of Colorado, Boulder, Colorado 80309, USA

M. I. GILMOR
Canadian Institute of Steel Construction, 201 Consumers Road, Suite 300, Willowdale, Ontario M2J 4G8, Canada

A. GRANSTRÖM
AG Konstruktioner, Björns Backe 3, S-182 35 Danderyd, Sweden

S. A. IOANNIDES
Structural Affiliates International, Inc., 2416 Hillsboro Road, Nashville, Tennessee 37212, USA

J. JANSS
Centre de Recherches de l'Industrie des Fabrications Métalliques, Liège, Belgium

J. P. JASPART
Institut du Génie Civil, Université de Liège, 6 Quai Banning, 4000 Liège, Belgium

P. A. KIRBY
Department of Civil and Structural Engineering, The University, Mappin Street, Sheffield S1 3JD, UK

N. KISHI
Muroran Institute of Technology, Muroran 050, Japan

M.-S. KOO
Lehrstuhl für Stahlbau, RWTH Aachen, Mies-von-der-Rohe-Strasse 1, D-5100 Aachen, Federal Republic of Germany

J. KORDJAK
Metal Structures Research and Design Centre 'Mostostal', Ul. Krucza 16/22, 00-926 Warsaw, Poland

S. D. LINDSEY
Stanley D. Lindsey & Associates Ltd, 1906 West End Road, Nashville, Tennessee 37205, USA

J. LOPETEGUI
Lehrstuhl für Stahlbau, RWTH Aachen, Mies-von-der-Rohe-Strasse 1, D-5100 Aachen, Federal Republic of Germany

R. Maquoi
Institut du Génie Civil, Université de Liège, 6 Quai Banning, 4000 Liège, Belgium

E. Martinez Romero
E. Martinez Romero SA, Av. Nuevo Leon 54—2nd floor, 06140 Mexico DF, Mexico

K. G. Matsuoka
Muroran Institute of Technology, Muroran 050, Japan

F. M. Mazzolani
Engineering Faculty, Istituto di Tecnica delle Costruzioni, University of Naples, Piazzale V. Tecchio, I-80125 Naples, Italy

R. Melchers
Department of Civil Engineering and Surveying, University of Newcastle, Newcastle, New South Wales 2308, Australia

D. A. Nethercot
Department of Civil and Structural Engineering, The University, Mappin Street, Sheffield S1 3JD, UK

S. G. Nomachi
College of Industrial Technology, Nihon University, Izumi, Narashino 275, Japan

A. Picard
Département de Génie Civil, Université Laval, Ste Foy, Québec G1K 7P4, Canada

C. Poggi
Department of Structural Engineering, Politecnico di Milano, Piazza Leonardo da Vinci 31, I-20133, Milan, Italy

J. B. Radziminski
Department of Civil Engineering, University of South Carolina, Columbia, South Carolina 29208, USA

G. Sedlacek
Lehrstuhl für Stahlbau, RWTH Aachen, Mies-von-der-Rohe-Strasse 1, D-5100 Aachen, Federal Republic of Germany

W. A. SIBAI
IREM/DGC, Swiss Federal Institute of Technology, CH-1015 Lausanne, Switzerland

H. H. SNIJDER
TNO Institute for Building Materials and Structures, PO Box 49, 2600 AA Delft, The Netherlands

J. W. B. STARK
TNO Institute for Building Materials and Structures, PO Box 49, 2600 AA Delft, The Netherlands

Ch. STUTZKI
Lehrstuhl für Stahlbau, RWTH Aachen, Mies-von-der-Rohe-Strasse 1, D-5100 Aachen, Federal Republic of Germany

F. TSCHEMMERNEGG
Universität Innsbruck, Technikerstrasse 13, A-6020 Innsbruck, Austria

J. WARDENIER
Delft University of Technology, Stevinweg 1, PO Box 5048, 2600 GA Delft, The Netherlands

R. ZANDONINI
Department of Engineering, University of Trento, Mesiani di Povo, I-38050 Trento, Italy

P. ZANON
Department of Engineering, University of Trento, Mesiani di Povo, I-38050 Trento, Italy

INTRODUCTORY NOTES

Reidar Bjorhovde
Department of Civil Engineering
University of Pittsburgh
Pittsburgh, Pennsylvania, U.S.A.

Jacques Brozzetti
C.T.I.C.M.
Saint Rémy lès Chevreuse, France

and

André Colson
Laboratoire de Mécanique et Technologie
Ecole Normale Supérieure
Université Paris VI
Cachan, France

When the idea of an International Workshop on Connections and the Behaviour, Strength and Design of Steel Structures was first conceived, it reflected the recognition that whereas much research and development work has been conducted on steel structures in general, and on the miscellaneous components and connections in particular, relatively little attention has been paid to the increasingly important aspects of the unified, complete structure. Part of this deficiency is clearly related to the complexity of the structures, and the fact that most researchers have devoted their attention to determining the characteristics of the individual elements. It is also related to the analytical and experimental problems that have been encountered by investigators of topics that are highly non-linear and three-dimensional in nature, for which neither computational nor testing facilities of adequate capacity have been available until but in recent years.

The advent of very powerful computers, coupled with the development of sophisticated testing and data acquisition equipment, have made for the completion of research studies of increasingly more complicated member, connection and, to a certain extent, overall frame response problems. The former difficulties of being capable of solving the problems, given the proper tools, are therefore now being removed at an accelerating pace. However, this poses an array of problems of a different nature for the researchers, but especially for design standard writers and their users, the design engineers. Specifically, the volume of information is becoming large to the point where it is no longer realistically possible to keep abreast of all new developments. Secondly, methods and procedures vary substantially from one locale to the next, with the result that interpretation of the findings and how they may apply to a particular case will be a dubious undertaking. Finally, with studies being carried out in universities, research laboratories and companies around the world, some amount of duplication of research is essentially unavoidable.

On the background of the problems that have been outlined in the preceding, the idea of an intensive workshop was conceived. It was felt that with the gathering of a group of researchers with extensive experience in the study of steel structures and their connections, to present and discuss their own ongoing research project as well as those of others, it would be feasible to arrive at an assessment of the state-of-the-art. In addition, having established the status of current knowledge, an important function of the workshop would be to address the criteria and topics of future research, as well as to discuss suitable vehicles for gathering and maintaining a coherent data base for connections. The latter activity would be particularly important for designers and code writers, to serve as a source of reliable information on connections.

To achieve the most in-depth, intensive discussions of the problems, it was decided to limit the workshop attendance to a relatively small group of internationaly recognized researchers, code writers and design engineers with extensive background in the study of connections and steel structures. It was felt that this was essential to the success of the workshop, on the premise that the attendees would not need to be educated by each other in the traditional technical conference format, but rather would be able to have truly in-depth exchanges of opinions, methods, procedures, results, and future directions. For the same reasons the wokshop program was organized such that the discussion time could be maximized, with primary emphasis on the informal exchange of ideas and understanding. The proceedings reflect the importance of the discussions and the evaluations of research needs and data collection efforts, to the effect that individual chapters address these items following the written contributions of the invitees for each technical session.

It was decided to organize the workshop into technical sessions with the following primary subject coverage :

Session N°1 : Local Analysis of Joints
Session N°2 : Modeling of Load-Deflection Behaviour
Session N°3 : Methods of Frame Analysis
Session N°4 : Frame Stability and Simplified Methods
Session N°5 : Design Requirements and Codes
Session N°6 : Data Base Organization and Summary

The following comments highlight the topics that were to be discussed within each session, along with the anticipated outcome.

The session on local analysis of joints was intended for a review of current theoretical and experimental research dealing with the local behaviour of actual connections under load. Local behaviour was defined as the detailed and full-range response of the connection, including all connecting parts, mechanical fasteners, welds, and so on. Such studies are essential for the understanding of the failure mechanisms of the connection, as well as for identifying the main load and deformation controlling parameters. The three primary characteristics of non-linear behaviour can also be identified, namely, stiffness, ultimate strength, and ductility.

The expected outcome of the local analysis session was the acquisition of information on the most important numerical and experimental research work on local connection behaviour, including load transfer mechanisms and failure modes. It would also establish familiarity with the types of connections that are used in actual structures around the world, and facilitate correlation efforts.

Due to the complexity of connection strength and behaviour, it was decided to divide the session on modeling of load-deflection behaviour into two sub-sessions. Their topics of coverage and anticipated outcome are discussed in the following.

The first sub-session on modeling was aimed at mathematical modeling with parameter identification, as well as comparison with test results. Specifically, rather than dealing with the local conditions, as were the topics of the first session, the discussions and papers of this modeling session were intended to reflect the overall connection response. Special emphasis was therefore placed on developing an inventory of connection moment-rotation mathematical models. The latter may have been based on curve-fitting techniques or statistical analyses of test results ; it may also have been possible to derive M-Φ-equations that are associated with the physical characteristics of the connections. Further, in recognition of the non-linear response of all realistic connections, contributors were asked to consider the global characterization of semi-rigid joint behaviour, including for column footings. Finally, the correlation between connection model and test results is important, and it was also intended to obtain an evaluation of the effects of the sense of loading, i.e. monotonic and cyclic load effects would be significant considerations.

The expected outcome of the first modeling sub-session was the development of an inventory of mathematical models of semi-rigid connections, including a collection of moment-rotation (or similar) curves and the attendant primary variables for strength, behaviour and ductility.

The second sub-session on connection modeling was aimed at the classification problem, with an introduction to the data base concept. Thus, it was felt that sufficient experimental and other information is currently available for several connections in various countries. The expected outcome of the session might be the development of a classification system for beam-to-column connections, including considerations such as initial stiffness, ultimate strength, and rotation capacity at service and ultimate load levels.

The session dealing with the subject of frame analysis was intended for a presentation of the methods of structural analysis which specifically include the constitutive law of the connections. Procedures are currently available which incorporate first and second order geometric and material effects for the connections, the structural members, and the entire frame, but it was felt that it would be important to establish exactly what the individual solutions can accomplish, and how well they correlate with tests (if any) and other analytical solutions.

The anticipated outcome of the frame analysis session was the establishment of a repertory of computer programs, including definitions of the primary assumptions and variables. It was also expected that some frame test results would be presented, to lend further support to the theoretical research results.

Recognizing the unique problems of frame and member stability, the session addressing the topics of frame stability and simplified methods was intended for dealing with : (1) the influence of semi-rigid connection characteristics on the stability of members and frames, including serviceability and ultimate limit state considerations, and (2) the development of simplified methods of calculation for column and frame stability, as well as for member forces. It was felt that the simplified solutions might be more adaptable to practical usage, such as would be considered by design engineers.

The expected outcome of the frame stability session was the presentation of stability solution procedures for members and frames, taking into account known stiffness characteristics of the connections. The computer solutions that were thought of would also be expected to facilitate the computations of member stress resultants and displacements throughout the frame. It was also thought to be possible that these results might indicate approaches that could take into account second order (P-Delta) effects, without necessarly having to go through complete second order frame solutions.

The technical session addressing design requirements and codes was aimed at dealing with subjects such as the status of incorporating semi-rigid connection and frame response concepts into design codes as well as practice. This would necessarily have to include some considerations of cyclic loading, such as may be encountered under seismic conditions, as well as material requirements. A particular group of concerns would be those associated with serviceability and ultimate limit states, the questions of story and overall lateral deflections (drift) of frames, and the format of limit states design criteria for structures with semi-rigid connections. This should also consider the questions of necessary and available rotation capacities.

The outcome of the discussions of code requirements was envisioned to be an assessment of the state-of-the-art of code development in various countries, along with indications of what is perceived to be needed for practical design criteria. In particular, it was known that actual structural designs had been performed by at least one company in the United States and possibly others elsewhere ; it would be of significant interest to have the assessments of such designers of the current knowledge and future needs.

In addition to providing a final review of all of the papers and the respective discussions of the various technical sessions, the final session was intended as the outlet for two primary groups of presentations, namely, (1) a discussion and evaluation of a potential data base organization and (2) the reports of the session reporters and research reporters. The former subject was of significant interest, especially to the effect that it was hoped that a proper organizational structure would facilitate future exchanges of information, as well as avoid any duplication of work. The research reporters and their function were considered crucial to the workshop, as these reports would focus only on the work that needs to be undertaken, to remove current deficiencies of available data and to add to data bases as well as methods of analysis and design. It is noted that the use of research reporters constitutes the first time such assignments have been utilized in a technical conference setting ; it might be considered by other groups for similar activities in the future.

SESSION N°1

LOCAL ANALYSIS OF JOINTS

BOLTED AND WELDED CONNECTIONS IN STEEL STRUCTURES

Jan STARK
Eindhoven University of Technology, Eindhoven, The Netherlands

Frans BIJLAARD
TNO Institute for Building Materials and Structures, Delft, The Netherlands

Gerhard SEDLACEK
University of Aachen, Aachen, West Germany

ABSTRACT

On European (Eurocode 3) as well as on world level (ISO/TC 167) work is going on to come to international harmonisation of design procedures for steel structures. Both activities focus on a presentation of design codes in a limit state design (LRFD) format. A brief review will be given of the design rules for bolts, as now proposed in the drafts of the international documents. To validate the strength functions and to determine suitable values for the model factors, a statistical re-evaluation of available test information is carried out. Preliminary results of this study are presented and compared with the proposed new design rules.

INTRODUCTION

The 1984 draft of Eurocode 3 is now under revision on the basis of the comments from the EEC member states. The final draft is planned for the end of 1988. The European Commission has recognized the need for background information that confirms and explains the rules presented in EC3. The authors of this paper are charged with the task to provide such information for the rules on fasteners and joints. The results of this study are also relevant for the work on an international standard in ISO/TC 167. The conversion of traditional working stress design methods for connections into a limit state design format was not easy. It was found that many of the rules in existing national standards were based on engineering judgement, more than on a consistent evaluation of experimental evidence. The following procedure is used for the determination of the design rules on fasteners in EC3. Based on observation of actual behaviour in tests and on theoretical considerations a "design model" is selected. Then by statistical interpretation of all available test data i.e. regression analysis the efficiency of the model is checked. Eventually the design model has to be adapted until the correlation of the theoretical values and the test data is sufficient. Then through manipulation with the statistical data of the test population, the model factor γ_{Rd} can be determined. Dividing the strength function by the model factor γ_{Rd} gives the design expression.

A first attempt for this evaluation, based on a procedure proposed by the EC Coordination Group [1, 2], is reported in [3]. The results showed clearly that the procedure was not adequate for the purpose of checking strength functions and had to be adapted.

In this paper a revised evaluation procedure for the same strength functions as in [3] is presented.

DESIGN MODELS FOR BOLTS

Bolts in a joint can be loaded in tension, shear or a combination of tension and shear.

<u>Design model for a bolt loaded in tension</u>

If a bolt is subjected to increasing tensile load fracture will occur in the threaded portion. By definition the stress area of the bolt A_s is the cross sectional area of a smooth cylinder which, subjected to tension, has the same failure load as the threaded part of a bolt from the same

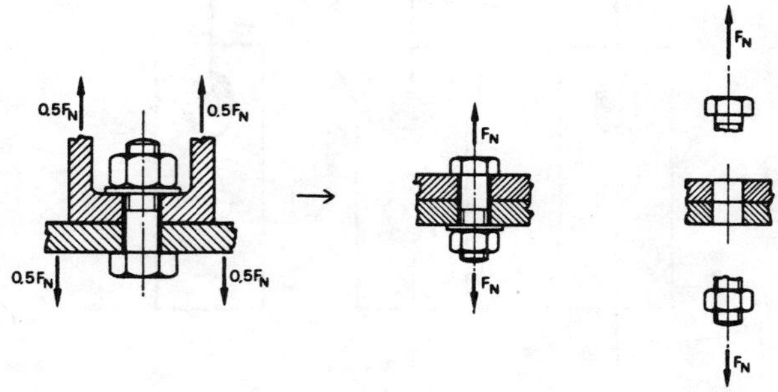

Fig. 1 Bolt loaded in tension

material. The formula for A_s, given in Fig. 2, is internationally standardised.

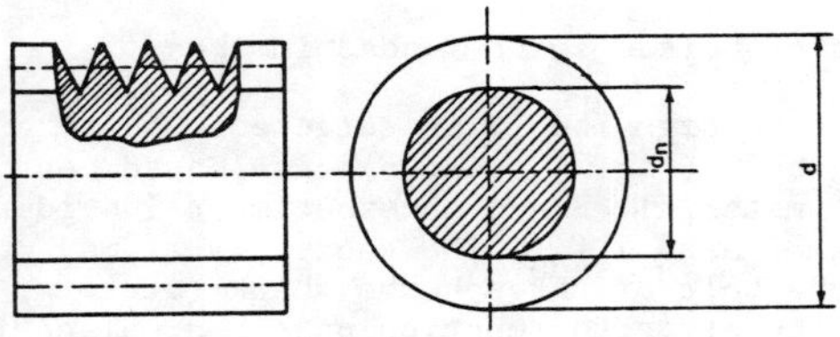

$$A_s = \frac{\pi}{4} \left(\frac{d + 3\, d_n}{4}\right)^2$$

Fig. 2 Definition of stress area

It is evident that, based on the definition of A_s, the strength function for a bolt in tension will be taken as:

$$F_t = f_{ub} A_s$$

where f_{ub} is the ultimate tensile strength of the bolt material.

In a joint so called prying forces may occur due to flexibility of the connected parts. This is illustrated in Fig. 3.

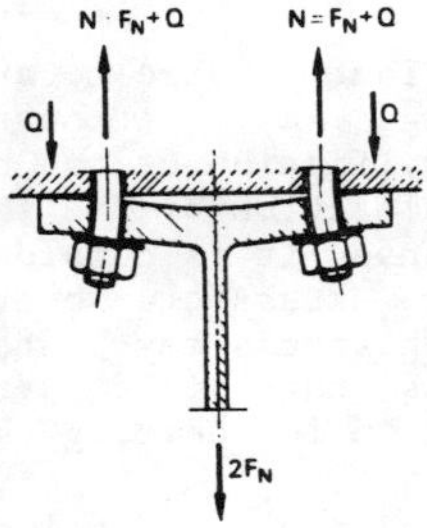

In conventional design methods the occurence of prying forces is sometimes taken into account in the design expression for F_t. Principally this is not correct and therefore in EC3 it is suggested to account for prying forces on the loading side of the safety verification.

Fig. 3 Prying forces Q in a T-stub joint

Design models for bolts loaded in shear

The maximum resistance of a joint loaded in shear may correspond to four different failure modes as illustrated in Fig. 4.

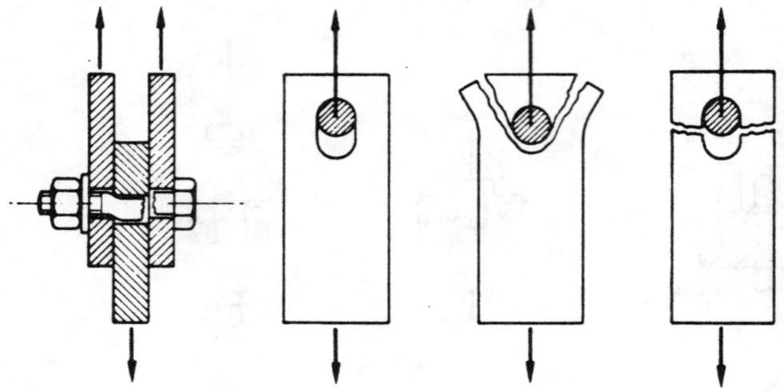

shear failure of bolt | hole elongation | shear failure of plate | tension failure of plate

Fig. 4 Possible failure modes of bolted joint loaded in shear

For each of these failure modes, a "design model" is selected. The strength function for shear failure of the bolt (Fig. 4, mode a) is based on the pure shear strength of the bolt material. Based on experimental evidence, the pure shear strength τ_{ub} is taken as 0.7 f_{ub}. The shear plane may pass through the threaded portion of the bolt or through the shank (see Fig. 5). Dependent on the case considered the strength function per shear plane is:

$F_v = 0.7\ f_{ub} A_s$ or $0.7\ f_{ub}\ A$

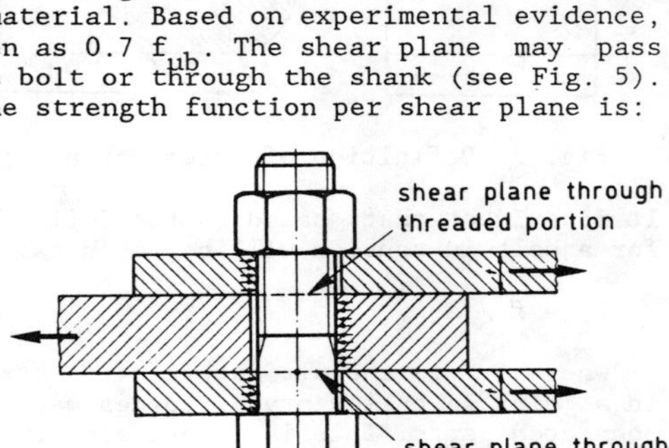

Fig. 5 Bolt loaded in shear

If the end distance of the bolt hole is sufficiently large to avoid occurence of the shear mode of failure shown in Fig. 4c, the bearing pressure between the bolt and the plate will cause ovalization of the hole under increasing load. Although even a significant ovalization of the hole is not an ultimate limit state in the strict sense, it is considered necessary to give a limit for the mean bearing stress to control deformation. According to [4] the mean ultimate bearing stress may be taken as 3 f_y if the plates are in compression, where f_y is the yield stress level of the plate material. The strength function for bearing (hole elongation) will be:

$F_b = 3\ f_y\ d\ t$

where d is the shank diameter and t is the plate thickness.

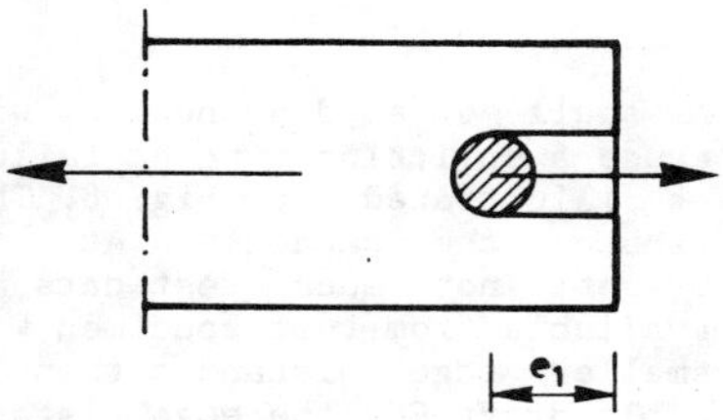

In EC3 it is assumed that this value is also applicable for plates in tension if the end and edge distances are sufficiently large. In case of smaller end distances the shear mode (Fig. 4, mode c) will occur. To determine the main parameters of the strength function for this case, the simplified model shown in Fig. 6 is used.

Fig. 6 Simplified model for shear failure of the plate

Assume that the length of the longitudinal shear planes is approximately 0.9 times the end distance e_1. According to the Von Mises yield criterion, the design shear stress is $\tau_y = 0.6\ f_y$. The design load will then be:

$$F_{yb} = 2 * 0.9\ e_1\ t * 0.6\ f_y$$

$$= 1.1 \frac{e_1}{d} f_y\ d\ t$$

$$= \alpha\ f_y\ d\ t$$

where α is a function of $\frac{e_1}{d}$.

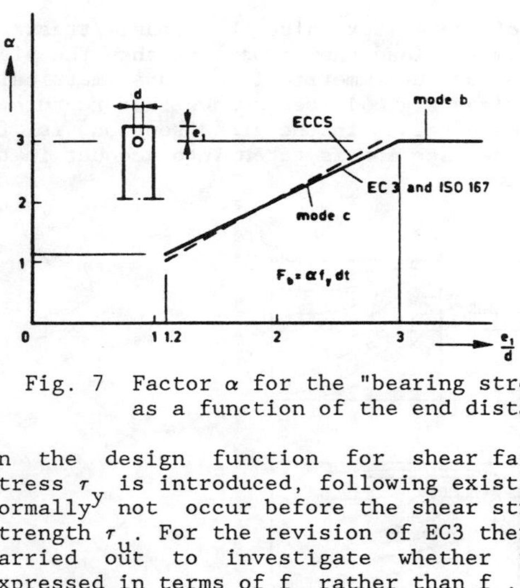

Fig. 7 Factor α for the "bearing strength" as a function of the end distance

So the design function takes the same form as for hole elongation with the "bearing stress" $3\ f_y$ replaced by $\alpha\ f_y$. This design function is used in draft EC3 where α is taken as $\alpha = \frac{e_1}{d} \leq 3$ (see Fig. 7). It is important to realise that the design function for bearing in fact covers two different failure modes.

In the design function for shear failure of the plate, the yield shear stress τ_y is introduced, following existing practice. However, failure will normally not occur before the shear stress is equal to the ultimate shear strength τ_u. For the revision of EC3 therefore comparison studies will be carried out to investigate whether the strength function for F_b can be expressed in terms of f_u rather than f_y. Shear failure of the plate can also occur when the pitch P between two bolts in a row is relatively small. To account for this failure mode, the factor α in the strength function for bearing is also made dependent of $\frac{P}{d}$ (see Table 1).

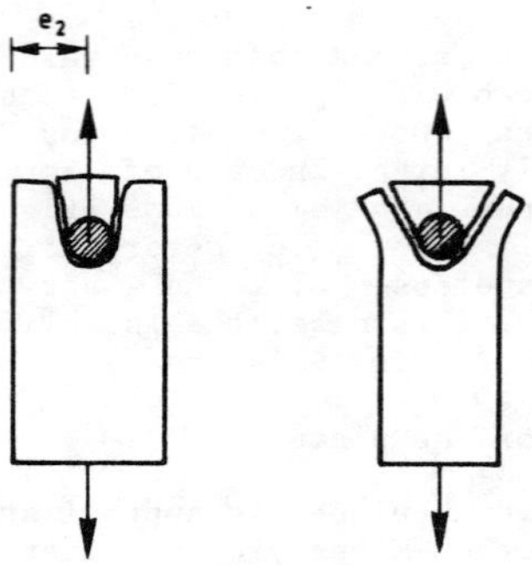

A small edge distance e_2 will cause a splitting type of failure as illustrated in Fig. 8. This reduces the capacity. At the present not much test data are available from test specimen with smaller edge distances then 1.5 d. In draft EC3 the edge distance is therefore limited to a minimum of 1.5 d.

Fig. 8 Illustration of influence of edge distance on shear failure mode of the plate

The strength function for tensile failure of the connected parts (Fig. 4, mode d) is based on the assumption that failure occurs if the net section A_n is fully stressed upto the tensile material strength f_u. So the strength function is:

$$R_t = A_n f_u$$

To prevent excessive deformations of the member, also the mean stress in the gross cross-section at ultimate load should be less than the yield strength, with $\gamma_M = 1$. In the case of unsymmetrical or unsymmetrically connected members such as angles (see Fig. 9), tests show that rupture at an edge may start well before the mean stress in the gross section is f_u. In that case a reduced effective net section is taken into account in the strength function, as illustrated in Fig. 9.

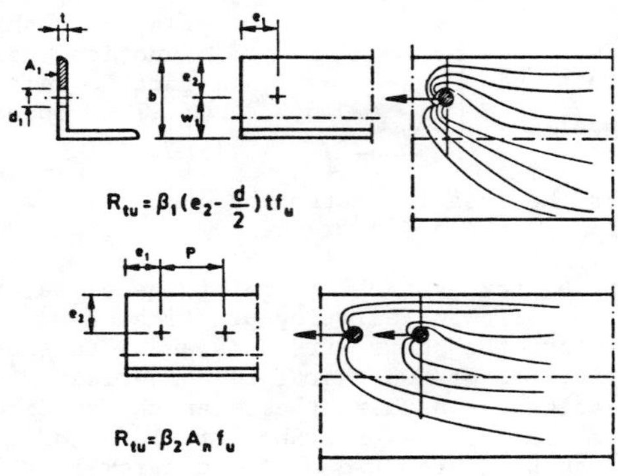

Fig. 9 Illustration of reduced effective net section for an angle connected on one leg only

Design model for bolts under the combined action of tension and shear

If the shear plane passes through the threaded portion of the bolt (Fig. 5), failure will most certainly occur over the shear plane. The strength function is based on a simple interaction formula:

$$\left(\frac{T}{F_t^*}\right)^2 + \left(\frac{V}{F_v^*}\right)^2 \leq 1$$

where F_t^* and F_v^* are the design capacities with respect to tension and shear and should be determined for the stress area A_s (see Table 1), T is the normal force and V is the shear force acting on the bolt.

If the shear plane passes through the shank, dependent on the ratio of tension and shear, failure may either occur in the shank (shear plane) or the threaded part. In the interaction formula F_t^* and F_v^* may be determined for the cross section A of the shank, but additionally the tensile capacity of the threaded part should be checked with:

$$T \leq \frac{f_{ub} A_s}{\gamma_{Rd}}$$

In Table 1 the strength functions according to some recently published international documents are summarised in comparison with the strength functions assumed for the evaluation of test results.

Capacities	ECCS TC 10 [5]	Draft ISO 167 [6]	Draft EC3 [7]	Based on evaluation of test results
Tensile capacity F_t^*	smallest of: $0.8 f_{yb} A_s$; $\frac{0.8 f_{ub} A_s}{1.25}$	$\frac{f_{ub} A_s}{\gamma_r}$	$\frac{f_{ub} A_s}{1.25}$	$\frac{f_{ub} A_s}{\gamma_{Rd}}$
Shear capacity F_v^* - Thread	$\frac{0.7 f_{ub} A_s}{1.25}$	$\frac{0.6 f_{ub} A_s}{\gamma_r}$	$\frac{0.7 f_{ub} A_s}{1.25}$	$\frac{0.7 f_{ub} A_s}{\gamma_{Rd}}$
- Shank	$\frac{0.7 f_{ub} A}{1.25}$	$\frac{0.6 f_{ub} A}{\gamma_r}$	$\frac{0.7 f_{ub} A}{1.25}$	$\frac{0.7 f_{ub} A}{\gamma_{Rd}}$
Bearing capacity F_b^*	$\alpha f_y d t$; $\alpha = 1.25 \frac{e_1}{d} - 0.5$; $\alpha = 1.25 \frac{p}{d} - 1.75$; $\alpha \leq 3$	$\alpha f_y d t$; $\alpha = \frac{e}{d}$; $\alpha = \frac{p}{d} - 0.5$; $\alpha \leq 3$	$\alpha f_y d t$; $\alpha = \frac{e}{d}$; $\alpha = \frac{p}{d} - 0.75$; $\alpha \leq 3$	$\frac{\alpha f_u d t}{\gamma_{Rd}}$; value of α to be determined from test
Tensile of capacity of connected members R_t^* Plate	$A f_y$ or $\frac{A_n f_u}{1.25}$	$\frac{A f_y}{\gamma}$ or $\frac{A_n f_u}{\gamma}$	$A f_y$ or $\frac{A_n f_u}{1.25}$	$\frac{A f_y}{\gamma_{Rd}}$ or $\frac{A_n f_u}{\gamma_{Rd}}$
Angle, 1 bolt	$2(e_2 - \frac{1}{2} d)tf_y$	not covered	$2.2(e_2 - \frac{1}{2} d)tf_y$	$\frac{\beta_1 (e_2 - \frac{1}{2} d)tf_u}{\gamma_{Rd}}$
Angle, 2 or more bolts	$\beta A_n f_y$		$\beta A_n f_y$	$\frac{\beta_n A_n f_u}{\gamma_{Rd}}$

Table 1 Summary of design rules for bolted connections

EVALUATION OF TEST RESULTS

To validate the strength functions and to determine suitable values for the model factors a re-evaluation of available test information is carried out. Test results from different sources are systematically stored in a data base at the University of Aachen. The total collection of test results on bolts amounts at the present more than 1900. By a joint CEB-ECCS-coordination group, a uniform evaluation procedure for test results is developed, which is independent of the material and the structure considered [1, 2]. This procedure is adopted by the Eurocode organisation. But as mentioned in the introduction this procedure had to be adapted and the revised procedure is described in [8].

For the case that for each bolt and eventually the connected plate the actual material strength and the geometric properties are known, the procedure as used in [8] is described. For every specimen i the experimental value r_{vi} is compared with the value r_{ti}, calculated with the relevant strength function (without γ_{Rd}) and taking into account the actual properties of the specimen. For a chosen test population, the values of r_{vi} and r_{ti} can be plotted in a diagram (Fig. 10).

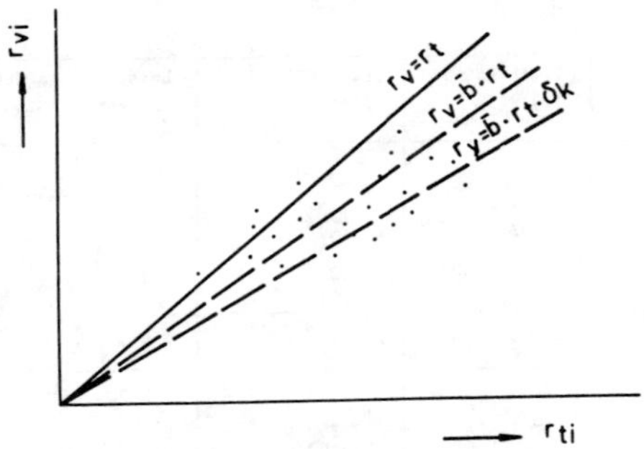

Fig. 10 Sample mean and correction for deviation

If the strength function is exact and complete, all the points (r_{vi}, r_{ti}) lie on the bisector of the angle between the axes of the diagram and the correlation coefficient $\rho_t = 1$. However, in general the points will scatter. So the mean value $\bar{r}_v$ and the standard deviation s_{rv} of the test values r_{vi} respectively the calculated values r_{ti} are determined:

$$\text{Tests}: \bar{r}_v = \frac{1}{n} \cdot \Sigma r_{vi}; \; s_{rv} = \sqrt{\frac{1}{n} \cdot (\Sigma r^2_{vi} - n \cdot \bar{r}_v^2)}$$

$$\text{Theory}: \bar{r}_t = \frac{1}{n} \cdot \Sigma r_{ti}; \; s_{rt} = \sqrt{\frac{1}{n} \cdot (\Sigma r^2_{ti} - n \cdot \bar{r}_t^2)}$$

The correlation coefficient $\hat{\rho}_t$ then follows from:

$$\hat{\rho}_t = \frac{\Sigma r_{vi} \cdot r_{ti} - n \cdot \bar{r}_v \cdot \bar{r}_t}{n \cdot s_{rv} \cdot s_{rt}}$$

If the value of $\hat{\rho}_t$ is greater than 0.9 the correlation is considered sufficient. Then a mean value correction is carried out. The straight line is determined, which intersects the origin and represents the mean values of the tests.

$$r_v = \bar{b}\; r_t$$

The coefficient of this line follows from:

$$\bar{b} = \frac{1}{n} \sum_{i=1}^{n} \frac{r_{vi}}{r_{ti}}$$

The formula for this line of mean values is then multiplied by a coefficient δ_k to find the characteristic line (5% fractile). The value of δ_k can be calculated with the statistical data of the population [8]. Finally with good approximation the model factor γ_{Rd} can be calculated:

$$\frac{1.03}{\bar{b}\,\delta_k}$$

Not all test reports provide the data on actual material and geometric properties of test specimen as required for the straight forward procedure described before. Two additional cases can be distinguished:

- Simple values of the material strengths in the sample are given, but these can not be related exclusively to individual test specimen;
- Only the characteristic values of the material strength are given.

The procedure has been modified for these cases. Instead of the strength of each specimen, the mean value of the sample is used and the standard deviation of the coefficient δ_k is increased properly. Comparison calculations show that the modified procedure gives satisfactory results.

PRELIMINARY RESULTS OF THE EVALUATION FOR BOLTS

Only as an indication some preliminary results are given. At present for the evaluation all relevant test data in the data base are used without any critical review of the test reports. In Fig. 11 the r_{vi} - r_{ti} diagram is given for shear tests on 8.8 bolts, where the shear plane passes throuth the threaded protion. Fig. 12 shows the ratio between r_{vi} and r_{ti} for the same type of test, but for different bolt grades.

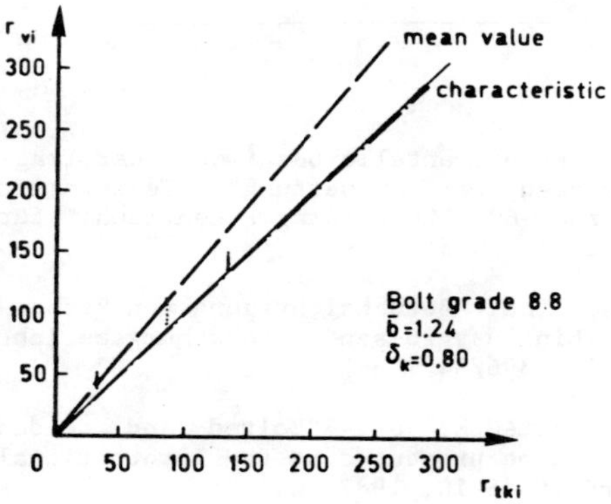

Fig. 11

Shear failure of 8.8 bolts
Shear plane through threaded portion

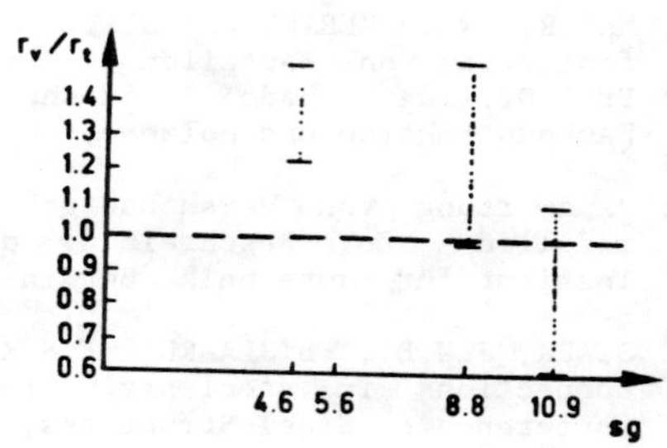

Fig. 12

Shear failure of bolts of different grades with shear plane through threaded portion

This figure clearly shows the tendency that higher grades have lower shear capacities. Based on this information it should be considered to reduce for 10.9 bolts the ultimate shear strength of 0.7 f_u, as assumed in the strength function (Table 1)..

CONCLUDING REMARKS

For welded connections the same procedure will be followed in evaluation of test results. At the moment about 510 test results are available mainly from the international test serie. An example is shown in Fig. 13. The design strength was based on:

$$\frac{F}{\Sigma\ a\ \ell} \leq \alpha\ f_y$$

where F is the tensile force on the connection, $\Sigma\ a\ \ell$ is the sum of the throat thickness times the length of the welds and α depends on the steel grade (α = 0.85 for Fe 360; α = 0.80 for Fe 420; α = 0.71 for Fe 510).

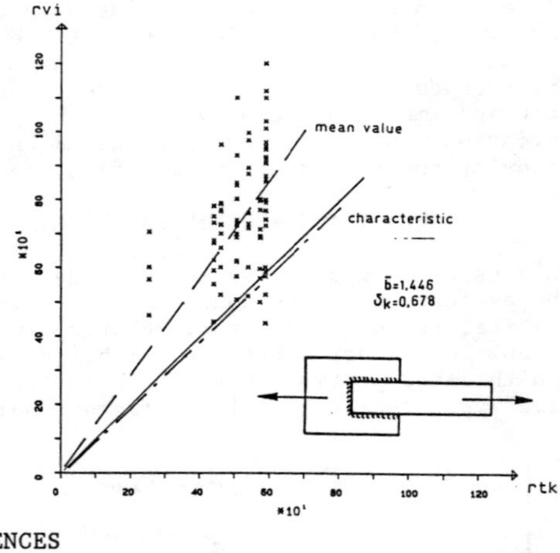

Fig. 13 Weld failure

REFERENCES

1. MAIER, W., KERSKEN-BRADLEY, M., "Experimentelle Bestimmung der Tragfähigkeit von Bauteilen - Grundideen der Auswertung" Festschrift Prof.Dr.-Ing. Bodo Heimeshoff zum 60. Geburtstag, Lehrstuhl für Baukonstruktion und Holzbau, TU München, 1986.

2. "Auswertung von Versuchsergebnissen bei Berücksichtigung von Rechenmodellen - Zur Beschreibung der Einflussgrössen", Forschungsbericht Institut für Bautechnik, Berlin IV 1-5-396/84.

3. STARK, J.W.B., BIJLAARD, F.S.K., SEDLACEK, G., "Bolted and welded connections in steel structures". To be presented at the International Conference on Steel Structures, Cardiff, 8-10, 1987.

4. "European Recommendations for Steel Construction", European Convention for Constructional Steelwork (ECCS), Brussels, March 1978, ECCS-EG 77-2E.

5. "European Recommendations for bolted connections: in structural steelwork", European Convention for Constructional Steelwork, Technical

Committee 10: Bolted and Welded Connections, Publication No 38, Fourth Edition, March 1985.

6. "Steel Structures, Materials and Design", Draft Document ISO/TC/167/SC1 Doc. N114E, Norges Byggstandardiseringsrad, Oslo, January 1986.

7. "Eurocode No 3: Common Unified Rules for Steel Structures", Commission of the European Communities, Brussels 1984, Report EUR 8849 DE, EN, FR.

8. "Procedure to determine the characteristic strength functions from tests". Proposal for Part 1 of the background documents for the design rules in Eurocode 3.

STRENGTH OF CONNECTIONS IN BENDING COMPARED TO ULTIMATE BEHAVIOUR RATHER THAN ELASTO-PLASTIC THEORY

Anders Granström
AG Konstruktioner,
Björns backe 3,
S-182 35 Danderyd,
SWEDEN

ABSTRACT

In the design of structures, large structural elements such as beams and columns are generally designed according to elastic or elasto-plastic theory based on the yield strength measured on small tensile specimens. When it comes to connections these theories are often extremely conservative, particularly where bending is the governing failure mode.

In this paper the reason for the great strength of connections is discussed, and a few test results are presented that illustrate the magnitude of the strength increase in relation to elasto-plastic theory.

INTRODUCTION

The yield limit or the 0.2% limit is generally used as a design criterion for beams and columns, the strength of which are not governed by elastic instability. That procedure is justified, since only moderate straining is necessary to eliminate even great residual stresses created by the rolling and welding processes.

It is well known that steel, due to strain hardening, exhibits a greater strength than corresponds to complete yield of a cross section. However, strain hardening generally starts at rather great strain values, see table 1 |1|. In a beam there is little use of this extra strength since it is paired with excessive yielding in large areas and large deflection, which may be unacceptable even in an ultimate state condition.

In connections, however, excessive straining is local and affects the over-all deformations in the large structure to a much smaller extent. Therefore strain hardening should be considered, and may offer a great contribution to the design strength of a connection. Actually, it was found in |2| that the ultimate strength in bending of a detail may be

TABLE 1
Results from tensile & compressive tests on various structural steels

SS Steel No	Plate/ beam mm	Tensile tests R_{eL} N/mm²	R_{eH} N/mm²	R_m N/mm²	Strain hard. starts at %	Compressive tests R_{eL} N/mm²	R_{eH} N/mm²	Strain hard. starts at %
1412	HE200A	297	306	469	1.5	306	317	1.2
1412	Pl 15	259	270	426	1.3	312	329	1.2
2624	Pl 12	755	805	800	2.8	767	792	

considerably greater than UTS. The reason for this behaviour is that the true fracture stress in a tensile test, due to contraction, fig 1, is about twice as great as UTS. This fact is also shown in fig 2 |3| for a common structural steel, designated *I*, and a cold finished steel, designated *II*.

With plates in bending the contraction of the tensile zone is restrained by the compressive zone. Therefore the potential factor of strength increase should be about 2 related to UTS, and about 2 to 3 related to yield, depending on the type of steel.

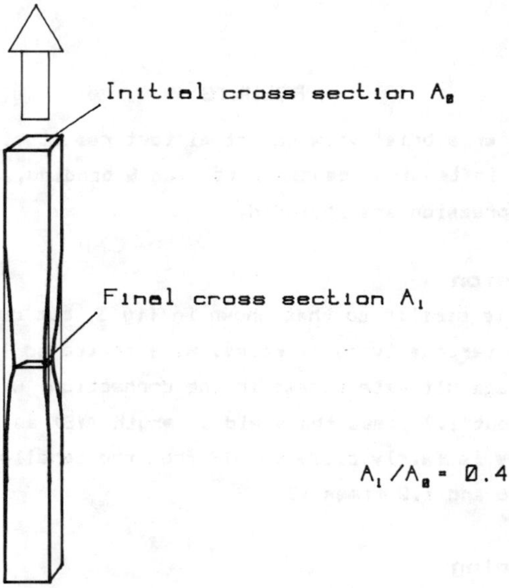

Figure 1. Tensile test at failure, where the fracture stress calculated on the initial area A_0 is slightly less than UTS

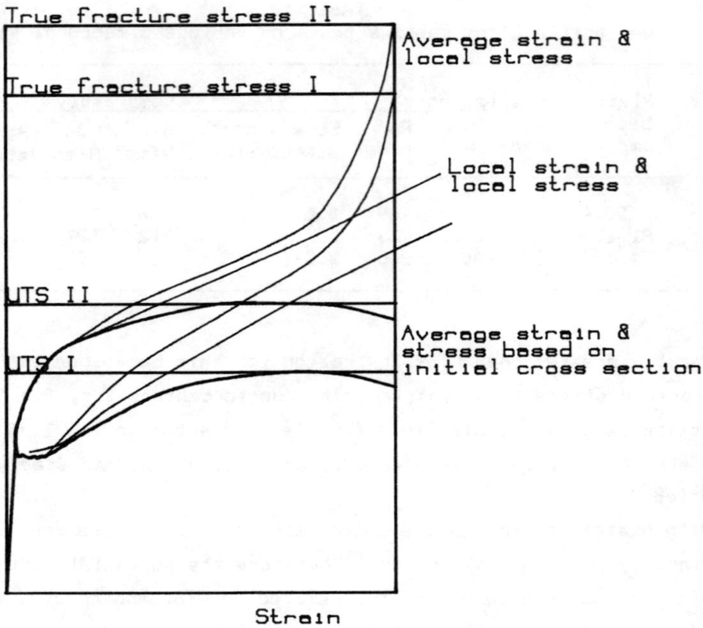

Figure 2. Principal stress-strain relationships according to |3|

RESULTS

In this chapter a brief view on actual test results is given, including details in tension, bending, tension & bending, and compression. Also welds in compression are included.

Details in tension

In |2| details similar to that shown in fig 3, cut out from larger details, and with various types of welds, were tested to failure in tension. The average ultimate stress in the connection, with little variation, was about 1.7 times the yield strength (YS) as measured in a tensile test. This is fairly close to UTS from the tensile tests, which varied between 1.6 and 1.8 times YS.

Details in bending

Also in |2|, details similar to that shown in fig 3 with various types of welds. The loading span was in most cases 300 or 500mm. Most details were loaded to large deformations without fracture. Only in one case, with giant fillet welds (a=17mm) one of the horizontal members failed, in brittle fracture at 2.1 times YS. Many of the other details were loaded up to 2.6 times YP without fracture.

Another investigation |4| with thick plate in bending showed an ultimate stress of 2.1 times YS, see fig4.

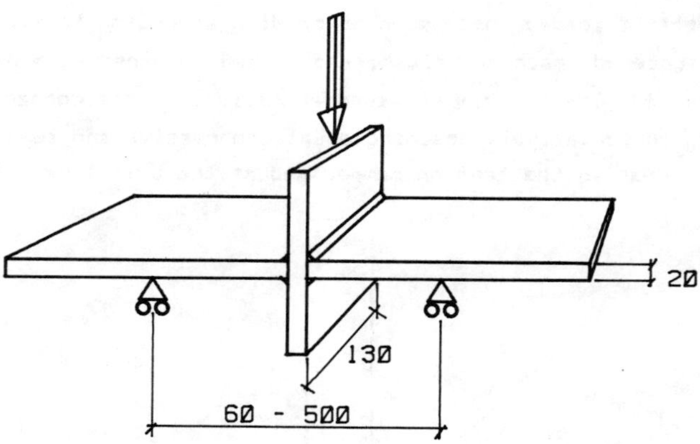

Figure 3. Mild steel details tested in bending

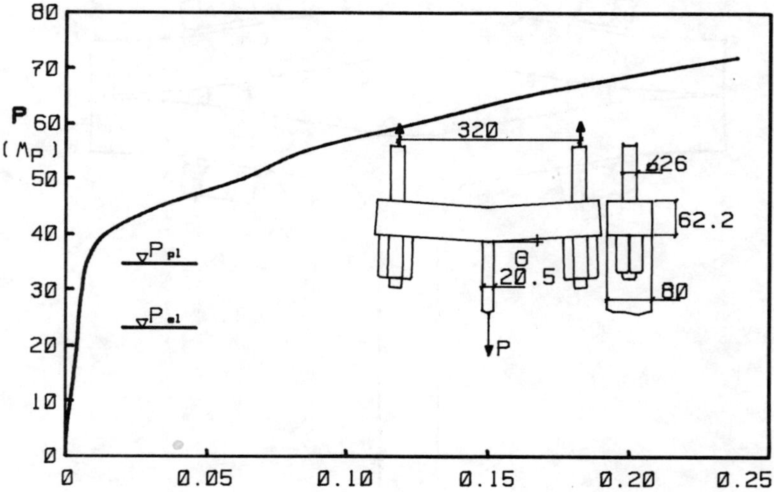

Figure 4. Bending test according to |4|

Even in bolted end-plate connections |5|, where the strains close to the tension member are considerably greater than at the yield line near the bolts, average ultimate bending stresses of 3.5 times YS for mild steel, and 1.8 times YS (and UTS) for ductile EHS, were evaluated neglecting the different magnitude of strain at the two yield lines.

Details in tension & bending

The details loaded in tension or bending according to fig 3 were later straightened, machined flush, and bolted together as end-plate connections |6|. The maximum stresses at failure of the connections were calculated, conservatively assuming equal compressive and tensile stress and equal stress in the tension member and at the bolt line, with minor

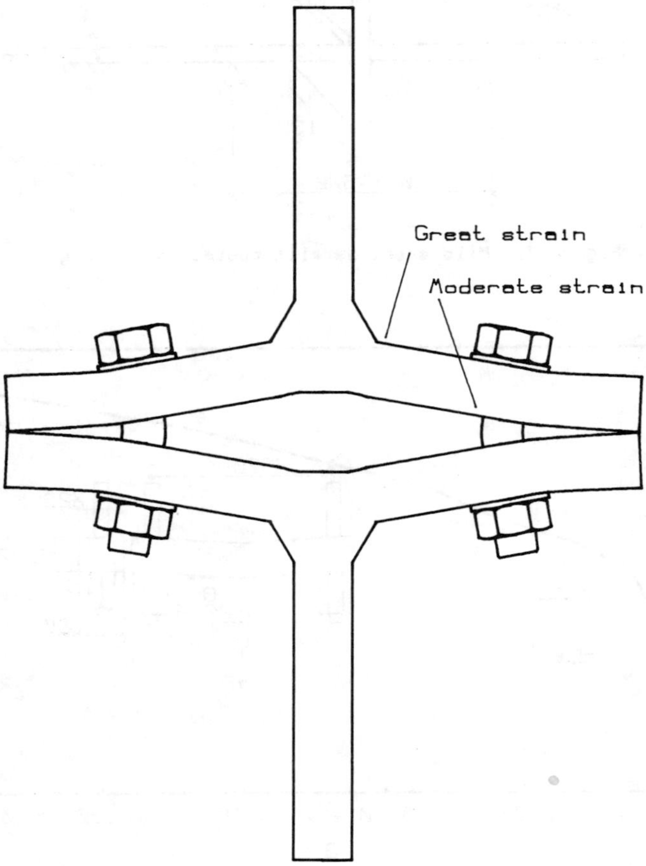

Figure 5. Uneven strain distribution in bolted end-plate connection

corrections for geometric distortion during the test and different YS of the end plate, fig 6. In spite of the great mean tensile stress, also shown in fig 6, the maximum stresses are of the magnitude of 3 - quite compatible with the ultimate stresses in pure bending.

Considering the mode of fracture and the magnitude of the ultimate stress, a correction for shear stresses is obviously not useful. Factors that should be considered should rather be large strain concentration, heavy welding, and thick plate.

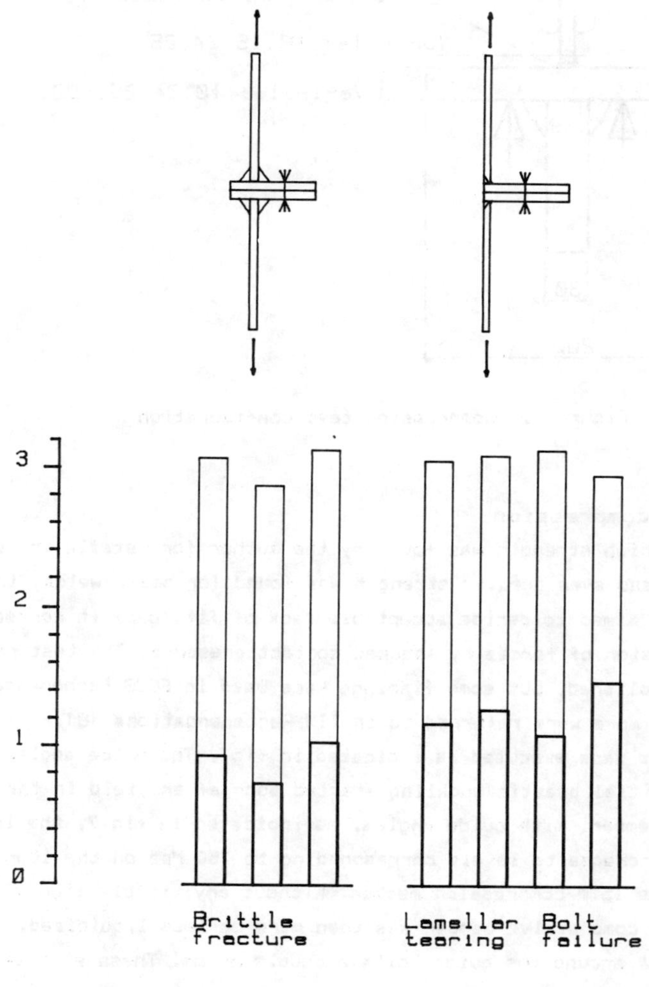

Figure 6. Mean stresses and maximum stresses related to yield strength

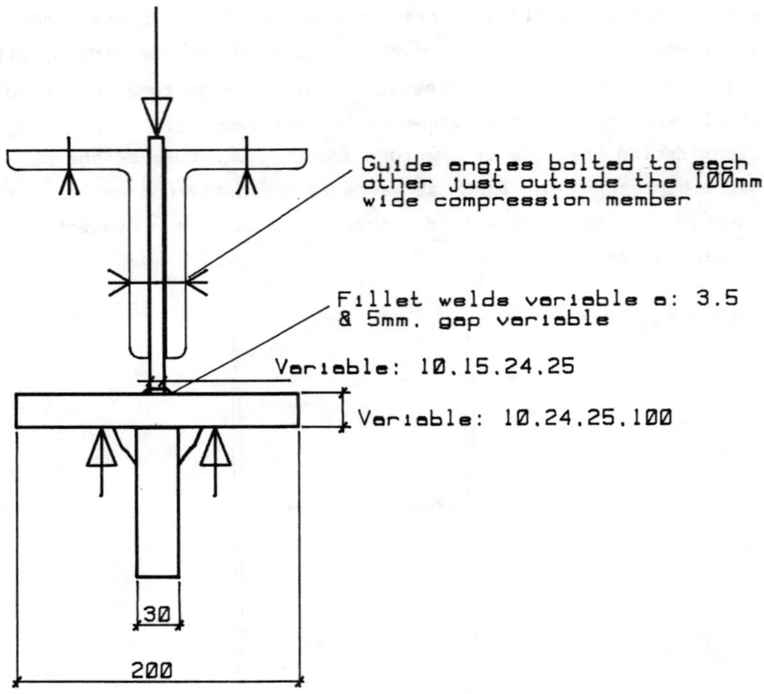

Figure 7. Compression test configuration

Details in compression

Equally high strength was found by the author for details in pure compression, and even greater strength was found for small welds, in an investigation aimed to decide acceptable lack of fit (gap) in connections with transmission of forces by assumed contact pressure. The test results were never published, but some findings were used in ECCS Recommendations |7| and a few more were referred to in IIW Recommendations |8|.

The tests were executed as indicated in fig7. The guide angles were used since initial plastic buckling started soon after yield in the thin compressive member. With guide angles, as indicated in fig 7, the load was possible to increase to levels corresponding to 750 Mpa on the 10mm & 800 Mpa on the 15mm compression member. Without any visible sign of fracture, the compressive member was then more or less liquidized, "floating out" around the guide bolts a couple of mm. These stresses correspond to about 3 times YS of the mild steel compression members.

In the case of the 3.5mm fillet welds, the maximum stress reached occasionally 2000 Mpa calculated as mean compressive stresses over the throat sections, which corresponds to 4 times the specified UTS for tensile specimens. Still the gap was not completely closed, and the welds sustained at subsequent tensile loading the specified UTS.

ACKNOWLEDGEMENTS

The author wish to thank the Swedish Institute of Steel Construction, which financially supported most of the investigations referred to in this paper, as well as the work with the paper itself.

The author also wish to thank the Swedish Council for Building Research, which financially made it possible for the author to go to the "workshop" in Cachan to present and discuss the results.

REFERENCES

1. Unpublished comercial investigation made by the author.

2. Granström, A., Methods of Assessing the Risk for Lamellar Tearing. The relevance of laboratory tests in assessing the strength of large structures, publ. 58, Swedish Institute of Steel Construction, Stockholm 1977.

3. Baehre, R., Historik - Material - Produkter. Kompendium i Stålbyggnad del I, Institutionen för Brobyggnad/Stålbyggnad, Royal Institute of Technology, Stockholm 1971.

4. Höglund, T., Om BSK och Detaljutformning av Stålkonstruktioner. FoU bygg/Väg- och vattenbyggaren, 1983, no 10, pp. 11-15, ISSN 0042-2177.

5. Granström, A., Bolted End-plate Connections II. EHS Steel & Asymmetric Bolt Locations, report 86:5, Swedish Institute of Steel Construction, Stockholm, September 1984, ISBN 91-85644-37-4.

6. Granström, A., The Strength of Bolted End-plate Connections. Ultimate Design Method & Lamellar Tearing Consideration, report 15:13, Swedish Institute of Steel Construction, Stockholm, February 1979, ISBN 91-85644-03-X.

7. ECCS Commission IX, Recommendations for Steel Constructions, ECCS-EG-76-1E, Vol. I, June 1976, p 153.

8. IIW Commission XV, Design Rules for Welded Joints with Gaps Loaded in Compression,, Welding in the World, 1977, 15, No 3/4, pp 73-78, ISSN 0043-2288.

EXPERIMENTAL STUDY OF THE NON-LINEAR BEHAVIOUR OF BEAM-TO-COLUMN BOLTED JOINTS

J. Janss
Centre de Recherchesde l'Industrie des Fabrications Métalliques (CRIF)
Liège
BELGIUM

J.P. Jaspart
and
R. Maquoi
M.S.M., Institut du Génie Civil, Université de Liège
BELGIUM

ABSTRACT

The paper is aimed to compare the behaviour of several types strong axis of beam-to-column bolted joints, which are commonly used in practice.

INTRODUCTION

An experimental investigation was recently performed at the University of Liège, with the financial support of IRSIA and ARBED [1]. Three types of connections between a beam and a column flange (strong axis connection) were considered : i) end plates, ii) double web cleat, and iii) flange cleats (fig. 1). All the connections are bolted ones with bolts of quality 10.9 (H.S. bolts), preloaded to 0.8 times the material yield stress or, in some cases, with a lower torque moment (controlled hand tightening). The basic data are listed in table 1 ; all the mechanical and geometrical properties of specimens were measured but are not listed here. The testing arrangement is shown in figure 2. The load P is increased step by step either up to collapse of the connection, or up to a maximum deflection (20 cm) of the cantilever end. The deflections are measured at the ends of the column and cantilever beam, as well as the slopes of beam and column axis in the joint. From these measurements, the relative rotations of the connection, of the joint and that due to the shear flexibility of the column web are deduced (fig. 3).

Test n°	Column C	Beam B	Relative Stiffness B/C	Connection (fig. 2)	Loading	Axial force N	Bolt tight.	Clea-rance d (mm)
01	HEB160	IPE200	0,78	a	S	300kN	C	-
02	HEB160	IPE200	0,78	b	S	-	C	0
03	HEB160	IPE200	0,78	c	S	-	C	15
04	HEB160	IPE200	0,78	a	S	700kN	C	-
05	HEB160	IPE200	0,78	b	S	-	C	15
06	HEB160	IPE200	0,78	c	S	-	M	15
07	HEB160	IPE200	0,78	a	S	-	C	-
08	HEB160	IPE200	0,78	b	Cyc	-	C	0
09	HEB160	IPE200	0,78	c	S	-	C	15
010	HEB160	IPE300	3,35	a	S	-	C	-
011	HEB160	IPE300	3,35	b	S	-	C	15
012	HEB160	IPE300	3,35	c	S	-	C	15
013	IPE240	IPE200	0,50	a	S	-	C	-
014	IPE300	IPE200	0,23	a	S	-	C	-
015	HEB160	IPE200	0,78	a	Cyc	-	C	-
016	HEB160	IPE200	0,78	b	S	-	M	15
017	HEB160	IPE200	0,78	c	Cyc	-	C	15
018	HEB160	IPE300	3,35	a	Cyc	-	C	-
019	HEB160	IPE300	3,35	b	Cyc	-	C	0
020	HEB160	IPE300	3,35	c	Cyc	-	C	15
021	IPE240	IPE200	0,50	a	Cyc	-	C	-
022	IPE300	IPE200	0,23	a	Cyc	-	C	-

-: The axial force is negligible or d has no meaning
S : static loading ; Cyc : cyclic loading
C : preloading to 0.8 f_y ; M : controlled hand tightening

Table 1 - Basic data for the test specimens

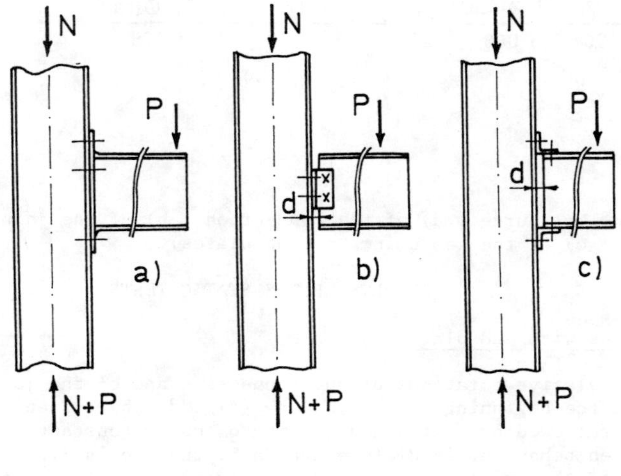

Figure 1 - Types of joints

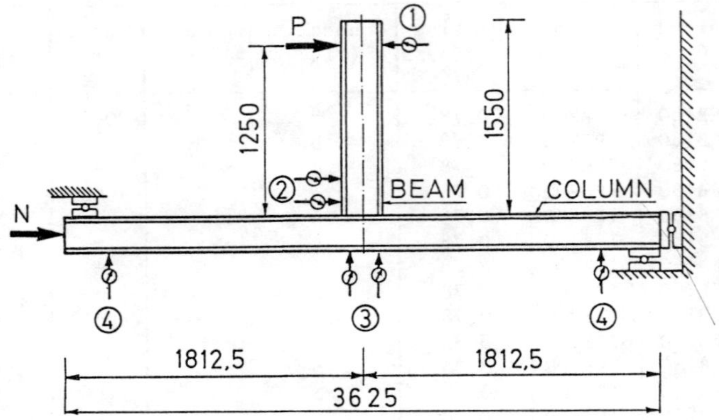

Figure 2 - Testing arrangement

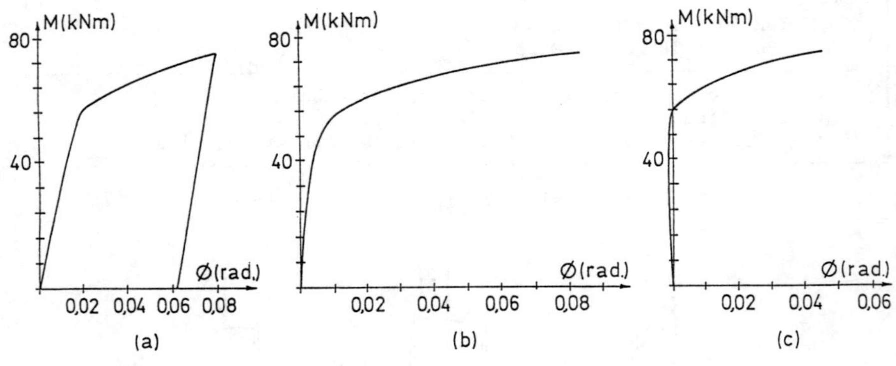

Figure 3 - M-ϕ Curve : a) of the connection ; b) of the joint ; c) of the web column - (end plates).

RESULTS OF THE STATIC TESTS

Connections with end plates.

The relative rotations of the connection and of the joint are nearly linear at the beginning of the loading (fig. 3.a,b). Then occurs yielding, which is followed by a strain hardening of quasi constant stiffness. The column web behaviour is different at all; the web is first nearly infinitely rigid, then yields rather suddenly (fig. 3c). From the test results the influence of the main parameters can be summarized as follows :

a) The axial force has no sensible effect on the yield shear force in the range $\sigma < 130$ MPa , i.e. $\sigma = 0.5\ f_y$, while the yielding moment of the column flange is affected so that the linear portion of the M-ϕ curve is reduced ; this observation is in accordance with Packer's and Morris' conclusions [2] but in contradiction with Zoetemeijer' and Munter' ones [3];

b) The initial stiffness of the joint seems just be little influenced by the axial force ;

c) It is necessary to distinguish between both behaviours of the column web and of the connection and to modelize them separately, as it appears clearly from figures 4. a-c. In test 07, the column web and the connection contribute the rotation in similar proportions, while, in test 010, the deformation is mainly due to the shear flexibility of the column web, which is weaker than the connection. It can also be observed that the use of the full elastic strength of the beam would imply a marked non linear behaviour of the joint.

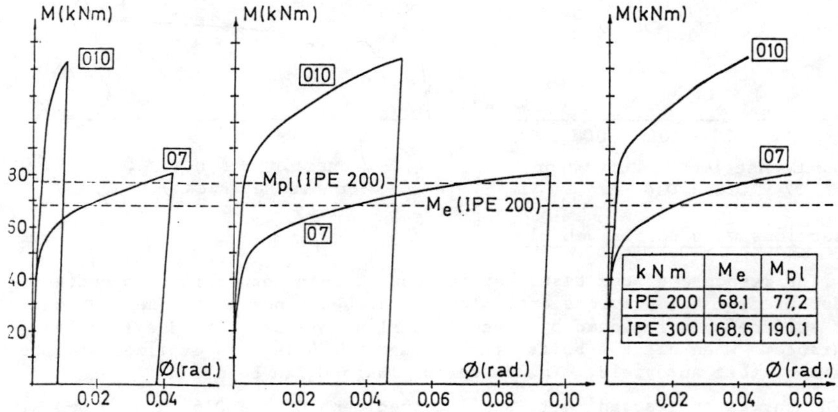

(a) (b) (c)

Figure 4 - M-ϕ Curve : a) of the connection ; b) of the joint ; c) of the web column - (end plates).

d) The initial stiffness in tests 07 and 014 - specimens fitted with same end plates - is shown dependent on the dimensions of the column flange (fig. 5). The more marked non linear behaviour of specimen 014 is due to a weaker column flange and to the yielding of the compressed zone of the web column. With respect to similar stiffness, specimen 013 benefits from the stronger rigidity of IPE240 column flange. Unlike test 07, the collapse of the connections 013 and 014 is reached, due to yielding in the compressed part of the column web, and the associated rigidity becomes nearly zero.

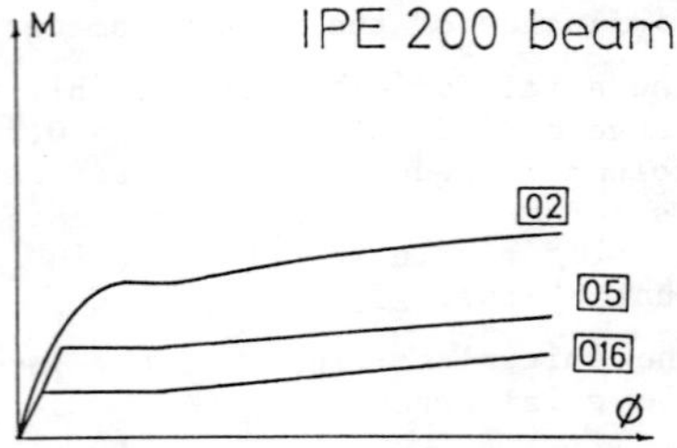

Fig.6-Effect of bolt tightening and of beam depth on connection M-ϕ curves (double web cleat)

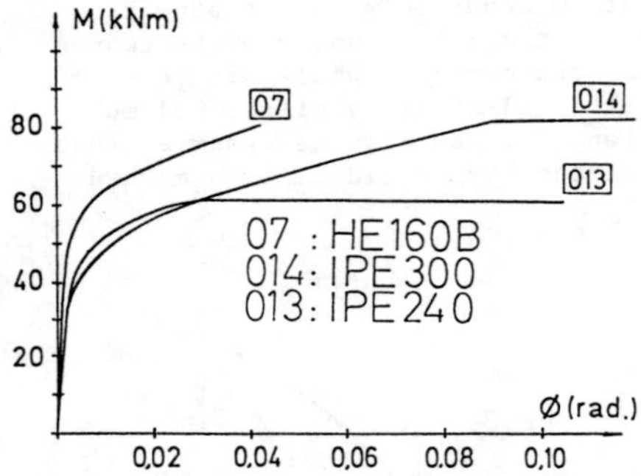

Fig.5- Connection M-ϕ curves for tests 07 - 013 - 014.

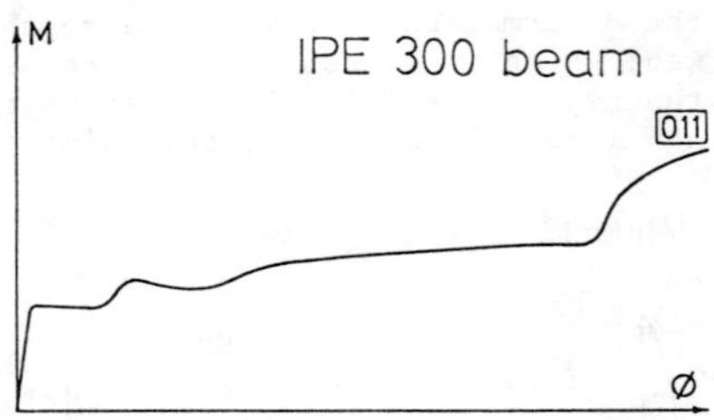

Fig.7- Connection M-ϕ curve for test 011 (double web cleat).

Connections with double web cleat.

The response of the tests carried out on such connections is quite different from the previous one. The linear behaviour with a nearly constant stiffness is followed by successive slips of the bolts due to hole clearance. When all the bolts are bearing, the holes are ovalized and the cleats distort and yield. Following conclusions can be drawn :

a) The curves corresponding to similar specimens 05 and 016 (d = 15 mm) are plotted in figure 6. Only the bolt tightening is different : preload to 0.8 f_y for specimen 05 and controlled hand tightening for specimen 016. The lèvel of tightening does not alterate neither the initial stiffness, nor the behaviour after slip ; only the bending moment associated to the first slip varies and does it proportionnally to the preload in the bolts (fig. 6).

b) By reducing the clearance d (fig. 1) from 15 mm to zero (specimen 02), the transmission of the applied bending moment is modified with the result of a much higher initial stiffness (fig. 6). Indeed compression is transmitted by direct bearing between beam-and column flanges, while tension is supported by friction, then by shear after slip occurs, and last by bearing pressure. The non linearity before slip is due to distorsion and yielding of the angles in the tensile zone, and last by the slip initiation in the upper bolts in the beam.

c) A deeper beam (IPE300) allows the use of deeper angles with four bolts (specimen 011), with the result of an initial stiffness nearly equal to that of specimen 010 (connection with end plates). The clearance d=15mm allows the beam-and column flanges bearing and explains the increase of rigidity at the end of the test (fig. 7).

Connections with flange cleats.

The M-ϕ curves dealing with tests n° 03 - 06 - 012 are shown in figure 8. It can observed that :

a) The initial stiffness of tests 03 and 06 is not affected by the level of tightening and that the ultimate strength is quasi independent of this level. However, the slip is initiated earlier for test 06.

b) With a deeper beam, (test 012), the initial stiffness does not change significantly ; so for the overall shape of the M-ϕ curve.

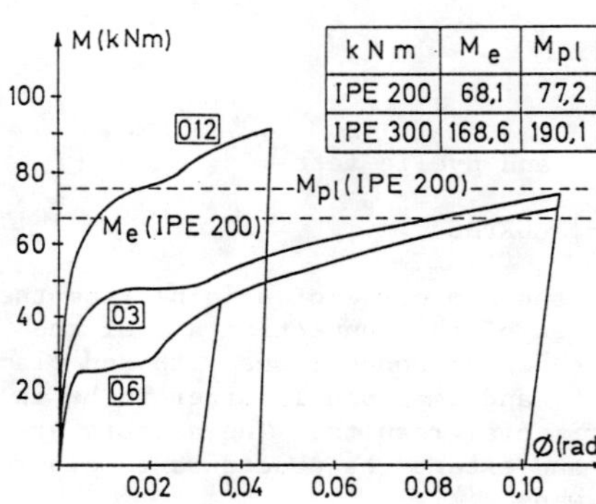

kNm	M_e	M_{pl}
IPE 200	68,1	77,2
IPE 300	168,6	190,1

Fig. 8 - M-ϕ curves for tests 03 - 06 - 012 (flange cleats).

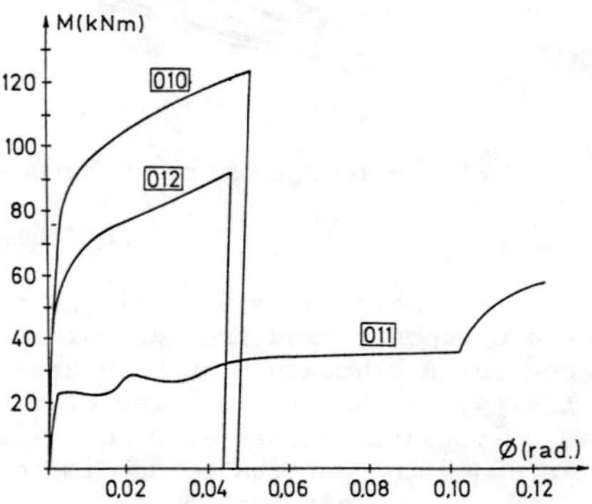

Fig. 9 - Comparison of M-ϕ curves for 3 types of connections.

Comparison between different types of connections.

Figure 9 shows for instance the M-ϕ curves associated respectively to the three types of connections for a joint between a HE160B column and an IPE300 beam.

CYCLIC TESTS

The cyclic tests were performed in accordance with a standard procedure [5]. At the present time, the results have not yet been fully investigated. Therefore only a cyclic test is compared to a static one, both being carried out on identical specimens (fig. 10).

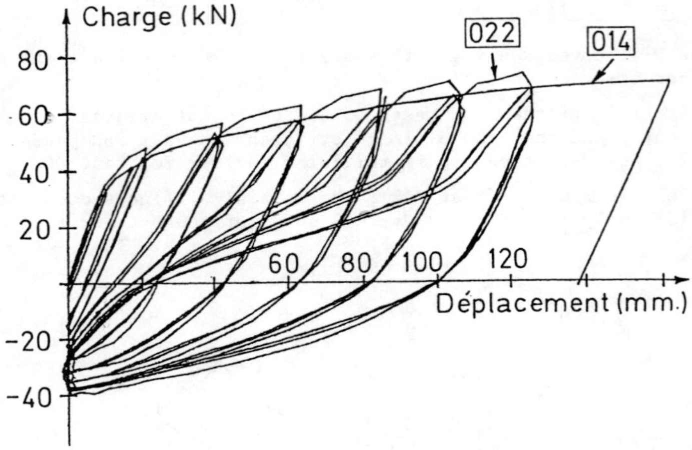

Figure 10 - Comparison between static and cyclic tests.

MODELLING OF M-ϕ CURVES

An ambitious goal would be to define the M-ϕ curve of a joint from the whole geometry and the mechanical properties of the connection and of the connected elements. Such an attempt was made for connections with end plates [4] but has to be somewhat modified [6] and improved in order to be in a quite satisfactory agreement with experimental results. The authors are intended to contribute such improvements and extend the procedure to other kinds of connections and to shear column webs.

REFERENCES

1. Jaspart, J.P., Essais sur assemblages poutre-colonne d'axe fort. CRIF, Bruxelles (in preparation).

2. Packer, J.A. and Morris, L.J., A Limit State Design for the Tension Region of Bolted Beam-to-Column Connections. The Structural Engineer, Vol. 55, 10, October 1977, pp. 446-458.

3. Zoetemeijer, P. and Munter, H., Influence of an Axial Load in the Column on the Behaviour of an Unstiffened Beam to Column End Plate Connection. Stevin Laboratory Report N° 6-84-1, Delft University of Technology, January 1984.

4. Yee, Y.L. and Melchers, R.E., Moment-Rotation Curves for Bolted Connections. Jl. Struc. Engrg., A.S.C.E. Vol.112, 3, March 1986,pp.615-635.

5. E.C.C.S., Recommended Testing Procedure for Assessing the Behaviour of Structural Steel Elements under Cyclic Loads. E.C.C.S. Publication n° 45, 1986.

6. Maquoi, R. and Jaspart,J.P., Moment Rotation Curves for Bolted Connections - Discussion of [4]. (to appear in Jl. Struct. Engrg.).

PREDICTION OF MOMENT-ROTATION BEHAVIOR OF SEMI-RIGID BEAM-TO-COLUMN CONNECTIONS

James B. Radziminski
Department of Civil Engineering
University of South Carolina
Columbia, SC 29208
USA

and

Atorod Azizinamini
Construction Technology Laboratories, Inc.
Skokie, IL 60077
USA

ABSTRACT

This paper presents the results of a study of the static behavior of semi-rigid steel beam-to-column connections. From the experimental investigation, the geometric parameters which most significantly affect the moment-rotation performance of the connections have been evaluated. An analytical model was developed to predict the complete non-linear static moment-rotation behavior of the connections. The model was found to compare reasonably well with the test data for the range of connection stiffnesses examined.

INTRODUCTION

The analysis of steel frame building structures with fully rigid beam-to-column connections is a straightforward one, and buildings utilizing such framing systems have been used extensively in design practice. More recently, increasing interest has been expressed in the use of semi-rigid (partially restrained) connections for building frames, spurred by potential economies that may be realized from simpler connection details, and by "tuning" the connections (i.e., adjusting connection stiffnesses) to optimize the distribution of moments in the connected members.

Semi-rigid connections are characterized by a decidedly non-linear moment-rotation ($M-\phi$) behavior which, in turn, can contribute significantly to the performance of the complete structure, including strength limits and drift considerations [1-4]. In certain instances, definition of the complete $M-\phi$ relationship may not be necessary, and an estimate of the initial stiffness of the connection (initial slope of the $M-\phi$ curve) will suffice. For example, under service gravity live load and wind load fluctuations, corresponding to a nominally elastic response of the structure, frame analyses using linearized connection stiffnesses may be adequate [5,6]. Under extreme loading conditions, however, such as those imposed in a seismic event, use of the full non-linear $M-\phi$ curve (complete hysteresis loop) is required to assess structural performance.

In the study reported herein, the static moment-rotation behavior of connections consisting of bolted top and seat flange angles, and double web angles, was investigated. The objectives of the study have been to experimentally determine the geometric parameters which most significantly affect connection performance, and, from these data, to develop analytical models to predict the static $M-\phi$ behavior. Models used to predict the initial stiffness of the connections are reported elsewhere [7]. In this paper, a semi-empirical model for predicting the complete non-linear M-ϕ curve of the connections is presented.

EXPERIMENTAL INVESTIGATION

Description of Test Program

The test members consisted of a pair of beam sections attached to a centrally positioned stub column. The connection elements were comprised of top and seat angles bolted to the flanges of the beams and supporting column, together with double web angles bolted to the beam web and to the column flanges. The top and seat angles were the same for a given test, and the web angles were centered on the beam. ASTM A36 steel was used for the members and connection elements; all fasteners were 3/4-inch or 7/8-inch diameter ASTM A325 heavy hex, high strength bolts. A325 hardened washers were used under the turned elements.

Two beam sizes, W14X38 and W8X21, were used in the test program. For the 14-inch sections, the overall beam length was 20 feet, and for the 8-inch sections, 12 feet, so that the span-to-depth ratio was slightly less than 20 in each case. The stub column for the W14X38 beams was a W12X96 section; a W12X58 column section was used with the W8X21 beams. Heavy column sections were selected to minimize column panel zone deformation, thereby confining the deformations to the connection elements. Details of the W14X38 and W8X21 connections are shown in Figs. 1 and 2, respectively.

A pair of duplicate specimens was tested simultaneously using the framing arrangement shown in Fig. 3. A 55 kip, hydraulically actuated ram was used to apply load to the test members through the central stub column. In all of the tests, the controlled input variables were the rate and magnitude of actuator displacement. Complete details of the test procedures, including specimen preparation and types of instrumentation used, are reported in Ref. [8].

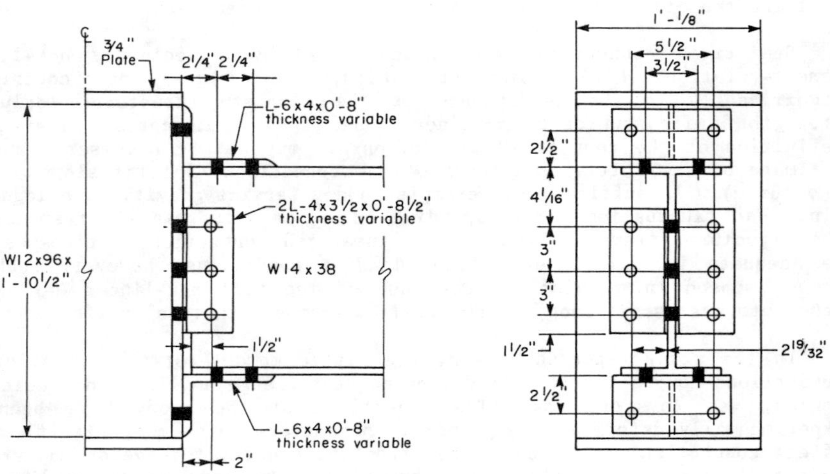

Figure 1. Details of Connection for W14X38 Beam

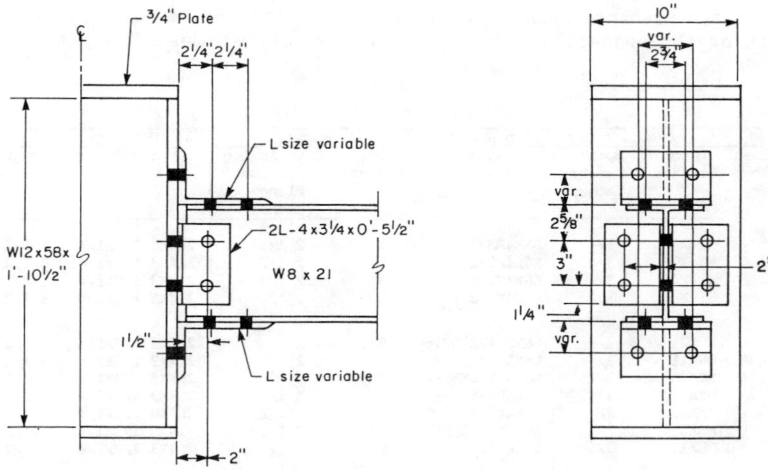

Figure 2. Details of Connection for W8X21 Beam

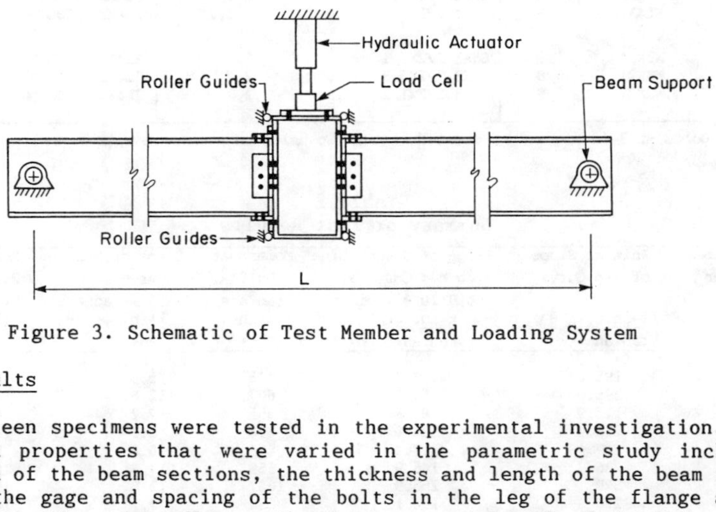

Figure 3. Schematic of Test Member and Loading System

Test Results

Eighteen specimens were tested in the experimental investigation. The geometric properties that were varied in the parametric study included: the depth of the beam sections, the thickness and length of the beam flange angles, the gage and spacing of the bolts in the leg of the flange angles connected to the column flange, the thickness and length of the web angles, and the bolt diameter. Details of the geometries of the individual test specimens are reported in Table 1.

A summary of the test results is presented in Table 2. Table 2 includes the initial stiffness of the connection, the slope and intercept moment of a secant line from the origin intersecting the M-ϕ curve at a rotation of $4X10^{-3}$ radians, and the moment and tangent slope at $24X10^{-3}$ radians. The latter slope, although not intended to represent a final, constant slope for a specific connection, does allow comparisons to be made among the various connections, as well as quantifying the degree of softening of a connection relative to its initial stiffness. Similarly, the reported secant slope and moment are indicative of the relative behavior of the connections during the early part of the loading cycle.

Table 1
Schedule of Test Specimens

Specimen Number	Beam Section	Bolt Diameter (inches)	Top and Bottom Flange Angles: Angle	Length (inches)	Gage in Leg on Column Flange (inches)	Web Angles: Angle	Length (inches)
14S1	W14X38	3/4	L6X4X3/8	8	2 1/2	2L4X3 1/2X1/4	8 1/2
14S2	W14X38	3/4	L6X4X1/2	8	2 1/2	2L4x3 1/2x1/4	8 1/2
14S3	W14X38	3/4	L6X4X3/8	8	2 1/2	2L4X3 1/2X1/4	5 1/2*
14S4	W14X38	3/4	L6X4X3/8	8	2 1/2	2L4X3 1/2X3/8	8 1/2
8S1	W8X21	3/4	L6X3 1/2X5/16	6	2	2L4X3 1/2X1/4	5 1/2
8S2	W8X21	3/4	L6X3 1/2X3/8	6	2	2L4X3 1/2X1/4	5 1/2
8S3	W8X21	3/4	L6X3 1/2X5/16	8	2	2L4X3 1/2X1/4	5 1/2
8S4	W8X21	3/4	L6X6X3/8	6	4 1/2	2L4X3 1/2X1/4	5 1/2
8S5	W8X21	3/4	L6X4X3/8	8	2 1/2	2L4X3 1/2X1/4	5 1/2
8S6	W8X21	3/4	L6X4X5/16	6	2 1/2	2L4X3 1/2X1/4	5 1/2
8S7	W8X21	3/4	L6X4X3/8	6	2 1/2	2L4X3 1/2X1/4	5 1/2
14S5	W14X38	7/8	L6X4X3/8	8	2 1/2	2L4X3 1/2X1/4	8 1/2
14S6	W14X38	7/8	L6X4X1/2	8	2 1/2	2L4X3 1/2X1/4	8 1/2
14S8	W14X38	7/8	L6X4X5/8	8	2 1/2	2L4X3 1/2X1/4	8 1/2
14S9	W14X38	7/8	L6X4X1/2	8	2 1/2	2L4X3 1/2X1/4	8 1/2
8S8	W8X21	7/8	L6X3 1/2X5/16	6	2	2L4X3 1/2X1/4	5 1/2
8S9	W8X21	7/8	L6X3 1/2X3/8	6	2	2L4X3 1/2X1/4	5 1/2
8S10	W8X21	7/8	L6X3 1/2X1/2	6	2	2L4X3 1/2X1/4	5 1/2

*Two Bolts at 3-inch spacing, mounted in top two holes on stub column, Fig. 1.

Table 2
Summary of Test Results

Specimen Number	Initial Slope of M-ϕ Curve $\left(\frac{\text{k-in.}}{\text{radian}} \times 10^{-3}\right)$	Slope of Secant Line to M-ϕ Curve at 4.0×10^{-3} radians $\left(\frac{\text{k-in.}}{\text{radian}} \times 10^{-3}\right)$	Moment at 4.0×10^{-3} radians (k-in.)	Slope of M-ϕ Curve at 24×10^{-3} radians $\left(\frac{\text{k-in.}}{\text{radian}} \times 10^{-3}\right)$	Moment at 24×10^{-3} radians (k-in.)
14S1	195.0	108.7	435	5.8	688
14S2	295.0	151.8	607	12.6	(947)*
14S3	115.9	88.8	355	7.2	652
14S4	221.9	124.0	496	8.3	822
14S5	247	117.4	467	10.8	763
14S6	286	146.9	589	10.3	1053
14S8	579	256.9	1027	15.0	1561
14S9	258	154.1	624	9.3	1024
8S1	66.7	44.3	177	4.1	329
8S2	123.4	69.0	276	1.5	(384)**
8S3	104.7	64.3	257	4.0	422
8S4	15.3	14.4	57.5	2.2	165
8S5	76.7	47.9	191.5	2.7	337
8S6	39.5	30.0	120	3.2	244
8S7	48.0	40.8	163	3.2	381
8S8	70	44.4	183	3.8	380
8S9	104	54.5	226	4.4	423
8S10	427	104.1	434	6.9	(634)***

* Major slip at 12×10^{-3} and 20×10^{-3} radians
** Major slip at 16×10^{-3} radians
*** Major Slip at 5.3×10^{-3} radians

The complete moment-rotation curves for several of the test connections are shown in Figs. 4 and 5. In all of the tests, the connections exhibited an M-ϕ response that became non-linear early in the loading sequence. This is attributed, primarily, to local yielding and eventual plastic hinge formation at each toe of the fillet in the angle attached to the tension flange of the beam. Another hinge developed in the vicinity of the bolt line on the leg of the flange angle attached to the column, together with progressive plastic hinging in the outstanding legs of the web angles. With the exception of three specimens that exhibited slip in the connection angles, all of the test connections were able to develop continually increasing moments through the full range of rotations imposed during the tests. During the latter period of loading, a nearly constant or only very gradually decreasing positive slope of the moment-rotation curve was exhibited by all but one of the specimens.

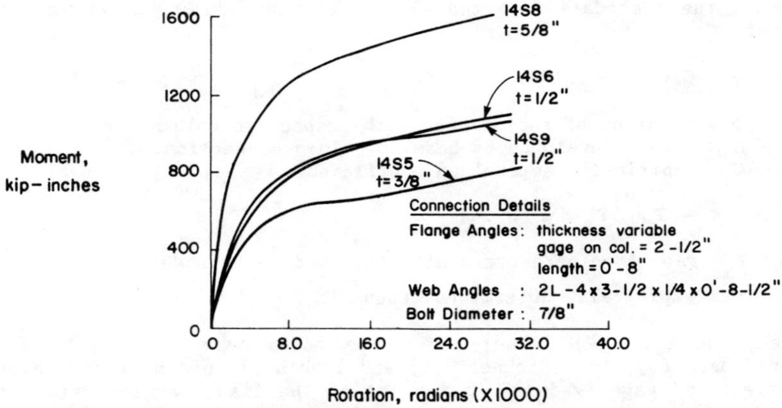

Figure 4. Typical Moment-Rotation Curves for W14X38 Specimens

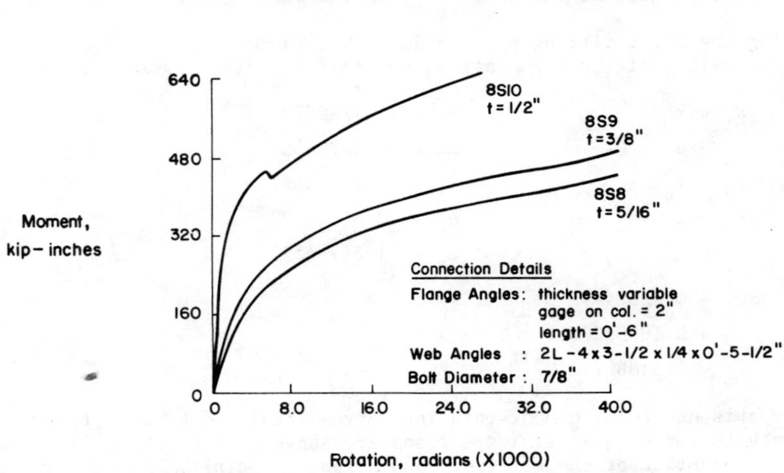

Figure 5. Typical Moment-Rotation Curves for W8X21 Specimens

From the experimental investigation [7,8], it was found that, of the geometric parameters studied, those that most significantly affect the initial stiffness and moment-rotation performance of the connections are: The depth of the beam section to which the connections were framed, the thickness of the flange angles, and the gage (including the effect of bolt diameter) in the leg of the flange angle attached to the column. Variations in the length of the flange angles, and the length and thickness of the web angles, had a less pronounced effect on connection response.

PREDICTION OF MOMENT-ROTATION BEHAVIOR

In 1969, Sommer [9] developed a technique to predict the moment-rotation curve for flexible type connections. The use of Sommer's method can lead to predictions of a negative M-ϕ slope. To overcome this problem, a modified procedure was developed [7]; the analytical model, which uses the test data from the parametric study described above, is of the form:

$$\phi = C_1(KM) + C_2(KM)^3 + C_3(KM)^5 \qquad [1]$$

where: ϕ = rotation of end of beam with respect to column face
M = moment developed by beam-to-column connection
C_i = empirically determined coefficients

and $$K = P_1^{\alpha_1} P_2^{\alpha_2} \cdot \cdot \cdot P_n^{\alpha_n}$$

where: P_i = geometric parameters affecting M-ϕ relationship

α_i = empirically determined exponents

The parameters (P) selected for the model were the depth of the supported beam (d), the thickness (t) and length (l) of the flange angles, the "effective" gage ($g-d_b/2$) in the leg of the flange angles attached to the column flange (where g is the gage from heel of angle and d_b is the bolt diameter), and the thickness (t_c) of the web angles.

Using the curve-fitting process described in Refs. [7,8], the following numerical values of the exponents α_i and coefficients C_i were obtained:

$P_1 = t$ $\qquad \alpha_1 = -1.1280877$

$P_2 = d$ $\qquad \alpha_2 = -1.2870455$

$P_3 = t_c$ $\qquad \alpha_3 = -0.41454097$

$P_4 = l$ $\qquad \alpha_4 = -0.69412158$

$P_5 = g-d_b/2$ $\qquad \alpha_5 = 1.34994572$

and $C_1 = 0.2232429X10^{-4}$

$C_2 = 0.1850728X10^{-7}$

$C_3 = 0.3188976X10^{-11}$

Comparisons of the moment-rotation curves predicted by Eq. [1] with the test results for several test specimens are shown in Figs. 6,7. In general, with the exception of the stiffest connections in both the W14X38 and W8X21 test series, the model offered reasonable approximations to the moment-rotation curves of the connections up to the limits of rotation examined in the test program. For the stiffest connections of each beam size, the model underestimates the moments developed at the larger connection rotations. It is recommended, therefore, that Eq. [1] be restricted to connections in which the flange angle thickness is of the same order as the thickness of the beam flange; use of this expression should not be extrapolated to predicting the M-ϕ response of significantly stiffer connections nor to connections consisting of top and seat angles only.

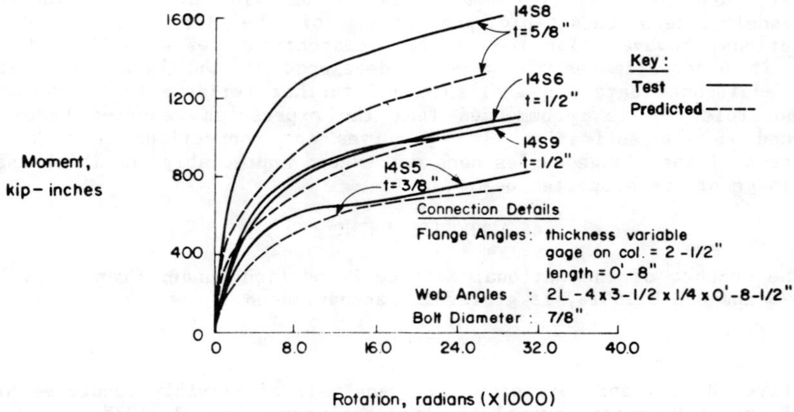

Figure 6. Predicted Moment-Rotation Curves for W14X38 Specimens

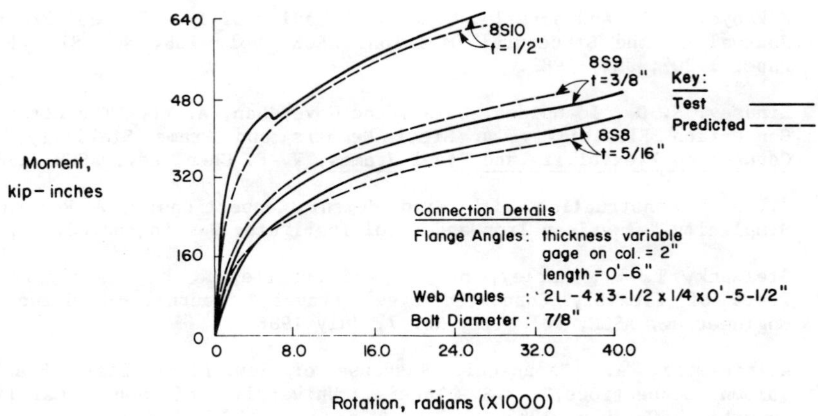

Figure 7. Predicted Moment-Rotation Curves for W8X21 Specimens

SUMMARY, CONCLUSIONS

A test program was conducted to evaluate the static moment-rotation performance of semi-rigid beam-to-column connections consisting of top and seat flange angles, and double web angles. Initial stiffnesses and complete M-ϕ curves were quantified for connections of varying geometry.

From the test program it was found that, for the type of connection studied, the geometric parameters that most significantly affect moment-rotation behavior are: the depth of the beam section to which the connection elements are framed, the thickness of the flange angles, and the effective gage in the leg of the flange angles attached to the column.

Using the test results, a parametric model was developed to predict the complete non-linear static moment-rotation behavior of the connections. The model offers reasonable predictions of the M-ϕ response of the connections; however, for the stiffest connections tested with each beam size, it underestimates the moments developed at the larger rotations. Until additional data are available and further refinements to the model are possible, it is recommended that the expression reported herein be confined to the estimation of M-ϕ curves for connections in which the thickness of the flange angles does not exceed appreciably the thickness of the flange of the supported beam.

ACKNOWLEDGEMENT

The support of the National Science Foundation, under Grant Nos. PFR-7923520 and CEE-8115014, is gratefully acknowledged.

REFERENCES

1. Frye, M. J., and Morris, G. A., "Analysis of Flexibly-Connected Steel Frames," Canadian Journal of Civil Engineers, Vol. 2, 1975.

2. Akbari, F., "Drift of Flexibly Connected Frames," Computers and Structures, Vol. 15, No. 2, 1982.

3. Ackroyd, M. H. and Gerstle, K. H.., "Behavior of Type 2 Steel Frames," Journal of the Structural Division, ASCE, Vol. 108, No. ST7, Proc. Paper 17207, July 1982.

4. Lindsey, S. D., Ioannides, S. A., and Goverdhan, A. V., "The Effect of Connection Flexibility on Steel Members and Frame Stability," in Connection Flexibility and Steel Frames, W.-F. Chen, ed., ASCE, 1985.

5. "Type 2 Construction with Wind Moment Connections: A Return to Simplicity," American Iron and Steel Institute, Washington DC, 1976.

6. Stelmack, T. W., Marley, M. J., and Gerstle, K. H., "Analysis and Tests of Flexibly Connected Steel Frames," Journal of Structural Engineering, ASCE, Vol. 112, No. 7, July 1986.

7. Azizinamini, A., "Monotonic Response of Semi-Rigid Steel Beam to Column Connections," M.S. Thesis, University of South Carolina, Columbia, SC, July, 1982.

8. Azizinamini, A., Bradburn, J. H., and Radziminski, J. B., "Static and Cyclic Behavior of Semi-Rigid Steel Beam-Column Connections," Structural Research Studies, Department of Civil Engineering, University of South Carolina, Columbia, SC, March 1985.

9. Sommer, W. H., "Behavior of Welded Header Plate Connections," M.S. Thesis, University of Toronto, Ontario, 1969.

EXPERIMENTAL ANALYSIS OF END PLATE CONNECTIONS

R. Zandonini and P. Zanon

Department of Engineering, University of Trento
Mesiano di Povo - I-38050 Trento, Italy

ABSTRACT

The prediction of joint rotational behaviour is a preliminary step in the analysis of semi-rigid frames. Mechanical models may be used and the behaviour of the whole joint obtained by "summing up" the contributions of the individual components. In order to develop simple models suitable to predict such component contributions results from ad hoc experimental research are needed. The paper presents an experimental technique adopted for that purpose. Results related to extended end plate connections are reported and their evaluation in order to characterize the moment rotation relationship discussed.

INTRODUCTION

The great majority of the investigations related to steel joints were, till recently, aimed at defining their ultimate strength capacity. The knowledge of the behaviour of the joints over its complete range were not of particular interest. Frame design was in fact based on the simple ideal models of rigid and pinned joints. Late research work, both experimental and numerical [1-5], contributed to a better understanding of the influence of joint flexibility on the performance of the whole frame. The possibility of considering the joint behaviour as a design parameter, allowing advantages in terms of cost effectiveness and/or structural reliability to be achieved, is indicated. The practicality of such an approach requires the capability of predicting the joint response.
This problem seems not yet solved. In effect, the number of governing mechanical and geometrical parameters (even for connections with simple detailing), suggests the suitability of an approach, which separate the contribution to the whole joint behaviour of its individual components (fig. 1): i.e. column panel zone (fig. 1b), column flanges (fig. 1c), connection elements (fig. 1d). The joint response is then obtained by "summing up" these contributions (fig. 1e). The feasibility of this procedure has already been established [6,7]. Moreover, such an approach has the advantage to favour a straight definition of the joint "rotation", a quantity which was given in the past a nonidentical meaning by different researchers.

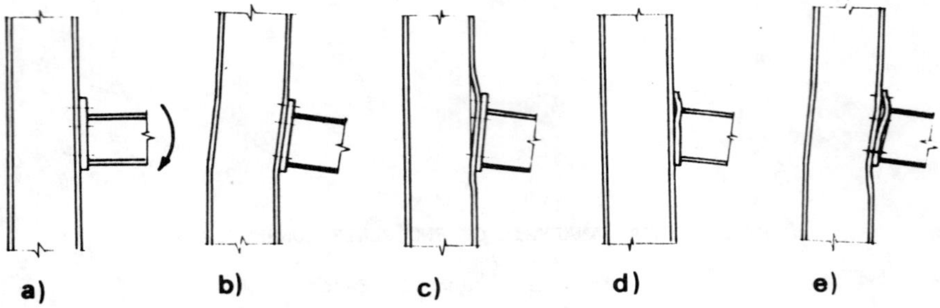

Fig. 1

The authors are conducting a research aimed at developing simple mechanical models, allowing for the prediction of the joint moment rotation curves. With reference to this philosophy by components, the first phase of the study is devoted to the behaviour of the sole connections [8]. This paper intends to highlight the experimental technique adopted and present a first series of results related to extended end-plate connections. Their evaluation is finally discussed in order to point out the parameters useful for the prediction and to single out the contribution of the different components (end plate and bolts). This data gives the preliminary information needed to select suitable mechanical models and represent the necessary benchmark to their validation.

ROTATION OF A CONNECTION AND COMPONENTS CONTRIBUTION

The rotation of the connection (fig. 1d) can be defined with reference to the end of the beam,as indicated in figure 2, under the assumption that this cross section remains plane. In reality, because of the yielding induced by the high contact pressures and/or by the interaction with the end plate in the tension zone, this assumption is generally not fulfilled.

Fig. 2

Nevertheless, it seems the most appropriate for its practicality, due to the simplicity and the consistency with the classical beam theory. In the same figure a criterion is graphically defined , which enables a singling out of the contribution to the connection rotation (ϕ_{cn}) of the deformability of the end plate (ϕ_p) and of the bolts (ϕ_b).

TESTS ON EXTENDED END PLATE CONNECTIONS

The geometrical characteristics of the connections tested are reported in figure 3. End plates with extension on one side or on two sides of the beam were considered. Within the same type of connection only the plate thickness was varied, allowing different ratios between the plate and bolts stiffness to be obtained.

The specimen was connected to a counterbeam with negligible deformability. Displacement transducers located as in figure 4 made it possible to determine: a) the connection rotation (four transducers A in fig. 4a); b) the rotation of beam cross section at 300 mm from the plane of the connection (two transducers B in fig. 4a); c) the rotation due to bolt deformation (four transducers C in fig. 4b). Finally, displacement transducers with the sensitive pin entering a concentric hole in the tension bolts (fig. 5) enabled a control on the initial tightening conditions and a following of the bolt axial deformation. Bolts were pretensioned up to a force equal to the 40% of the nominal tensile strength; this state of stress roughly approximates the snug tight condition.

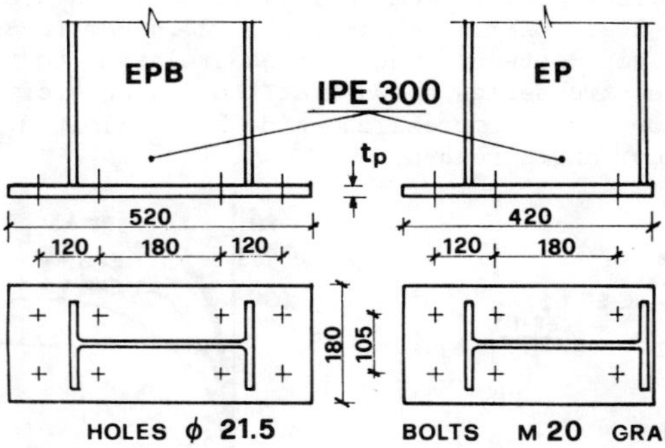

TEST	tp [mm]
EP 1-1 EPB 1-1	12
EP 1-2 EPB 1-2	15
EP 1-3 EPB 1-3	18
EP 1-4 EPB 1-4	22
EP 1-5 EPB 1-5	25

Fig. 3

The specimens were subject to a horizontal force applied to the upper end of the beam stub so to give rise in the connection to bending moment and shear. In order to obtain a realistic appraisal of the connection response, the load is increased by a series of subsequent loading and unloading cycles.

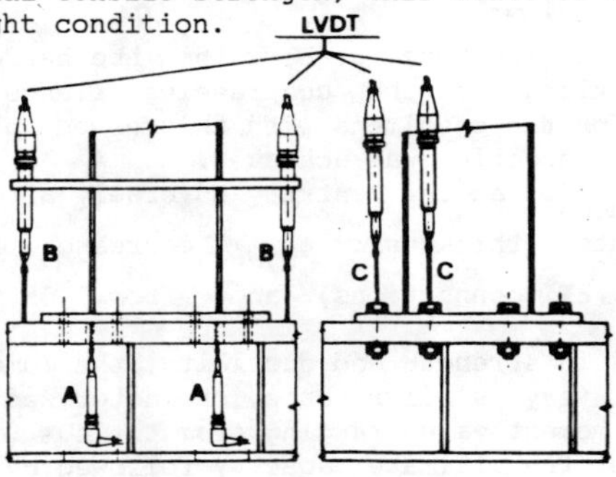

Fig. 4a Fig. 4b

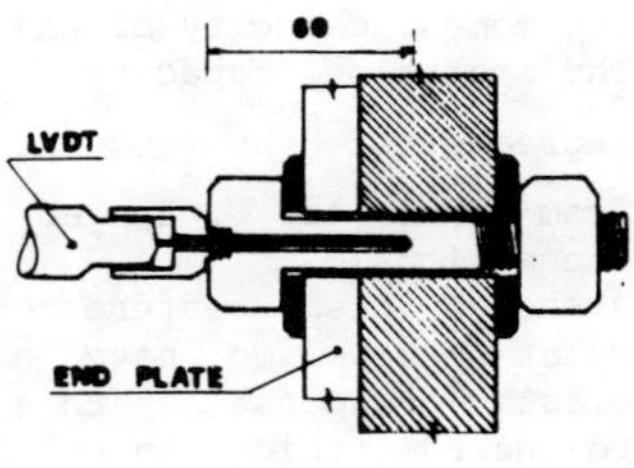

Fig. 5

EXPERIMENTAL RESULTS AND THEIR EVALUATION

The rotational behaviour of the overall connection is described by the moment - rotation characteristics. The figure 6 gather all the M - ϕ_{cn} relationships obtained for the specimens tested, grouped by type of connections. The average value $\bar{M}_{p,b}$ of the plastic moments of the beam is also reported, determined on the base of the experimental yield strength and the actual dimensions of the cross section. It may be remarked that all the connections showed a moment capacity greater than the plastic moment of the beam, though their ultimate strength increased with the end plate thickness from 181.5 kNm (EP1-1) to 225.7 kNm (EP1-5) for the EP tests and from 202.8 kNm (EPB1-2) to 227.5 kNm (EPB1-5) for the EPB tests. A comparison between the curves related to the corresponding specimens of the two series shows that the presence of the extension of the end plate on the compression side has a limited influence on the connection behaviour and strength.

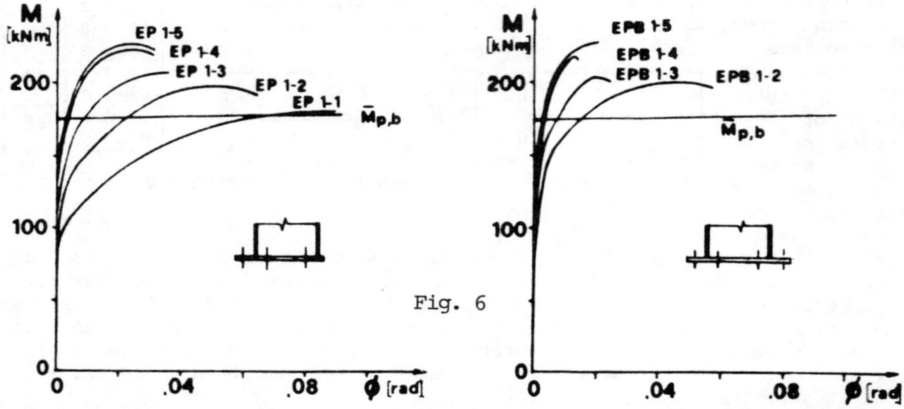

Fig. 6

The significant strains in the beam stub, within the hardening range, caused the inelastic buckling of the compressive flange (and of the adjacent zone of the web for the specimens with thicker end plate) before the moment capacity of the connection was achieved.

The rotation capacity $\phi_{u,cn}$ is at the contrary adversely affected by the increase of the end plate thickness: $\phi_{u,cn}$ decreases respectively from 0.08 to 0.02 rad (EP connections) and from 0.05 to 0.01 (EPB connections).

Although these differences in strength and ductility, the moment rotation relationships do have a very similar shape, characterized by a linear elastic response up to a moment value ranging from the 40% (thin plates) to the 60% (thick plates) of the ultimate capacity followed by a behaviour remarkably nonlinear. The shape and the extension of the curve in the latter range depends upon the connection component, which principally is affected by yielding. When the end plate is relatively thin, the connection rotation in the nonlinear range is mainly due to the elastic-plastic bending of this component (fig. 7): for the connection EP1-1 the end plate contributes more than the 90% of the overall rotation. The rotation capacity of the whole connection is then substantially affected by the local ductility of the tension zone of the end plate. This

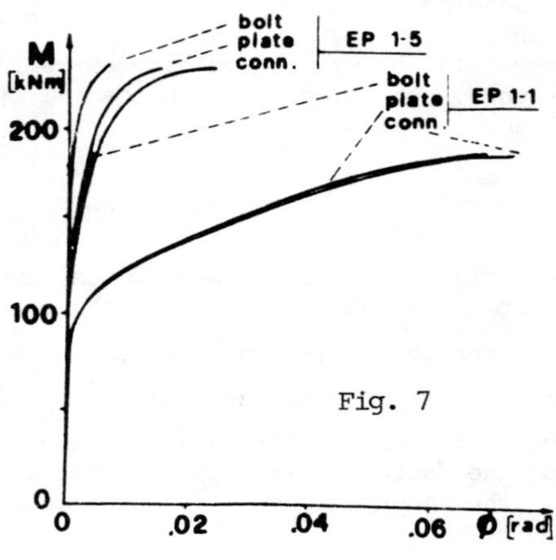

Fig. 7

ductility is generally very high, but may be considerably reduced as a consequence of the process of welding the beam flange and the plate, which modifies the material characteristics in a zone were important plastic strains should develop. The contribution of the bolts becomes more and more significant (up to about 35% for the test EP1-5) with the increase of the end plate thickness. The limited ductility, typical of high strength bolts, tends then to affect negatively the rotation capacity of the overall connection.

The measures related to the bolts elongation pointed out some interesting features of their behaviour; in particular, as it appears from the curves in figures 8 and 9, related to the average deformation histories of the bolts external and internal to the beam stub respectively for the connections EP1-1 (t_p = 12mm) and EP1-5 (t_p = 25mm),:

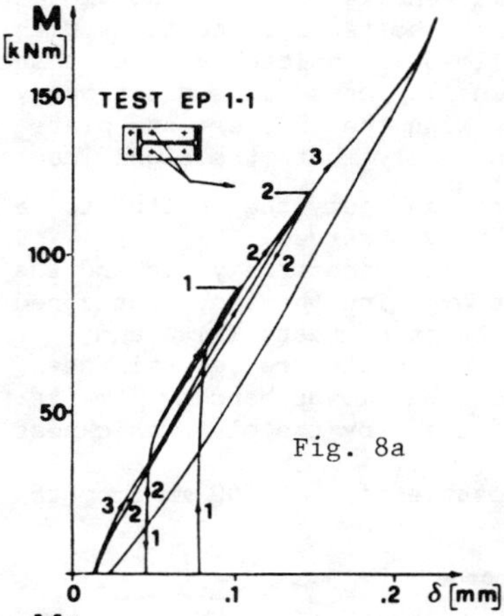

Fig. 8a

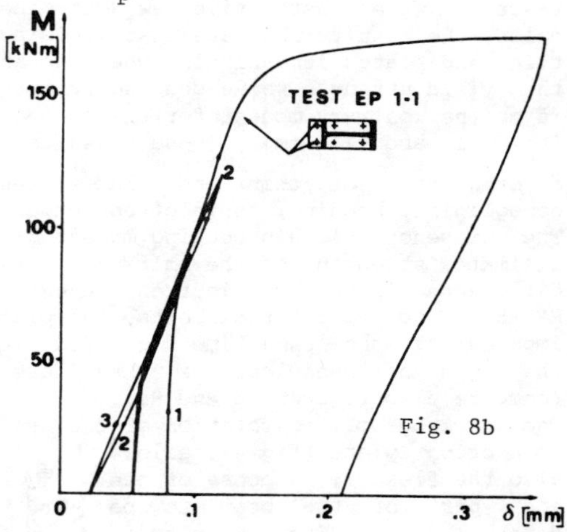

Fig. 8b

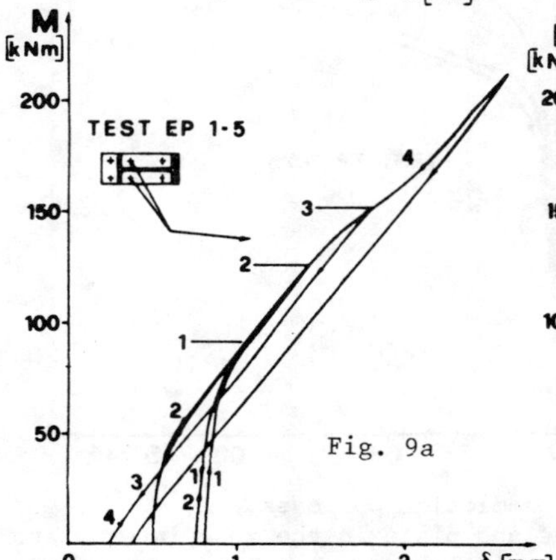

Fig. 9a

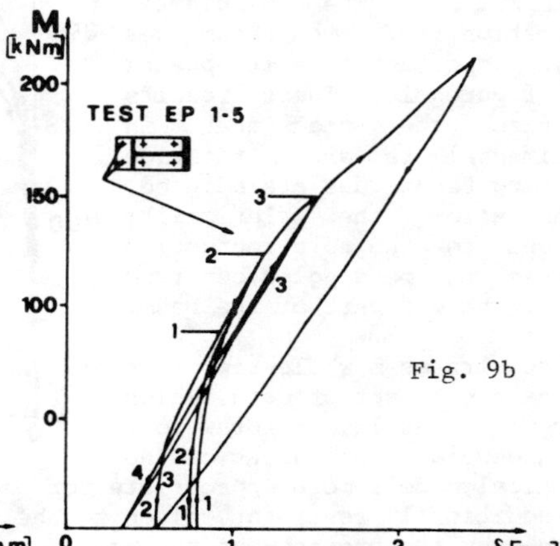

Fig. 9b

a) The initial preload significantly decreases, and often tends to vanish, since the very first loading cycles, though the bolt stress remains in the

elastic range well below the elastic limit. This loss of preloading seems then to depend principally upon the permanent deformation of the washers.
b) The bolts internal to the beam are substantially more stressed than the external bolts. However, the increase of the end plate thickness reduces this difference in the stress state.
c) The end plate thickness affects the type of the bolt stress state, as a comparison between the curves related to the external bolts (figg. 8b for the test EP1-1 and 9b for the test EP1-5) already suggest. The external bolts in the specimen EP1-1 experienced remarkable plastic deformations, while their behaviour in the specimen EP1-5 shows a limited nonlinearity, although the ultimate strength of the connection is attained under a higher applied moment. If an attempt is made to determine the axial force in the bolt by relating the deformation measured during the connection tests to the constitutive law determined via tensile tests on individual bolts, the results indicate that yielding in the bolts of connections with thin end plates is beginning when the axial load is noticeably lower than the yield strength value deduced from the tensile tests. This discrepancy is on the contrary modest for the connections with the thickest end plates (t_p= 21 and 25 mm). These results point out that the significant deformations of thin end plates tend to subject the bolts to a nonegligible bending, as in effect experimentally observed.
The presence of this bending moment greatly influences the yield and the ultimate strength of the bolt and therefore explains the above mentioned differences in the behaviour of the external bolts of tests EP1-1 and EP1-5. As to the internal bolts, the plate, due to its greater stiffness, imposes at the same time higher axial forces but lower bending moments; the type of behaviour is then less affected by the plate thickness (compare also figures 8a and 9a).
The knowledge of the rotation of the beam cross section at 300 mm from the connection plane (fig.4a) allows also the flexural response of the end part of the beam to be determined. The plastic deformation of the beam was always important, as it appears from figure 10. These results emphasize the need that, in experimental research, rotation measuring techniques are adopted, which allow the flexibility inherent to the sole connection response to be singled out from that of the end part of the beam.

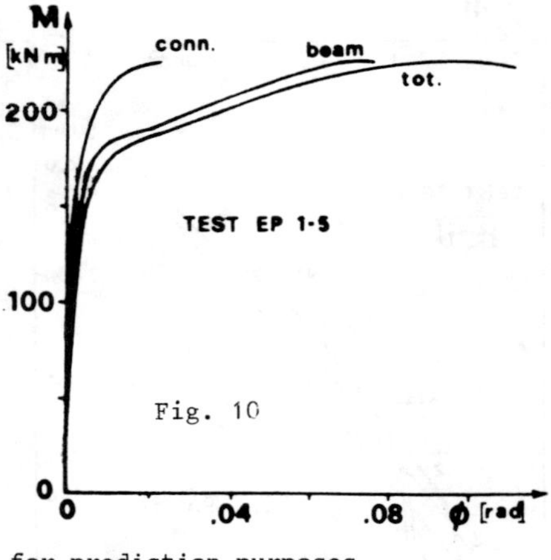

Fig. 10

The results as a whole represent a consistent set of data, which makes it possible to draw some first general conclusions on the mechanical models more appropriate for prediction purposes.
The additional restraint, given to the end plate in the zone internal to the beam by the presence of the beam web, influences to a large extent the connection behaviour. Such a restraint causes a nonnegligible dissymmetry, with respect to the beam tension flange,both of the plate deformation and of the stress state of the same plate and of the bolts. Consequently, the models which symmetrize the plate and bolts behaviour,

such as the T stub model, may not always be able to predict the connection response with satisfactory accuracy. In particular, they cannot (at least they cannot directly) recognize the different stress state of the internal and external bolts. The complexity of the observed behaviour suggests that a greater accuracy can be achieved by models, which account in a more realistic way for the end plate behaviour, by dealing separately with the external and the internal part. The former part can be simply modelled as a cantilever element clamped at the beam flange and subject to the bolt reaction and to the possible prying forces. The part internal to the beam has to be considered as a plate with appropriate edge restraints. A first validation of this modelling procedure is given by the results related to the prediction of the connection plastic moment (see the folowing section). The plastic mechanism presented in figure 11 (and obtained by modifying the proposal of reference 9) allows this moment to be determined with an error lower than the 9%. For thick end plates the plastic mechanism of the internal part of the plate may not be activated, because it would correspond to a force in the bolt greater than its tensile strength. A further improvement may be obtained, by refining, on the base of the experimental data, a method proposed by one of the authors [10].

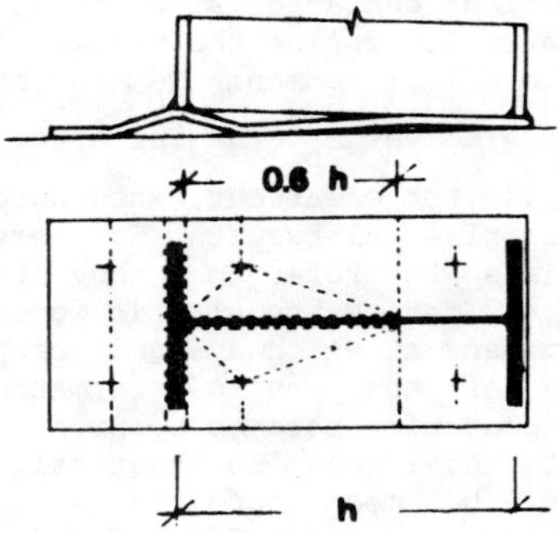

Fig. 11

MODELLING OF THE M - Φ CHARACTERISTICS

The use of simple mechanical models in order to predict the moment rotation relationships requires the definition of the parameters which can be regarded as "characteristics", i.e. suitable to describe the rotational behaviour of the connection. The model accuracy is then checked on the base of its capability of evaluating these parameters with satisfactory precision. The experimental results were therefore thoroughly analysed with the aim at defining the characteristic parameters for end plate connections.
Figure 12 shows the rotational response of a specimen to the first loading cycles. The effect of the loss of the preloading of the bolts on the overall connection behaviour is apparent. In the initial phase, two different stiffness values characterize the response related respectively to the conditions of plate in contact (k_i) and of plate separated from the flange of the counterbeam ($k_{i,red}$). The loading (and unloading) branch defined by the stiffness k_i shortens up to vanish after a few cycles. The stiffness $k_{i,red}$ (which remains almost constant for all the loading and unloading branches) characterizes completely the elastic branches of the subsequent cycles. The inelastic behaviour is highly nonlinear and shows a remarkable strain hardening; however, it seems possible to identify a characteristic value of the stiffness, k_{st}, defined as the mean value of the stiffnesses of all the inelastic parts of the loading branches (fig. 12).

The values of the stiffnesses k_i, $k_{i,red}$ and k_{st} determined for the connections tested are reported in table 1, where also the contributions of the end plate and of the bolts are singled out.

In order to describe completely the connection characteristic it is also necessary to define the values of the elastic limit moment M_e and of the plastic moment M_p. The latter definition is not clear-cut, when based on the experimental results. It seems appropriate to refer to a conventional value and define the plastic moment as the moment at which a significant increase of the plastic deformation of the connection becomes apparent. This moment level may be practically obtained by means of the stiffness parameters previously defined. Two boundary values of M_p ($M_{p,inf}$ and $M_{p,sup}$) may then be determined on the base respectively of the initial stiffness k_i and its reduced value $k_{i,red}$, following the procedure illustrated in figure 13. The two evaluations of the plastic moment differ only slightly for the connections tested; the mean value was then assumed and is reported in table 2. The table 2 gathers the values of M_e and M_p; the ultimate moments M_u of the connection, the plastic moment of the beam stub $M_{p,b}$ and the rotation ϕ_u corresponding to M_u are also reported.

The knowledge of the above mentioned parameters enables a modelling of the connection behaviour as illustrated in figure 14. However, representation with higher level of accuracy can also be used.

An analysis of tables 1 and 2 provides useful indications for the development of mechanical models suitable for predicting the response of the connection.

In table 1 the contributions to the connection stiffness supplied by the

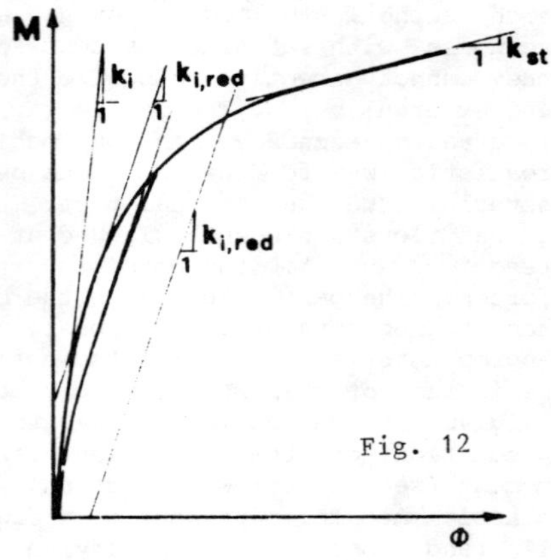

Fig. 12

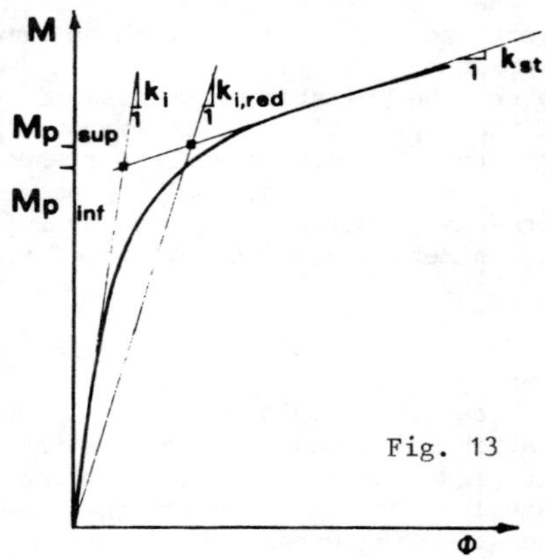

Fig. 13

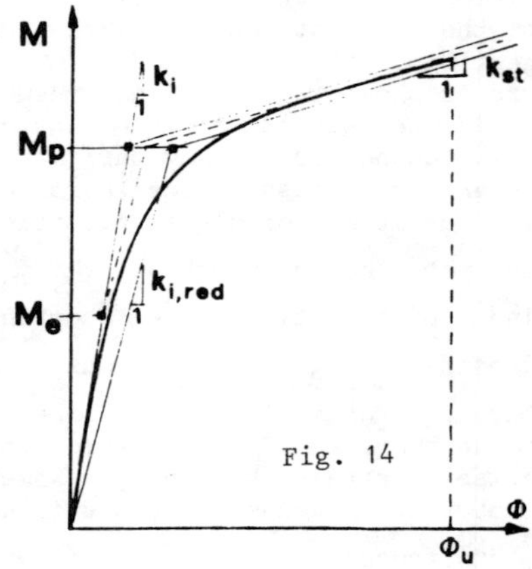

Fig. 14

end plate and by the bolts are singled out. These contributions were computed on the basis of the M - ϕ relationships determined for the single components (fig. 10). The overall stiffness, defined with reference to the moment-rotation curve of the whole connection, was verified to be in very good agreement with the value deduceable by considering the two component as arranged "in series". Values of k_i are mean values related to the cycles with maximum applied moment lower than the elastic limit moment. They are influenced by the bolt preloading and by initial contact conditions. As to the reduced stiffness $k_{i,red}$, it may be noticed that: a) the bolt contribution depends on plate stiffness and increases with this parameter; b) the increment of the plate contribution is less significant than expected on the basis of the variation in thickness, indicating an important influence of the plate restraint conditions, which change with the plate flexibility (also in relation to the bolts flexibility); c) in all cases the plate deformation contributes more to the joint flexibility; however the bolt contribution is never negligible. Finally, as to stiffness k_{st} it may be noticed that: a) no correlation exists between the bolt contribution and the plate thickness; b) plate deformation is fundamental for the assessment of the overall stiffness.
Elastic limit moments $M_{e,cn}$ reported in table 2 range between the 55% and the 80% of the plastic moment $M_{p,cn}$ and from the 40% and the 62% of the ultimate moment of the connection $M_{u,cn}$. Plastic moments are an almost constant percentage (between the 70% and the 80%) of the ultimate moment capacity and are close to the plastic moment of the beam.
Furthermore, if results are evaluated by component, as done for the stiffness parameters, the analysis of the M-ϕ curves enables a determination of the levels of moment, for which the onset of yield is achieved for the plate and for the bolts. Moment levels at which a sharp increase of the plastic deformation shows up for the two components can also be defined.

CONCLUDING REMARKS

The capability of assessing the contribution given by each joint component, is important for the development of methods suitable for the prediction of the joint response. An experimental procedure allowing the collection of the needed data for the setting up of such an approach with reference to the sole connection is proposed.
The procedure was applied in a series of tests on end plate connections; the response of each component (i.e. the end plate and the bolts) was singled out together with its influence on the connection behaviour.
The results evaluation made it possible to determine the main stiffness and strength parameters for the modelling of the connection behaviour. The parameters "characteristic" of the overall M - ϕ relationship can be obtained by means of simple mechanical models for the single components. The present phase of the research is devoted to the definition of such models

ACKNOWLEDGEMENTS

The research is made possible by a grant of the Italian Ministry of Education (M.P.I.).

REFERENCES

1. Davison J.B., Kirby P.A. and Nethercot D.A., Column Behaviour in PR Construction - Experimental Behaviour, ASCE Convention, New Orleans, September 1986.
2. Zandonini R., Steel Frames with Semi-Rigid Joints, Proc. 2nd Regional Colloquium on Stability of Steel Structures, Vol. I/2, Budapest, September 1986, pp. 373-380.
3. Anderson D., Bijlaard F.S.K., Nethercot D.A. and Zandonini R., Analysis and Design of Steel Frames with Semi-Rigid Connections, IABSE survey, to appear.
4. Cosenza E., De Luca A. and Faella C., Inelastic Buckling and Postcritical Behaviour of Steel Structures with Semi-Rigid Joints, Proc. 2nd Regional Colloquium on Stability of Steel Structures, Vol. I/2, Budapest, September 1986, pp. 279-285.
5. Bijlaard F.S.K. and Zoetemeijer P., Influence of Joint Characteristics on the Structural Response of Frames, In Steel Structures; Recent Research Advances and Their Applications to Design, ed. M.N. Pavlovic, Elsevier Applied Science Publishers, London, 1986, pp. 109-134.
6. Yee Y.L. and Melchers R.E., Moment-Rotation Curves for Bolted Connections, J. of Structural Engineering, ASCE, vol. 112, ST3, March 1986.
7. Klein H., Das elastisch - plastische Last-Verformungsverhalten M-Φ steifenloser, geschweisster Knoten fur die Berechnung von Stahlrahmen mit HEB-Stutzen, Doctoral Thesis, University of Innsbruck, 1985.
8. Zandonini R. and Zanon P., Semi-Rigid and Flexible Connections: an Experimental Investigation, Proc. of the Int. Conf. Steel Structures; Recent Research Advances and Their Applications to Design, Part I, Budva, Yugoslavia, September 1986, pp.273-282.
9. Surtees J.O. and Mann A.P., End Plate Connections in Plastically Designed Structures, Proc. Conf. on Joints in Structures, Vol.1, University of Sheffield, July 1970.
10. Zanon P., Bolted Connections: the Ultimate Limit State Behaviour of End Plates, Costruzioni Metalliche, No. 1, 1985.

TABLE 1

	k_i (1),(3)			$k_{i,red}$ (1)			k_{st} (1)		
TEST	Plate	Bolts	Conn.	Plate	Bolts	Conn.	Plate	Bolts	Conn.
EP 1-1	330	893	244	51	169	39	1.22	19.80	1.15
-2	353	850	249	64	146	45	1.60	10.00	1.38
-3	383	950	273	86	156	56	3.28	14.50	2.67
-4	408	872	277	99	186	65	5.20	13.50	3.76
-5	476	973	320	134	189	78	5.70	11.36	3.79
EPB 1-2 (2)	415	788	272	75	144	49	1.77	8.00	1.45
-3	380	748	252	128	166	72	3.67	10.60	2.72
-4	441	894	295	148	202	85	8.00	9.20	4.29
-5	452	745	281	172	216	96	7.40	11.50	4.52

(1) All stiffnesses k are expressed in (KNm/rad) x 10^{+3}

(2) EPB 1-1 is not reported because of a premature failure due to bolt stripping

(3) Values for tests EPB are obtained neglecting the effect of the initial bow of the end plate below the compression flange.

TABLE 2

	beam[1]	Connection [1]								
TEST	$M_{p,b}$	$M_{e,cn}$	$M_{e,cn}/M_{u,cn}$	$M_{e,cn}/M_{p,cn}$	$M_{p,cn}$	$M_{e,cn}/M_{u,cn}$	$M_p/M_{p,b}$	$M_{u,cn}$	$M_{e,cn}/M_{p,b}$	$\phi_{u,cn}$
EP 1-1	176.1	73.6	0.406	0.622	118.4	0.652	0.672	181.5	1.03	0.08
1-2	175.8	81.8	0.410	0.555	147.3	0.738	0.838	199.5	1.13	0.05
1-3	170.6	98.1	0.480	0.628	156.3	0.765	0.916	204.4	1.19	0.03
1-4	173.3	130.8	0.588	0.765	170.9	0.768	0.986	222.4	1.28	0.02
1-5	176.7	139.0	0.616	0.781	177.9	0.788	1.007	225.7	1.28	0.02
EPB 1-2	175.8	89.9	0.443	0.558	161.2	0.798	0.917	202.8	1.15	0.05
1-3	170.6	98.5	0.482	0.591	166.7	0.816	0.977	204.4	1.19	0.02
1-4	173.3	106.3	0.485	0.608	174.7	0.797	1.008	219.1	1.26	0.01
1-5	176.7	122.6	0.539	0.705	173.9	0.764	0.984	227.5	1.29	0.02

(1) Moments M are expressed in KNm, rotations ϕ are expressed in radiants.

BOLTED END-PLATE CONNECTIONS IN STEEL STRUCTURES

Henning Agerskov
Department of Structural Engineering
Technical University of Denmark
DK-2800 Lyngby
Denmark

ABSTRACT

High-strength bolted end-plate connections in steel structures have been studied in several investigations at the Technical University of Denmark. The present paper gives a short summary of the main results obtained in these investigations. In an analytical investigation of end-plate connections in beam sections, a design method was developed, and the basic equations of this method are given in the paper. Special attention was paid to studying the effect of prying action in the connection. Also T-stub connections and end-plate connections in circular sections were studied to clarify the strength and stiffness characteristics. For all three types of connections, experimental investigations were carried out to verify the analytical results obtained.

INTRODUCTION

In recent years, extensive and increasing use has been made of high-strength bolted end-plate connections in steel structures. The situation today is that this type of connection is presumably the most widely used beam-to-column connection in building structures.

However, full utilization of end-plate connections requires an understanding of the behaviour of the connection when subjected to external loading. Due to a number of complications, e.g. the effect of bolt pretension, prying action, and the many parameters governing the total behaviour of the connection, end-plate connections have been studied in a large number of investigations during the last 10-15 years.

The present paper gives a short summary of the analytical and experimental investigations of bolted end-plate connections, carried out at the Department of Structural Engineering of the Technical University of Denmark. End-plate connections in beam sections, as those shown in Fig. 1, as well as T-stub connections and end-plate connections in circular sections, were studied.

One of the problems investigated was the development of prying forces in end-plate or T-stub connections when subjected to tension.

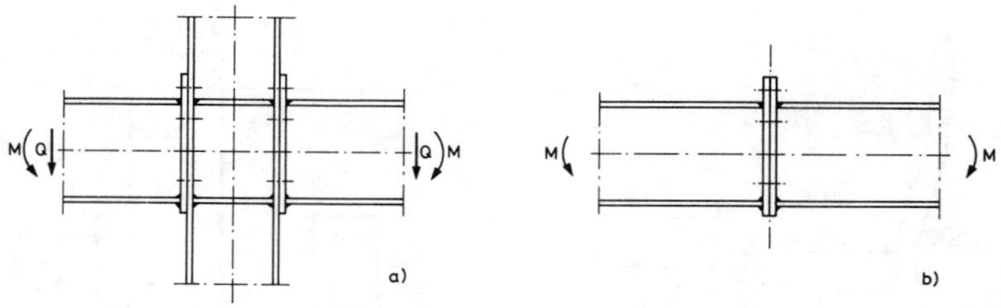

Figure 1. Beam-to-column and beam-to-beam end-plate connections.

Very divergent answers to the question of the size of the prying forces have been given over the years, ranging from a total negligence of prying forces to more than a 100 % increase in bolt forces due to prying action.

END-PLATE CONNECTIONS IN BEAM SECTIONS

In the investigation of end-plate connections in beam sections, a theory was developed that may be used to determine the yield load of the connection [1,2]. On the basis of this theory, the influence of the different variables affecting the behaviour of the connection was studied in an investigation in which a large number of connections were analysed on computer. This analysis showed that diagrams for direct use in the design could be established.

The purpose of the tests carried out in the experimental investigation was to study the strength and deformation characteristics of bolted end-plate connections. The experimental results show good agreement with the results obtained from the proposed design method.

Basic Equations of Proposed Design Method

In the following, a short survey is given of the basic equations resulting from the theory developed. With respect to the derivation of the equations and the basis of the theory, see Refs. [1,2].

The analytical model of the connection under consideration is shown in Fig. 2. The external tension in the beam flange is denoted by 2F and the bolt force by B. A prying force of magnitude Q is assumed to act at the end of the plate. Forces F and Q are tributary to each bolt. Fig. 3 shows the geometry of the bolts used in the connection.

The yield load of the connection is considered as the load at which the reduced yield moment is reached in the end-plate at the toe of the weld. Three basic cases are possible in the theory. In the first, separation takes place in the bolt lines before the reduced yield moment is reached. The situation shown in Fig. 2 corresponds to this case. In the second case, the reduced yield moment is reached before separation occurs at the bolt lines. The difference between the first and second case is that in the latter, a compressive force will exist between the plates in the bolt lines at the instant at which the reduced yield moment is reached. In the third case, which may arise when a relatively heavy end-plate is used, the bending deformation of the plate is so small that separation also occurs at the edges of the plate at the yield load, meaning that no prying forces exist.

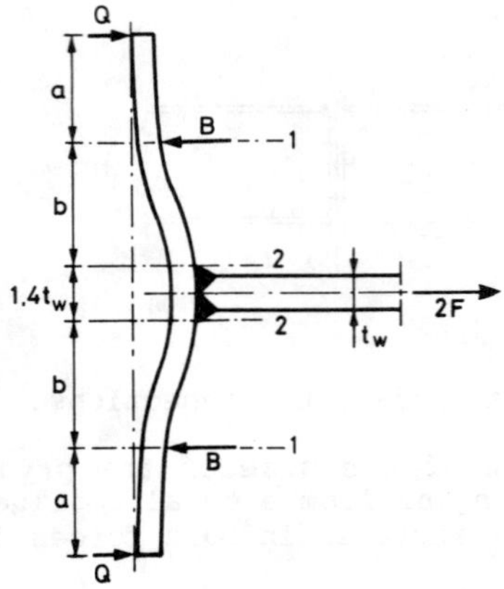

Figure 2. Analytical model of connection.

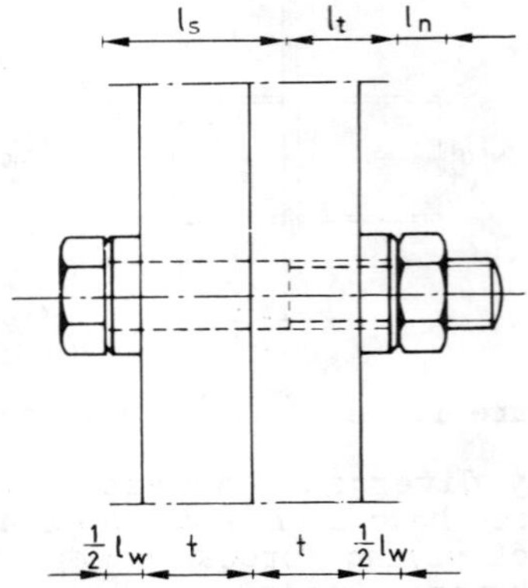

Figure 3. Geometry of bolts used in connection.

Fig. 4 shows the forces acting on the end-plate at the yield load of the connection in the three basic cases.

a. Separation before Yield Moment

The two basic equations are

$$Fb - Qa = \frac{1}{4}wt^2 \sqrt{\sigma_y^2 - 3\left(\frac{F}{wt}\right)^2} \tag{1}$$

and

$$\frac{1}{10}\frac{B_0 t}{A_s} + \frac{B - B_0'}{A_s}\left(\frac{1}{2}k_1 + k_4\right) = \frac{l^3}{wt^3}(F\alpha_1 - B\alpha_2) \tag{2}$$

Eq. 1 expresses the fact that the moment at the toe of the weld at the yield load must be equal to the moment corresponding to full plastification of the section. Eq. 2 states that the deflection in the bolt line of the middle surface of the end-plate must be the same in deformation of the plate as in deformation of the bolt.

According to Fig. 2, the distance from the edge of the end-plate to the bolt line is a and the distance from the bolt line to the toe of the weld is b. The thickness of the end-plate is t, the length of the end-plate tributary to each bolt is w, σ_y is the yield stress of the plate material, B_0 is the initial bolt tension, and B_0' is the bolt force at which separation occurs in the bolt lines. With $k_1 = l_s + 1.43\ l_t + 0.71\ l_n$, $k_2 = k_1 + 0.20\ l_n + 0.40\ l_w$, $k_3 = 0.40\ t$, and $k_4 = 0.10\ l_n + 0.20\ l_w$, B_0' is found from $B_0' = B_0(1 + k_3/k_2)$.

The lengths l_s, l_t, l_n, and l_w, referring to the bolts used in the connection, are shown in Fig. 3. In Eq. 2, A_s is the area of the bolt shaft, and α_1, α_2 are determined from $\alpha_1 = 3\alpha/2 - 2\alpha^3$ and $\alpha_2 = 6\alpha^2 - 8\alpha^3$, in which $\alpha = a/l$; and $l = 2(a + b)$.

According to Fig. 4(a), the prying force Q is $Q = B - F$. The unknowns, F, B, and Q can be determined from the above equations. The magnitude of the prying force may be given in dimensionless form by the ratio $\beta = 1 + Q/F$.

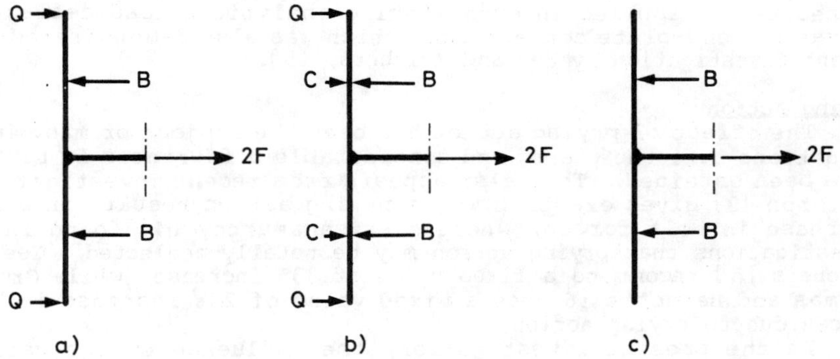

Figure 4. Forces acting on end-plate in the three basic cases: a) separation occurs at bolt lines before yield moment is reached; b) yield moment is reached before separation at bolt lines; and c) no prying action in connection.

b. Yield Moment before Separation

The two basic equations in this case are

$$Fb - Qa = \frac{1}{4}wt^2\sqrt{\sigma_y^2 - 3\left(\frac{F}{wt}\right)^2} \tag{3}$$

and

$$\frac{1}{10}\,\frac{B_0 - C}{A_s}t = \frac{l^3}{wt^3}[F\alpha_1 - (B - C)\alpha_2] \tag{4}$$

in which C is the compressive force between the end-plates at the bolt line, corresponding to the yield load of the connection. The relationship between the bolt force, B , and the plate compression, C, at the yield load of the connection is, [1,2]

$$C = B_0\,\frac{B_0' - B}{B_0' - B_0} \tag{5}$$

The prying force, Q, is Q = B - F - C. Insertion of C and Q in Eqs. 3 and 4 gives two equations, from which B and F may be determined. a, b, w, t, σ_y, B_0, B_0', A_s, l, α_1, and α_2 are as defined in the preceding section.

c. No Prying Action

In this case, the prying force is Q = 0, and thus $\beta = 1 + Q/F = 1.00$. The compressive force between the end-plates at the bolt line is C = 0. With the bolt force, B, equal to the external force, F , the yield load of the connection is determined from

$$Fb = \frac{1}{4}wt^2\sqrt{\sigma_y^2 - 3\left(\frac{F}{wt}\right)^2} \tag{6}$$

in which b, w, t, and σ_y are as defined in the preceding section.

The theory developed includes consideration of the deformations of the end-plates and the bolts. This means that the results obtained may be applied in establishing analytical load-deflection curves for end-plate connections, which was also demonstrated in a recent investigation by Yee and Melchers, [5].

Prying Action

The effect of prying action has been the subject of many investigations over the years, and considerable differences in results have been obtained. This also appears from recent investigations. Thornton [3] gives examples where prying action results in a 100% increase in bolt forces, whereas Krishnamurthy [4] found in his investigations that prying action may be totally neglected. Yee and Melchers [5] recommend a fixed value of 33% increase, while Grundy, Thomas and Bennetts [6] use a fixed value of 20% increase in bolt forces due to prying action.

In the present investigation, the influence of the various factors affecting the magnitude of the prying forces was studied. A computer analysis of a large number of connections (about 2800) was carried out, varying the different parameters within areas of practical interest. From this analysis it appears that, generally, an increase in the prying action in the connection, determined through the value of the ratio β, will result from an increase in: (1) the distance from the bolt line to the toe of the weld, b; (2) the bolt diameter, d_B; and (3) the yield stress of the bolt material, σ_{yB}, i.e., initial bolt tension. Generally, a reduction of the value of β will be the result of an increase in: (1) the distance from the edge of the end-plate to the bolt line, a; (2) the length of the end-plate tributary to each bolt, w; (3) the plate thickness, t; and (4) the yield stress of the plate material, σ_y.

On the basis of the results obtained in the computer analysis, combinations of the various parameters were investigated in order to establish diagrams giving the prying ratio as a direct function of certain combinations of the independent variables. It was ascertained that the prying ratio may be determined, with sufficient accuracy for practical purposes, from the values of the ratio b/a and a dimensionless quantity, γ, defined as

$$\gamma = \frac{wt^2\sigma_y}{bd_B^2\sigma_{yB}} \qquad (7)$$

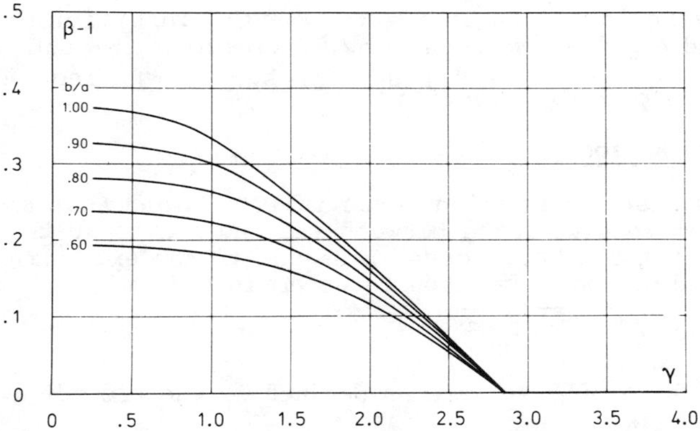

Figure 5. Prying ratio at yield load of connection.

The curves of the diagrams resulting from this investigation have been combined in a single diagram, shown in Fig. 5. The curves give the prying ratio at the yield load of a connection with specified values of b/a and γ. The curves have been drawn for values of $\gamma \geqq 0.25$. Smaller values of γ are of no practical interest. For $\gamma \geqq 2.85$, no prying appears in the connection, and $\beta = 1.00$.

Experimental Investigation

In the experimental part of the investigation of end-plate connections in beam sections, four series of tests were carried out. In these, high-strength bolted end-plate connections in rolled IPE- and HEB-sections were investigated. In all the tests, the bolt forces were determined by means of strain gages. The tests showed that the tensile force in the connection may, with sufficient accuracy for practical use, be considered as uniformly distributed over all the bolts in the tensile zone. Furthermore, a comparison of the test results with the results from the suggested design method showed good agreement between the values obtained, the deviation in bolt forces being, in most cases, only a few per cent.

The load-deflection behaviour of the connections was also studied in the investigation, and load-deflection curves are available from all the tests carried out. With respect to details of the experimental investigation and the results obtained, reference is made to Refs. [1,2].

T-STUB CONNECTIONS

The investigations of T-stub connections carried out at the Technical University of Denmark were part of the study of the tensile zone of end-plate connections. Two series of tests on high-strength bolted T-stub connections with either 4 or 8 bolts were performed. In the tests, the bolt forces, the load-deflection curves, the yield load, and the ultimate load of the connections, were determined and compared with corresponding analytical results.

One of the main aims of the experimental part of the study was to clarify whether end-plate connections may, with sufficient accuracy for practical purposes, be designed as T-connections, with the end-plate behaving as the T-stub flange and the beam flange as the T-stub web. The tests of the present investigation showed this to be the case. For further details of the results obtained in the investigations on T-stub connections, see Refs. [1,2,7].

END-PLATE CONNECTIONS IN CIRCULAR SECTIONS

Bolted connections in circular sections - solid round bars or tubes - may also advantageously be carried out as end-plate connections. In recent years, extensive use has been made of circular sections in the construction of steel lattice structures, e.g. in the form of towers (television, radio, and high voltage transmission towers), roof structures (space structures and lightweight plane trusses), and cranes (tower cranes and overhead travelling cranes).

The aim of this investigation has been to clarify the strength and stiffness characteristics of high-strength bolted end-plate connections in circular sections. The analytical and experimental inve-

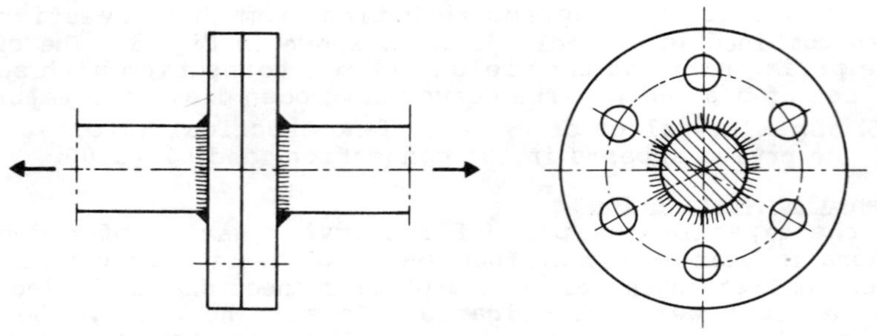

Figure 6. Bolted end-plate connection in circular section.

stigations carried out concentrated on connections of a type as shown in Fig. 6.

In the analytical investigation, a theory was developed that permits the analysis of bolted end-plate connections in circular sections, including determination of the yield load of the connection, bolt forces, the effect of prying action, etc.

In the experimental part of the study, a test series was carried out using bolted end-plate connections with varying round bar diameter and end-plate thickness. The stresses in the end-plate and the bolt forces were determined by means of strain gages.

A comparison of the results obtained from this test series with those obtained by means of the suggested design method shows little difference between the values obtained, the maximum deviation in bolt forces being only about 8%.

The yield load for the connection, as defined in the suggested design method, is also in this case considered to be the load at which the reduced yield moment is reached in the end-plate at the toe of the weld. The use of this definition of the yield load will result in some strength reserve due to strain hardening, but it also means that heavy plastic deformations of the end-plate will be avoided. In the connections with thin end-plates, the ultimate load was found to be more than 100% greater than the theoretical yield load. For the connections with heavier end-plates, the difference was found to be 25-75%.

It is the intention to continue the present investigation of end-plate connections in circular hollow sections.

More details about this investigation and the results obtained can be found in Refs. [8,9,10].

CONCLUSIONS

In the investigations carried out at the Technical University of Denmark various types of end-plate connections have been studied. For the three different connection types (end-plate connections in beam sections, T-stub connections, and end-plate connections in circular sections) analytical investigations have been performed, resulting in design methods that can be used to determine the strength, i.e. the yield load of the connection. In a number of investigations, yieldline methods have been applied, and these methods are useful in determining the ultimate load of end-plate connections.

However, they do not provide any information about the deformations in the connection. The analytical treatment of the present investigations includes consideration of the deformations of the individual connection elements. Thus, these methods can be applied in the analytical determination of load-deflection curves, as also demonstrated recently by Yee and Melchers [5]. Furthermore, the analytical methods of the present investigations may be useful in the study of dynamically loaded structures and of structures subjected to fatigue.

For all connection types studied, test series were carried out to verify the analytical results. Yield loads and ultimate loads were determined, as well as bolt forces and stresses in the end-plates. The development of prying action in the connections was studied, and the deformations of the connections were determined. The test results showed that the proposed design methods provide sufficient accuracy for practical purposes. The experimental results obtained concerning connection deformations may contribute to a data base of load-deflection curves, if established.

REFERENCES

1. Agerskov, H., High-Strength Bolted Connections Subject to Prying, Journ. of the Struct. Div., ASCE, Vol. 102, No. ST 1, Jan. 1976.

2. Agerskov, H., Analysis of Bolted Connections Subject to Prying, Journ. of the Struct. Div., ASCE, Vol. 103, No. ST 11, Nov. 1977.

3. Thornton, W.A., Prying Action - a General Treatment, Engineering Journal, AISC, 2nd Quarter, 1985.

4. Krishnamurthy, N., Loading and Evaluation of Splice-Plate Connections, ASCE Convention, Atlanta, Oct. 1979, Preprint No. 3696.

5. Yee, Y.L. and Melchers, R.E., Moment-Rotation Curves for Bolted Connections, Journ. of Struct. Engrg., ASCE, Vol. 112, No. 3, March 1986.

6. Grundy, P., Thomas, I.R. and Bennetts, I.D., Beam-to-Column Moment Connections, Journ. of the Struct. Div., ASCE, Vol. 106, No. ST 1, Jan. 1980.

7. Agerskov, H., Tests on High-Strength Bolted T-Stub Connections, Report No. R 106, Dept. of Struct. Engrg., Techn. Univ. of Denmark, 1979.

8. Agerskov, H. and Bjørnbak-Hansen, J., Bolted End Plate Connections in Round Bar Steel Structures, Report No. R 147, Dept. of Struct. Engrg., Techn. Univ. of Denmark, 1982.

9. Agerskov, H. and Bjørnbak-Hansen, J., Bolted Connections in Round Bar Steel Structures, Journ. of Struct. Engrg., ASCE, Vol. 111, No. 4, April 1985.

10. Bauingenieur, Kopfplattenverbindungen in zugbeanspruchten Rundstäben, Kurze Technische Berichte nach [9], Vol. 61, No. 6, June 1986.

STRENGTH AND BEHAVIOUR OF IN-PLANE WEAK AXIS JOINTS AND OF 3-D JOINTS

J. Janss
Centre de Recherches de l'Industrie des Fabrications Métalliques (CRIF)
Liege
BELGIUM

J.P. Jaspart
and
R. Maquoi
M.S.M., Institut du Génie Civil, Université de Liège
BELGIUM

ABSTRACT

A limited number of tests have been carried out on weak axis - and on 3-D beam-to-column connections. On base of the first results obtained, some conclusions are drawn.

INTRODUCTION

The behaviour of plane- or space frames may be analysed by means of computer programs. Some of the latter can account for the semi-rigidity of the beam-to-column joints. For beams (fig. 1.a) that are connected to the column flanges (strong axis connection), the information regarding the behaviour of the joints is available. It is largely missing when the beam (fig. 1.b) is connected to the column web (weak axis connection) [1] [2]. The question is still more intricate when 3-D joints (fig. 1.c) because of the interaction in the column web between both strong- and weak axis connections [3]. With a view to collect some data on weak axis - and 3-D connections, some tests were carried out at the University of Liège. It is only reported on those, the results of which have already been fully investigated.

The authors are indebted to IRSIA and ARBED for their financial assistance in this research work.

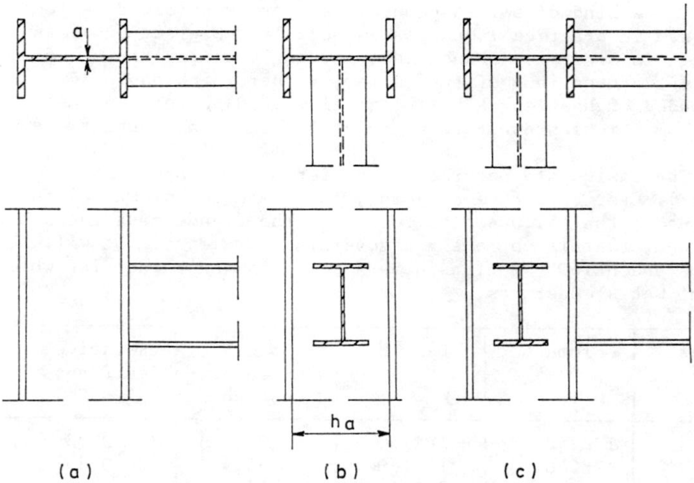

Figure 1 - Types of beam-to-column joints : (a) strong axis connection ; (b) weak axis connection ; (c) 3-D connection.

WEAK AXIS CONNECTIONS.

When defining the test specimens, three main parameters were accounted for : the type of connection, the relative beam-to-column stiffness and the slenderness h_a/a of the column web.

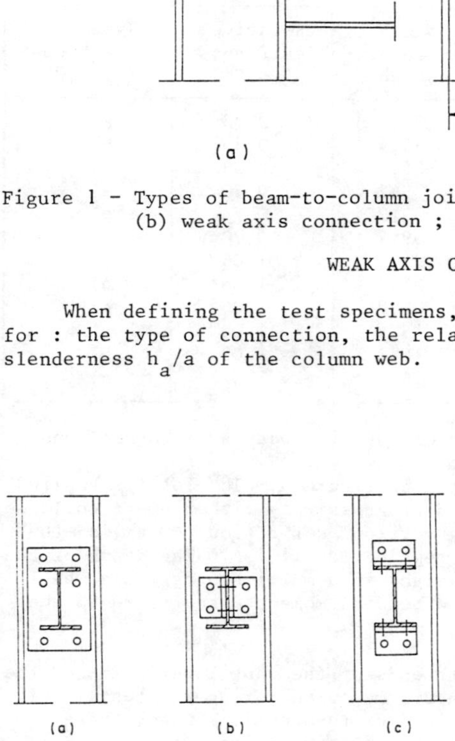

Figure 2 - Types of connections: (a) end plates ; (b) double web cleat (c) flange cleats

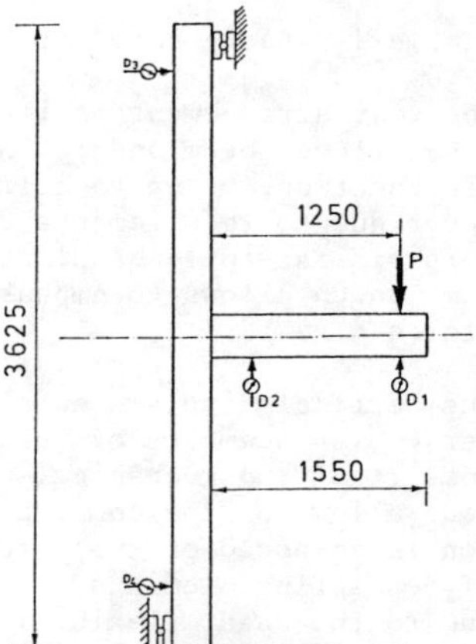

Figure 3 - Test arrangement.

Three kinds of bolted connections were considered because of their current use in practice : end plates, double web cleat and flange cleats (figure 2). In order to exhaust the carrying capacity of the column web and to assess the behaviour of the latter, rather thick angles or end plates and preloaded high strength bolts (quality 10.9) were used. All the mechanical and geometrical properties of columns, beams and connections were measured.

The basic data describing the set of test specimens are summarized in table 1. For each type of joint, four beam-to-column stiffness ratios are examined ; that allows also to change the slenderness ratio h_a/a of the column web, that is obviously a governing parameter. It will be noticed that HE180A and HE160B sections have a same weak axis inertia, while a very different web slenderness.

Test	Column C	Beam B	Column Web-slenderness h_a/a	Relative stiffness B/C	Type of connection (fig. 2)
A1	IPE240	IPE160	35,5	0,33	a)
A2	IPE240	IPE160	35,5	0,33	b)
A3	IPE240	IPE160	35,5	0,33	c)
A4	IPE300	IPE160	39,2	0,70	a)
A5	IPE300	IPE160	39,2	0,70	b)
A6	IPE300	IPE160	39,2	0,70	c)
A7	HEA180	IPE160	25,3	1,06	a)
A8	HEA180	IPE160	25,3	1,06	b)
A9	HEA180	IPE160	25,3	1,06	c)
A10	HEB160	IPE160	16,8	1,02	a)
A11	HEB160	IPE160	16,8	1,02	b)
A12	HEB160	IPE160	16,8	1,02	c)

Table 1 - Basic data for test specimens with weak axis connections

The test arrangement is illustrated in figure 3. The load is applied at the cantilever beam and is increased progressively either up to collapse of the connection, or up to a limiting vertical deflection (40 cm) of the cantilever due to requirements of the testing facilities. The depth of the column is similar to an usual storey height in a multi-storey frame while the beam length allows to emphasize the bending moment with regard to the shear force.

The vertical displacements at both ends of the cantilever beam and the transverse displacements of the column are measured during the tests. With the geometrical and mechanical properties of the sections used, these deflections allow for the computation of the relative joint rotation. This rotation is composed of the rotation of the connection proper (slip, distorsion of connecting elements, elongation of bolt shanks,...) and of the rotation due to the shear flexibility of the column web. Dedicated measurement set-ups allow to distinguish between both rotation components. For instance, figures 4 to 7 are dealing with the double web cleat connection A2; the rotation of the connection itself is divided into one part due to slip and another one resulting from the distorsion of angles.

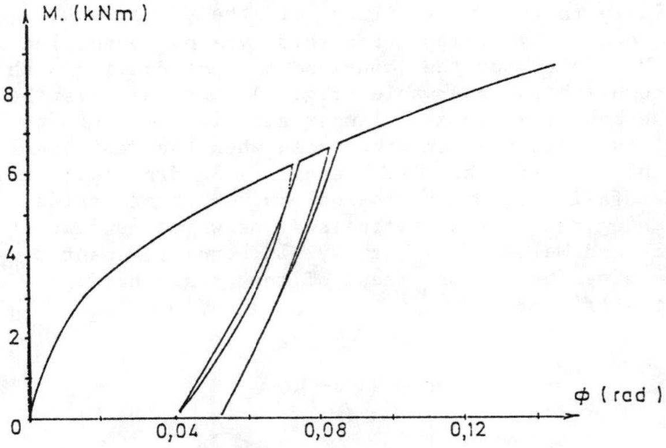

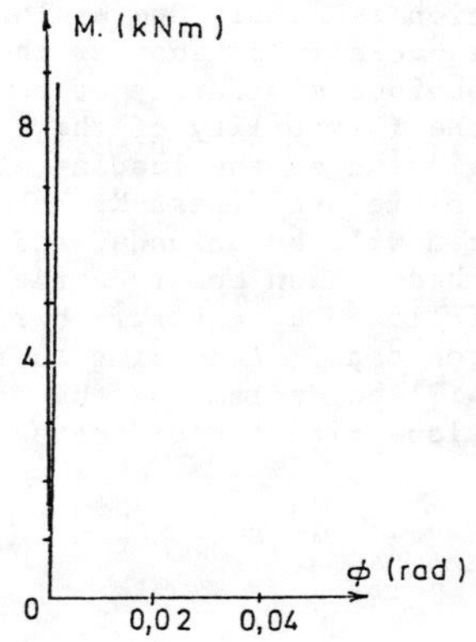

Figure 4 - Test A2: Joint Curve M-φ Figure 5-Test A2: Curve M-φ due to slip

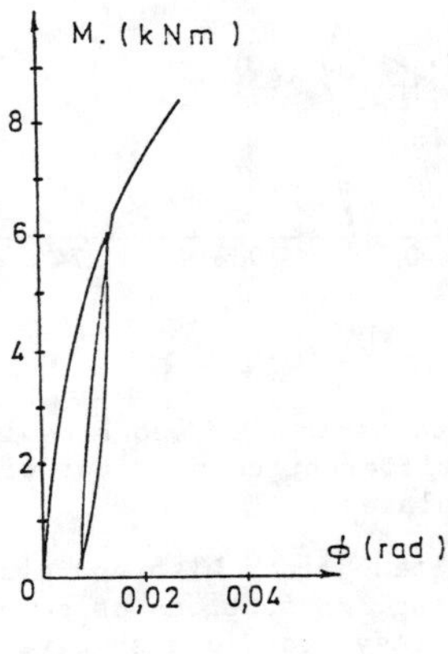

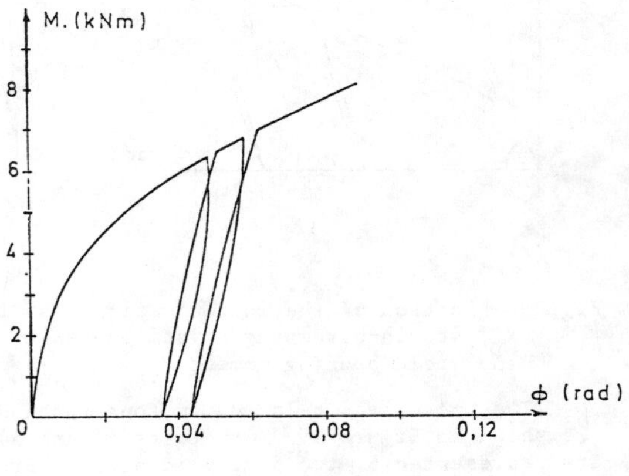

Figure 6 - Test A2: Curve M-φ due to angles distorsion

Figure 7 - Test A2: Curve M-φ due to the flexibility of the column web

The assessment of the results will be limited to the tests on joints with end plates, that are the sole to have been investigated in deep at the present time. A complete report will be prepared by the end of 1987 [5]. Because of their thickness (30 mm), the end plates are so stiff that they may be considered as not deformable. In addition, at the loading levels reached at collapse, the effect of the elongation of bolt shanks can be disregarded with respect to that of column web yielding. Thus the joint rotation is nearly due exclusively to the flexibility of the column web, especially as soon as the onset of yielding. For this type of connection an obvious similarity of the M-ϕ curve of the joint and of that dealing with the flexibility of the column web is observable (fig. 8). At the first beginning of the loading, the behaviour remains linear elastic and exhibits an initial stiffness K_i, that is reproduced and recovered when the test specimen will be unloaded and then loaded again, whichever the loading level reached. Then occurs a rather fast yielding of the column web, that is last followed by a strain hardening range, the stiffness K_p of which is quasi constant. According to Yee and Melchers [4], the yield bending moment M_{ye} will be defined as the one given by the intercept of the strain hardening slope with the ordinate axis (fig. 8).

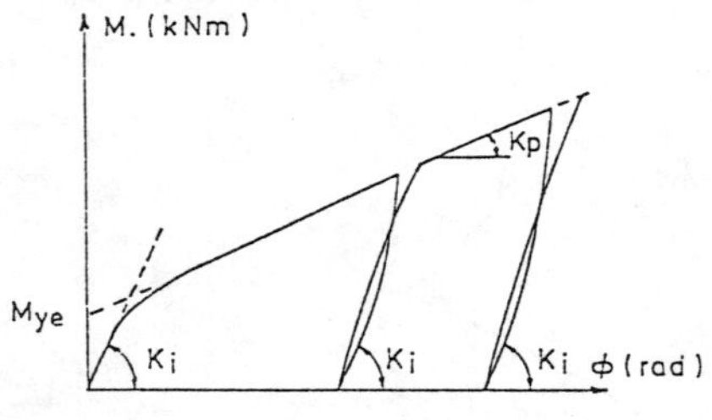

Fig.8-Definition of the initial stiffness K_i, strain-hardening K_p stiffness and yield bending moment M_{ye}.

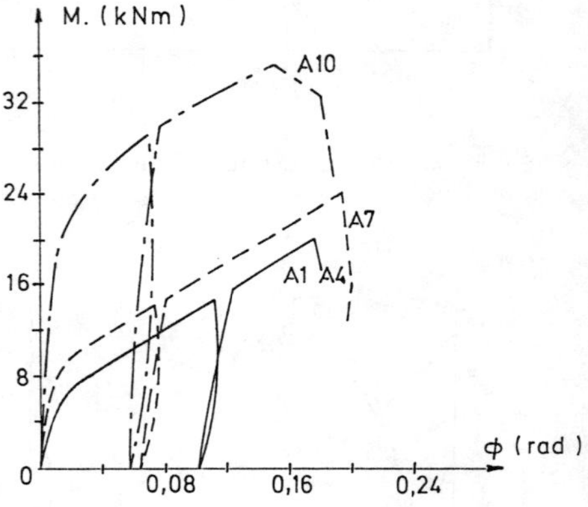

Fig.9-Comparison of web behaviour of different connections with end plates.

The M-ϕ curves related to four tests on bolted joints with end plates are shown in figure 9. Both tests A1 and A4 (using IPE sections as column) are represented by two nearly identical curves ; they provide a same value of the initial stiffness K_i, of the strain hardening stiffness K_p and of the yield bending moment M_{ye}. The strain hardening stiffness K_p looks like approximately constant for the four tests, i.e. whichever the slenderness h_a/a of the column web in the range 16-36. The initial stiffness K_i and the yield moment M_{ye} are shown to increase rapidly when the slenderness h_a/a decreases. The experimental values of M_{ye} are plotted in figure 10 against the web slenderness.

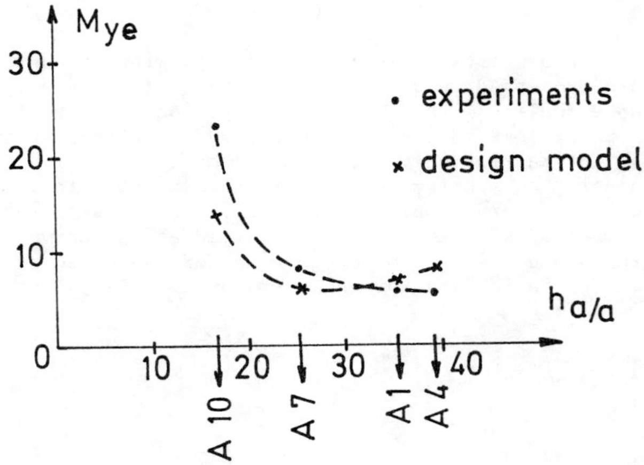

Fig.10 - Yield bending moment vs. slenderness of the column web.

Unlike for strong axis connections, no design model exists - at the authors' knowledge - for weak axis connections, which would provide the ultimate carrying capacity of the joint. By observing the yield lines in the column web, an attempt of establishing such a design model was made (fig.11). Though the effect of the web slenderness is qualitatively well reflected (fig. 10), some improvements of the model are shown necessary with a view to a better quantitative agreement. The consideration of more sophisticated yield lines patterns in the column web as well as the account of membrane force - due to the out-of-plane deflection of this web - in the yield criterion are in progress, so that a realistic design model can be hopefully expected soon.

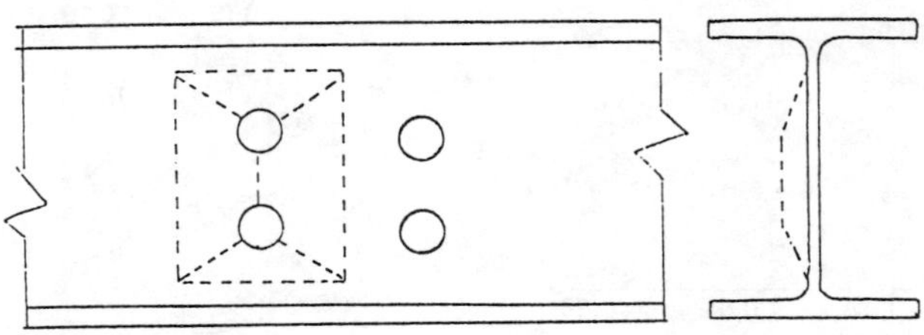

Figure 11 - Test A10 : Yield lines pattern

3-D CONNECTIONS

The results of tests on semi-rigid joints with strong axis connections (serie O) are presented in another paper [6]. It is briefly reported here-above on those dealing with weak axis connections (serie A). For a specified type of test specimen, both sets of results give two limiting points of an interaction curve (fig. 12), the aim of which is to show the carrying capacity of a column web involved in a 3-D joint. Any other point of the interaction curve would require tests on the associated 3-D specimen. A limited number of such tests were performed at the University of Liège;the web slenderness and the type of connection were selected as the governing parameters. The basic data for the test specimens are listed in table 2.

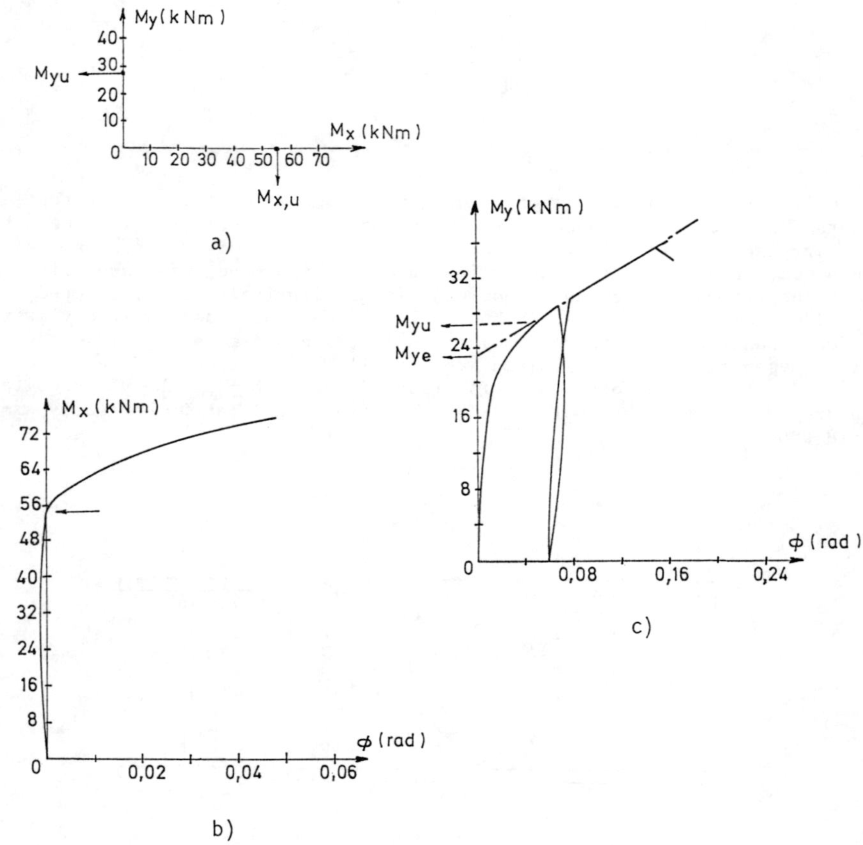

Figure 12 - (a) Test I1:Limiting points of the interaction curve, (b) Curve M-ϕ due to the shear flexibility (Test O4) of the column web ; (c) Curve M-ϕ of the connection (Test A10).

Test	Column	Weak axis beam P_1	Weak axis connection (fig. 2)	Strong axis beam P_2	Strong axis Connection (fig. 2)
I1	HEB160	IPE160	a	IPE200	a
I2	HEB160	IPE160	b	IPE200	a
I3	HEB160	IPE160	c	IPE200	a
I4	IPE300	IPE160	a	IPE200	a
I5	IPE300	IPE160	b	IPE200	a
I6	IPE300	IPE160	c	IPE200	a

Table 2 - Basic data for test specimens with 3-D connections

The test arrangements were quite similar to those used for the tests on joints with weak axis connections. The tests were performed by keeping constant the load P_2 at the end of the weak axis beam while the load P_1 at the end of the strong axis beam is increased progressively. The value of P_2 is associated to the onset of the bending moment M_{ye}, derived from the test on in-plane weak axis connection. Measurements are made similarly to those carried out for the testing of semi-rigid joints with strong axis connections (serie O) [6].

The behaviour of the column web, when a 3-D connection, can then be compared to that when strong axis connection only. Thus the M-ϕ curves due to the shear flexibility of the column web are plotted in figure 13 for test O7 (strong axis connection) and I1 (3-D connection). It is observable that the rotation remains very small till the onset of yielding in the web of the column fitted with a strong axis connection. The effect of shear deformation is much more marked when an additional loaded weak axis beam exists, because of a more progressive yielding of the web. However the influence of the weak axis beam tends to be masked when approaching the ultimate limit state.

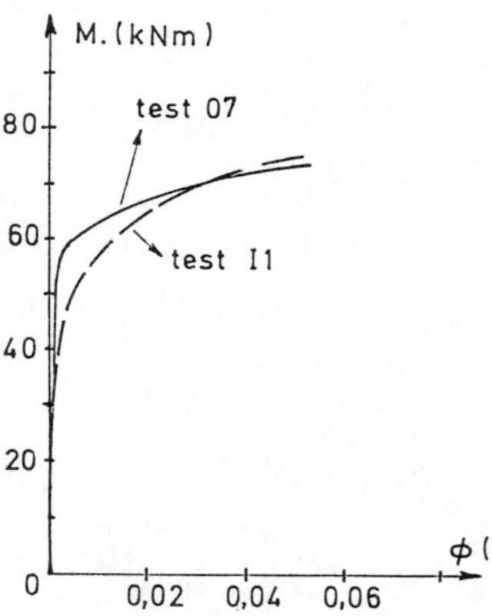

Figure 13 - Comparison of the shear flexibility of the column web between test I1 and test O7.

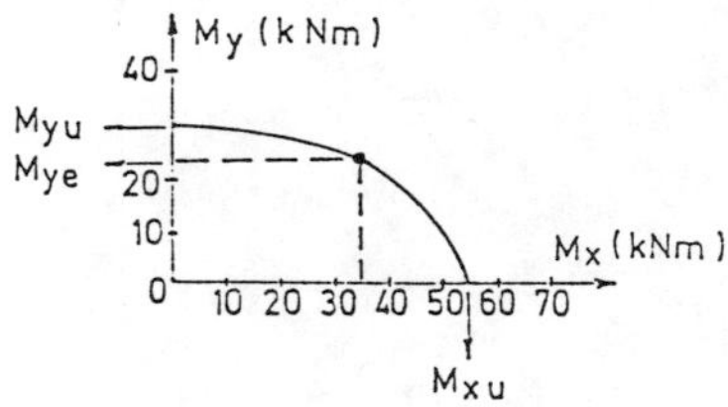

Figure 14 - Interaction curve for a 3-D connection

Such a test on a 3-D joint, carried out as indicated hereabove, enable to determine one additional point of the associated interaction curve and to sketch the shape of this latter (fig. 14). Obviously more experimental points are needed to assess more precisely this interaction curve and to suggest a design model for 3-D joints ; therefore additional tests will be performed soon in order to implement the information.

REFERENCES.

1. Rentschler, G.P. and Chen, W.F., Tests of Beam-to-Column Web Moment Connections. Jl. Struct. Div., A.S.C.E., ST5, Mai, 1980, 1005-1022.

2. Vircik, J., Marek, P. and Hudak, J., Limit States of Carrying Capacity Demonstration of the Reliability Conditions on a Selected Structural Detail. Design Limit States of Steel Structures - 1st International Correspondence Conference - Preliminary Report. Edited by J. Melcher, Brno, 1983.

3. Massonnet, Ch. and Save, M., Calcul Plastique des Constructions. Volume 1 Structures dépendant d'un paramètre, 3d edit., Nélissen Edit., Liège, 439-441.

4. Yee, Y.L. and Melchers, R.E., Moment-Rotation Curves for Bolted Connections. Jl.Struct. Engrg., A.S.C.E., Vol. 112, 3, March 1986, 615-635.

5. Jaspart, J.P., Essais sur assemblages poutre-colonne d'axe faible et tridimensionnels. C.R.I.F., Bruxelles (in preparation).

6. Janss, J., Jaspart, J.P. and Maquoi, R., Experimental Study of the Non Linear Behaviour of Beam-to-Column Bolted Joints, Proceedings, Workshop on Connections and the Behaviour, Strength and Design of Steel Structures, Cachan, May 25-28, 1987.

ANALYSIS OF JOINTS BETWEEN STRUCTURAL HOLLOW SECTIONS

J. Wardenier
Department of Civil Engineering
Delft University of Technology
Delft, The Netherlands

ABSTRACT

Structural hollow sections with various shapes are used in various combinations also with open sections. Furthermore, various joint configurations are possible. The large variety in sections, combinations and joint configurations make that many failure modes are observed depending on the geometry variables and loading.

A unified analysis is essential for a uniform presentation of the design rules. In this paper the criteria of failure, failure modes and the analysis are briefly discussed as a background for those not familiar with hollow section joints.

INTRODUCTION

Recent research in Europe, Japan, USA and Canada has added new evidence for the analysis of joints between hollow sections and combinations with open sections. In these programmes various types of axially loaded welded joints have been investigated. The main objective was to enlarge the range of validity of the IIW recommendations [1] and to provide test evidence to check the effective width criteria.

The IIW recommendations cover T, Y, X, N, K and KT joints for various member combinations (figures 1 and 2).

Since research and analyses are carried out in various countries, a major objective is to achieve consistent recommendations. It is important that the designer understands the behaviour, failure modes and the resulting design recommendations. This means also that similar failure modes for different joints should be treated in a similar way. Although there re-

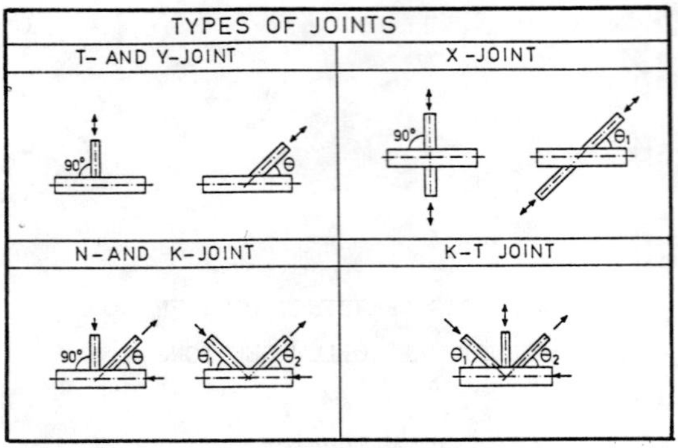

Figure 1. Types of joints

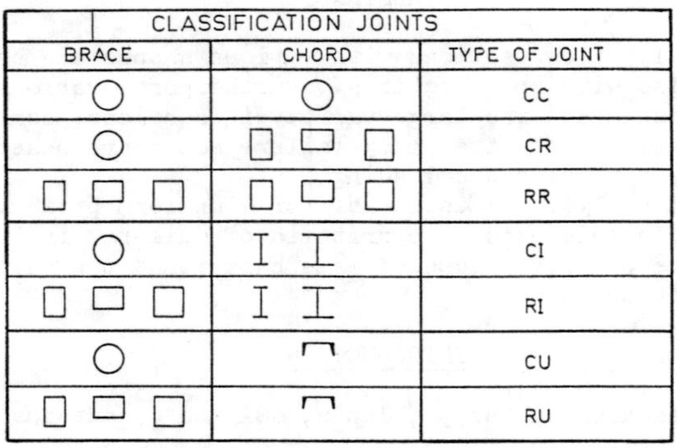

Figure 2. Member combinations

main areas which need further investigation, the current modified recommendations show a large degree of consistency.

CRITERIA OF FAILURE

The static strength can be characterized by the criteria shown in fig. 3, i.e.:

- ultimate load capacity (5)
- deformation criterion (2) or (3)
- visually observed cracks (4)

Because of the non-linear load deflection behaviour there is no international agreement about the determination of the yield load or regarding the deformation limit. For most types of joints the ultimate load is clearly defined.

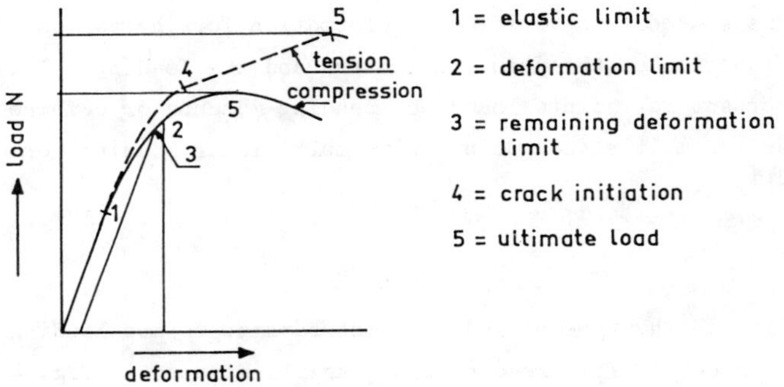

Figure 3. Failure criteria

For the design recommendations of the International Institute of Welding and also for those included in Eurocode 3, in principle the ultimate strength is taken as basis for the determination of the design strength (K, N, KT joints and other joints loaded in compression). The range of validity of the resulting formulae has been limited if the joint deflection at working load would exceed 1% of the chord diameter or chord width.

For joints loaded in tension in principle the same strength is given as for joints loaded in compression (buckling not governing). This means a strength which nearly agrees with that belonging to the kink in the load-deformation diagram (see figure 3). For joints which show excessive deformations at ultimate load, e.g. T and X joints between rectangular hollow sections with $\beta < 0.85$, the strength has been based on the analytical yield strength determined with a yield line model neglecting membrane and strain hardening effects. In this way the associated deformations are indirectly limited, however, in a conservative way.

Furthermore, in general the range of validity of the formulae has been limited to avoid crack initiation below the design load.

Various proposals exist for the determination of a deformation limit or for the determination of a yield load, but none has been accepted internationally. For example, Kurobane uses the kink in the load-deflection diagram if plotted on a log-log scale; Mang is using 80% of the load at which the maximum curvature in the load deflection diagram appears; Yura relates the deformation limit to the deformation of a member with a particular length based on offshore applications [7,8,9].

There is a need to describe the deformation for the various types of joints as a function of the joint parameters and the loading. This is especially important for joints loaded by bending moments or combined loadings. In particular for multiplanar joints with multiplanar loading more evidence is needed.

MODES OF FAILURE

Depending on the type of joint, joint parameters and loading conditions several basic modes of failure can occur, as illustrated in fig. 4 for joints between rectangular hollow sections:

a. plastic failure of the chord face or chord cross section

b. rupture of the braces from the chord (punching shear)

c. cracking of the braces near the joint

d. shear failure of the chord

e. local buckling in compressive areas of the joint members

f. weld failure

g. lamellar tearing

Many times failure occurs by combinations of the basic types mentioned above.

Due to the locally high peak stresses it is essential that weld failures are avoided. Thus the welds should be stronger than the connected members.

Lamellar tearing can be avoided by choosing approapriate steel qualities and welding processes, especially for heavy wall thicknesses.

Local buckling in the joint can be avoided by limitation of the validity range of the recommendations. This is important since the secondary bending moments in welded trusses of hollow sections are rather difficult to estimate. This is caused by the non-uniform stiffness along the perimeter at the intersection between chord and connecting brace.

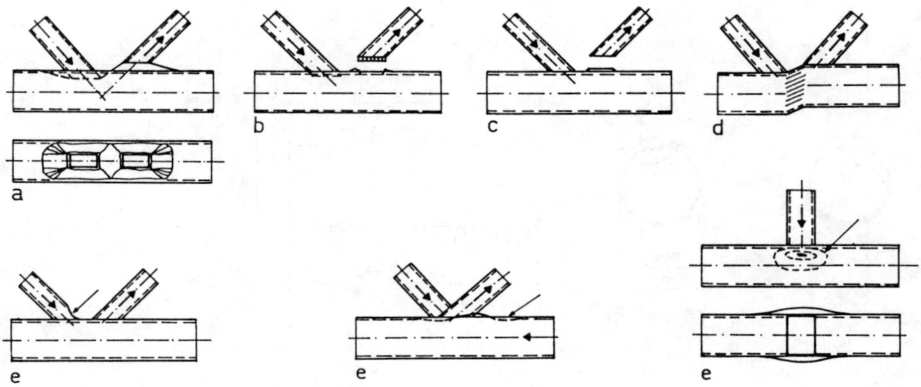

Figure 4. Modes of failure for rectangular hollow section joints

ANALYTICAL MODELS

For the analysis of the various failure modes analytical models are used to determine the governing joint parameters for the ultimate strength, e.g. :

a. Ring model (CC joints)

b. Punching shear model (CC-CR-RR joints)

c. Effective width model (CR-RR-RI-CI joints)

d. Yield line model (CR-RR joints)

e. Chord web bearing models (RR-RI-CI joints)

f. Model for shear failure of the chord (all K and N gap joints)

For indication some of these models are shown in figs. 5, 6 and 7. A detailed description of these models is given in [3, 6]. Due to the complex joint behaviour these models cannot fully describe the influence of all influencing parameters.

TEST EVIDENCE

All types of joints discussed in this paper have been extensively tested in many countries all over the world [3,5,6].
Based on the analytical models and the test evidence, the ultimate strength formulae have been determined.

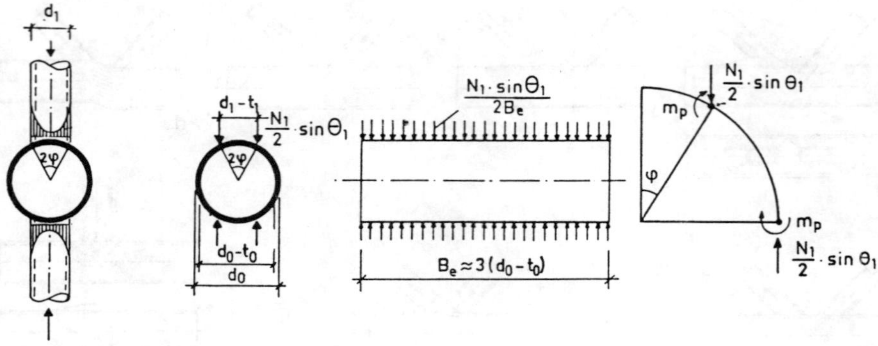

Figure 5. Ring model for CC joints

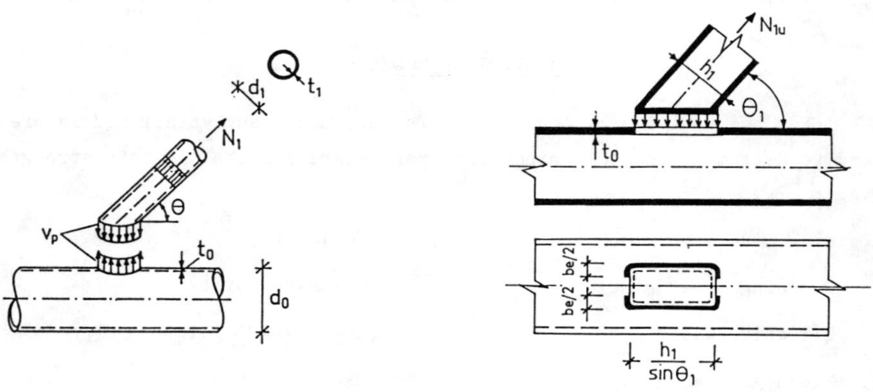

Figure 6. Punching shear model for CC and RR joints

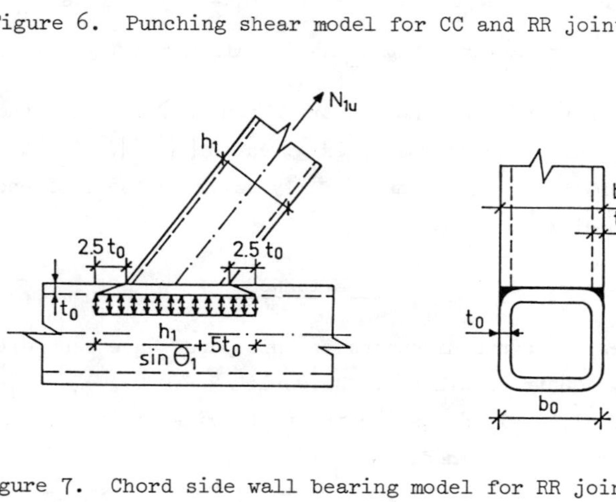

Figure 7. Chord side wall bearing model for RR joints

EVALUATION TO DESIGN RULES

Purely experimental design rules should, in principle, be avoided unless no other information is available and the limitations are clearly defined. However, only for particular failure modes pure analytical design rules can be used. For the more complex joints the design rules are of a semi-empirical nature.

The strength functions which are based on an analytical analysis and which describe a "yield" strength are not treated in a statistical manner. As far as these joints show a ductile behaviour the analytical strength function is adopted as design strength.

From the semi-empirical ultimate strength formulae characteristic strengths have been obtained using a semi-probabilistic approach [3,4,5,6,7 and 8]. All variables have been taken into account, e.g. scatter in test results, but also the variation in mechanical properties and dimensions and the fabrication tolerances. The resulting formulae are sometimes simplified for practical use.

The characteristic joint strength N_k, representing the strength that 95% of the joints are expected to reach, is divided by a partial joint factor γ_m to obtain the design strength $\hat{N}$. The partial joint safety factor γ_m adopted in the IIW and Cidect recommendations varies between 1 and 1.25 depending on the joint deformation capacity and the reserve in strength.

RECENT ADOPTED MODIFICATIONS TO THE IIW RECOMMENDATIONS

Recently, modifications have been adopted to the IIW recommendations. For joints between circular hollow sections the formula for the design strength of K joints has been modified, resulting in a somewhat lower strength for overlap joints with a high chord diameter to thickness ratio.

For joints with a rectangular section or an I section as chord, the range of validity has been extended, particularly the width to wall thickness ratio. Depending on the type of joint, plastic design sections or compact sections can now be used for the compression braces. For these joints also the effective width criteria have been modified. The updated version of the IIW rules will be published in the course of 1987/1988.

FUTURE RESEARCH

Subjects of further investigation are :

- multiplanar joints under multiplanar loading
- interaction between axial loading and bending moments
- beam to column connections
- stiffness and deformation aspects (behaviour and limits)
- local buckling behaviour
- influence weld configuration and weld size
- stress concentration factors (RR-CI-RI joints).

REFERENCES

1. IIW-XV-E, Design Recommendations for Hollow Section Joints - Predominantly statically loaded - IIW Doc XV-491-81/XIII-1003-81 (Revised version).

2. ECSC, Eurocode Nr. 3: Common Unified Rules for Steel Structures, Commission of the European Communities, Report EUR 8849.

3. Wardenier, J., Hollow Section Joints, Delft University Press, 1982.

4. Packer, J.A. and Kremer, J.S.M., Reliability of Tubular Joints in Offshore Structures. 4th International Conference on Struc. Safety and Reliability, Kobe, Japan, May 1985.

5. Packer, J.A., Birkemoe, P.C. and Tucker, W.J., Canadian Implementation of Cidect Monograph Nr. 6, University of Toronto, Dept. of Civil Engineering, Publ. Nr. 84-06.

6. Giddings, T.W. and Wardenier, J., The Strength and Behaviour of Statically Loaded Welded Connections in Structural Hollow Sections, Monograph Nr. 6, Cidect, 1986.

7. Koning, C.H.M. de and Wardenier, J., The Static Strength of Welded Joints between Structural Hollow Sections or between Structural Hollow Sections and H-sections. (Evaluation study). Parts 1, 2 and 3. TNO-IBBC and Stevin Reports 6-84-18/19/20.

8. Makino, Y. and Kurobane, Y. Strength and Deformation Capacity of Circular Tubular Joints. Proceedings of the International Meeting "Safety Criteria in Design of Tubular Structures", Tokyo, July 1986. Maruzen Co., Ltd., Tokyo.

9. Yura, J.A., Zettlemoyer, N. and Edwards, I.F. Ultimate Capacity Equations for Tubular Joints, Offshore Technology Conference Proceedings, Vol. 1, Nr. 3890, 1980.

STRENGTH OF WELDED CONNECTIONS WITH HOLLOW SECTION BRACINGS AND I OR H SECTION AS CHORD

J.M. ARIBERT
Structures Laboratory
Institut National des Sciences Appliquées (INSA)
35043 - Rennes Cedex - France

ABSTRACT

The topic relates to the failure by cracking of the bracing elements, considering three different types of connection. It is shown how to determine accurate design formulae towards this failure, using both experimental tests and theoretical modelling by plastic mechanisms.

INTRODUCTION

This paper deals with the ultimate strength of a particular type of semi-rigid welded connection which is used in Europe for the construction of lattice plane girders. This type of connection results from the association of a chord with an open section (I or H) and bracing members with hollow sections (rectangular or circular). Several failure modes have been observed for these welded connections, but one of these modes must be specially controled for design ; generally, it is called "failure by transverse cracking of the bracings", but in reality this failure mode includes two phenomena : the first one is a sudden fracture of the weld when the bracing member is in tension ; the second one is a local buckling of the end of the bracing member when this member is in compression. Concerning this transverse cracking failure, it is possible to obtain accurate design formulae by combining an empirical approach and a theoretical analysis based on plastic mechanisms. The aim of this paper is to give some ideas about a such procedure, essentially for the X type connection in tension ; however, a more general application of the procedure is briefly shown for two other types : the K type connection and the X type connection in compression.

As shown in figure 1, the stress distribution at the bottom of the bracing member is absolutely not uniform because the flexural stiffness of the chord's flange is stronger near the web than near its edges. In order to express this varying stiffness, most of the norms use the practical concept of "effective width b_e" ; consequently, the strength towards the cracking failure is calculated by :

$$N_{ci} = 2\,\sigma_{yi}\,t_i\,b_{ei} \qquad (1)$$

(i = 1 for compression and i = 2 for tension).

Generally, b_{ei} is given by :

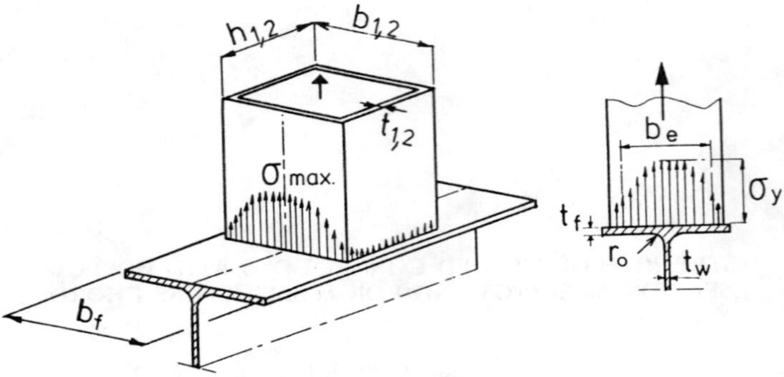

Figure 1. Geometry of the X type connection

$$b_{ei} = 2\, t_w + c_i\, t_f \qquad (2)$$

where t_w and t_f are the thicknesses respectively of the web and the flange of the chord. The constant c_i depends on the used norm ; for example, the French Standard NF - P - 22 - 255 [1] recommends the following values :

$$c_i = \frac{E}{k_i\, \sigma_{y,1,2}} \qquad (3)$$

with $k_1 = 60$ (compression) and $k_2 = 85$ (tension). It must be emphasized that the associated formulae **(1)**, **(2)** and **(3)** take into account a safety factor of 1.4 ; it is important to know this point for comparisons which might be made between the application of the previous regulation formula, or others, and experimental results.

In order to possess a better knowledge of the cracking failure, a specific experimental programme of about fifty connections of different types (X, K and N) with different sections (⊠, ϕ and H) has been carried out at the INSA of Rennes during the period 1981-1984. The studied parameters were the geometrical dimensions of the sections ($t_{1,2}$, $b_{1,2}$ and $h_{1,2}$ for the bracing members, t_w, t_f and b_f for the chord), the yield stresses of the members (σ_{yo}, $\sigma_{y1,2}$) and the angles $\theta_{1,2}$ between bracings and chord. A complete information about the data and results of all these tests may be found at the references [2, 3].

PROPOSED FORMULATION FOR X TYPE CONNECTION LOADED IN TENSION

Only from the analysis of the experimental results, we have been able to determine the main influence of the previous parameters on the ultimate load, and we have proposed a new design formula for the X type connections in tension :

$$N^*_{c2} = \frac{2\, t_2\, \sigma_{y2}}{\sin \theta_2} \left[(t_w + 2\, r_o) + 9.3\, \frac{\sigma_{yo}}{\sigma_{y2}} \left(t_f - \frac{t_2}{1.75} \right) \right] \qquad (4)$$

In figure 2, a comparison between experimental ultimate loads N_{u2} and the corresponding loads N^*_{c2} given by the formula **(4)** shows a satisfactory agreement

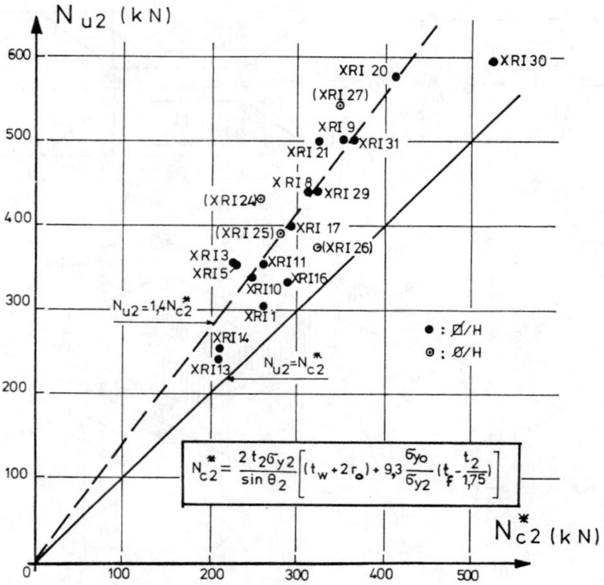

Figure 2. Comparison between experimental results and the formula **(4)**

if the safety factor 1.4 is taken into account ; more exactly, the mean relative deviation of the points towards the line $N_{u2} = 1.4\, N^*_{c2}$ is 9 % whereas it was 20 % with the formula **(1)**, **(2)** and **(3)**.

We have supplied the empirical formula **(4)** with a theoretical modelling which allows to calculate N_{u2} by means of a specific mechanism by yield lines. This mechanism is located in the chord's flange ; it is schematized at the figure 3 in the case of a rectangular hollow section with $\theta_2 = 90°$. The dimension k can be calculated by minimizing the internal work dissipated in the yield lines AC, AD and AB ; furthermore, the uniformly distributed load q_u is equal to the yield stress σ_{y2} multiplied by the thickness t_2 because the mechanism of the flange must be kinematicly consistent with the yield deformation of the bracing member. From this value of q_u, we can deduce the unknown length x of the loaded part which is parallel to the chord's web :

$$x = \frac{\sqrt{2}\, t_f^2\, \sigma_{yo}\, [2(b_f - m) + h_2/\sqrt{2}\,]}{4(b_2 - m)\, t_2\, \sigma_{y2}} - \frac{b_2 - m}{4} \quad ; \qquad (5)$$

then, if $0 \leqslant x \leqslant h_2/2$, the ultimate strength of the connection is calculated by :

$$N_{u2}^{(cal)} = 2\, t_2\, \sigma_{y2}\, (b_2 + 2x) \quad . \qquad (6)$$

If $x \geqslant h_2/2$, we have of course :

$$N_{u2}^{(cal)} = 2\, t_2\, \sigma_{y2}\, (b_2 + h_2) = N_{y2} \quad . \qquad (7)$$

A remark may be made about the distributed load q ; in reality, this load cannot be uniform along the length x because the stress distribution has an elasto-plastic

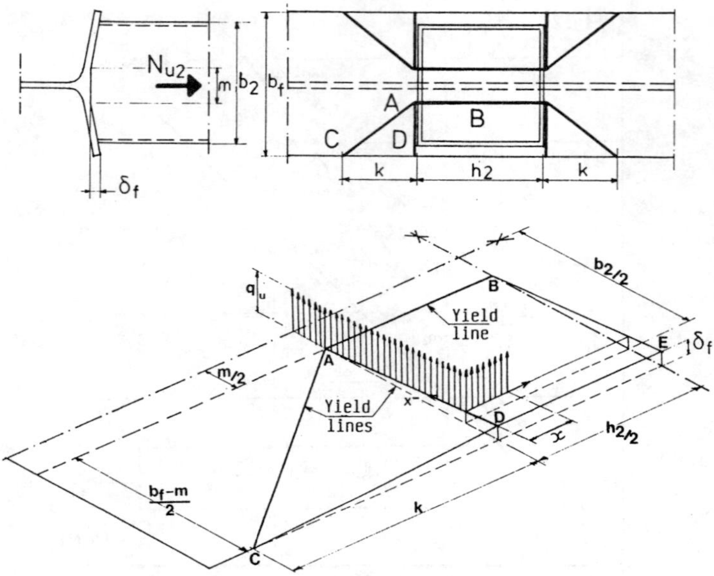

Figure 3. Yield pattern for the X type connection in tension

nature. But the hypothesis with q uniform is valid for a Limit Analysis because the vertical displacement along the length x is constant for the studied mechanism; in a way, x may be considered as an equivalent length for a rigid-plastic behaviour.

The previous modelling has been easy generalized **[4]** when the bracing member is not perpendicular to the chord ($\theta_2 \neq 90°$) ; the distribution of the load q is only different.

Many comparisons have been made between the calculation model and the experimental approach ; they confirm fully the good accuracy of the empirical formula **(4)** from a point of view both local and global. For example, the figure 4

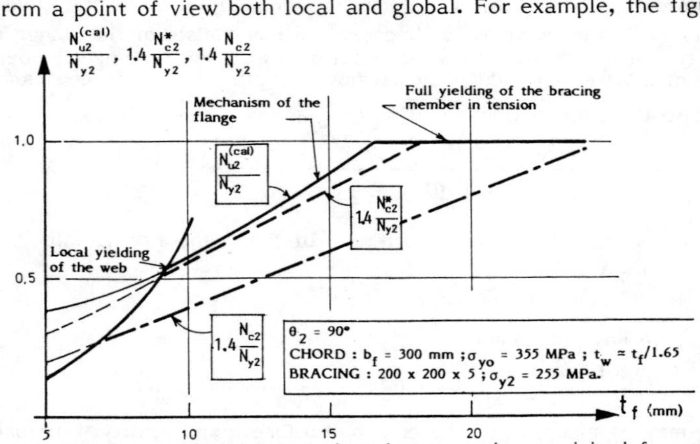

Figure 4. Example of local comparison between the empirical formulae and the calculation model

shows such a comparison for a given connection, but with a variable thickness t_f of the chord's flange ; we can see that the empirical formula **(4)** is clearly closer to the model results than the norm's formula **(1)**, **(2)** and **(3)**. In another way, the figure 5 deals with a global comparison for various manufactured profiles, up to a

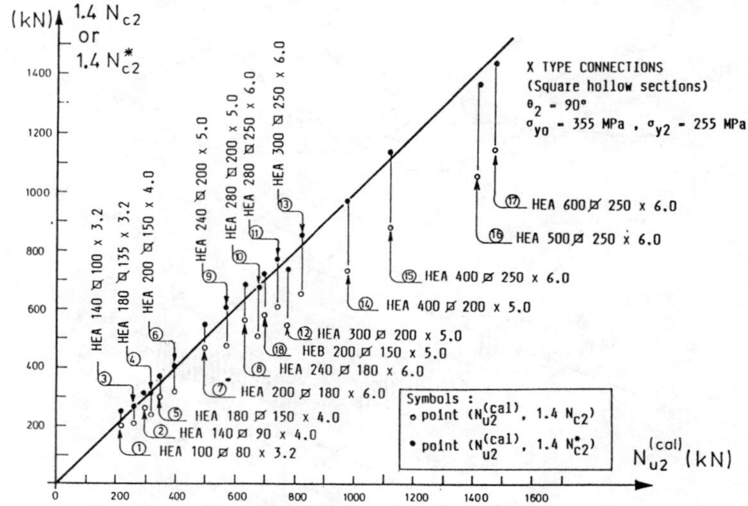

Figure 5. Global comparison between the empirical formulae and the calculation model

depth equal to 600 millimeters. The horizontal axis of the diagram relates to the ultimate load $N_{u2}^{(cal)}$ given by the theoretical model ; the vertical axis is meant to represent either the norm's load N_{c2} or the load N_{c2}^{*} given by the empirical formula **(4)**. Here again, the formula **(4)** is in perfect agreement with the model, for the whole field of application, whereas the norm's formula shows a real deviation.

GENERALIZATION TO OTHER TYPES OF CONNECTION

Hereafter we present only two generalizations, but a more complete study may be consulted at the reference **[4]**.

For the K type connections with overlap, experimental results have allowed to propose the following formula for the strength of the bracing member in tension :

$$N_{c2}^{*} = \frac{t_2\,\sigma_{y2}}{\sin\theta_2}\left[(t_w + 2\,r_o) + 9.3\,\frac{\sigma_{yo}}{\sigma_{y2}}\left(t_f - \frac{t_2}{1.75}\right)\right] + 8\,t_1\,t_2\,\sigma_{y2} + \frac{2\,h_r\,t_r\,\sigma_{yr}}{\sqrt{3}\,\sin\theta_2} \qquad \textbf{(8)}$$

This sophisticated formula includes three terms : the first one is equal to half the ultimate load of the bracing in tension if it is considered alone ; the second term relates to the strengthening due to the bracing in compression. And the third term results from another strengthening due to the effect of the overlap area $2\,h_r\,t_r$.

A model by yield lines has been found in good agreement with the formula **(8)**. As shown in figure 6, the main characteristics of this model are the following ones :

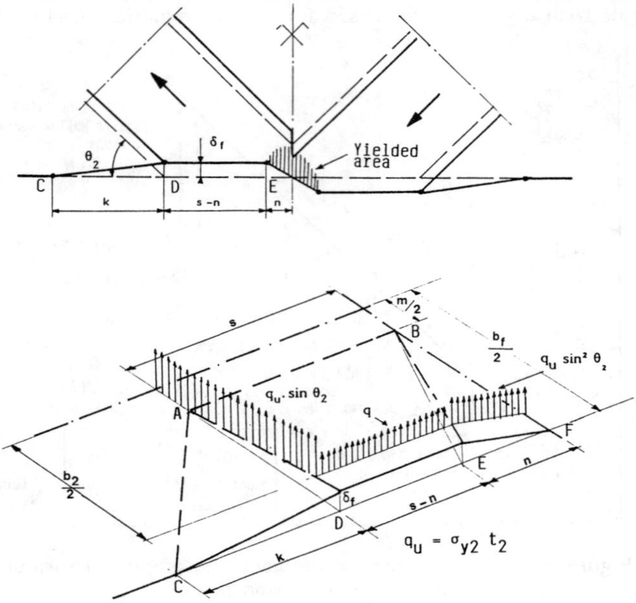

Figure 6. Yield pattern for the K type connection with overlap

1/ the exact position of the yield line BE (defined by the dimension n) is calculated by minimizing the ultimate load of the bracing member.

2/ The distributed load along the perimeter at the bottom of the bracing member is assumed to have three different constant values according to the part of this perimeter : the values $q_u \sin \theta_2$ and $q_u \sin^2 \theta_2$ are consistent with the yield deformation of the bracing member near the lines AD and BE ; as for the middle value q, its calculation results from the equality between the internal work and the external work of the mechanism.

3/ The relative displacement between the two bracing members is only possible if the overlap area is locally yielded by shearing.

The second generalization that we are going to present now, deals with the failure by buckling of the X type connection loaded in compression. Experimental results have led to the formula :

$$N^*_{c1} = 2 \, t_1 \, \sigma_{y1} \, [(t_w + 2 \, r_o) + 4.0 \, \frac{\sigma_{yo}}{\sigma_{y1}} (t_f + \frac{5}{3} t_1)] \qquad \textbf{(9)}$$

in which the influence of the angle parameter θ_1 is knowingly neglected on account of the nature of the failure.

The formula **(9)** has been checked by means of a mechanism including two parts, as shown in figure 7 : one part is in the chord's flange with a pattern like that of the

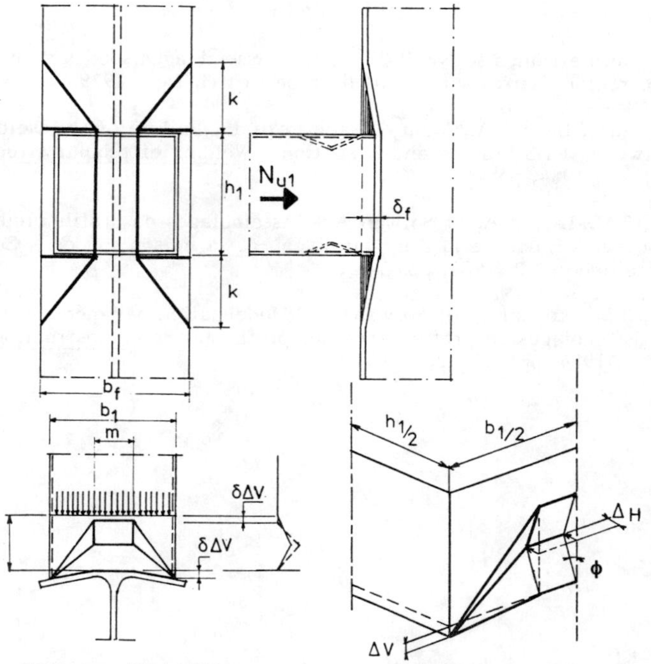

Mechanism with buckling

State of the bracing before buckling

Figure 7. Yield pattern for the X type connection in compression

connection in tension (fig. 3) ; the second part of the mechanism takes place near the bottom of the bracing member where second order geometrical effects must be considered for the calculation (so, at the figure 7, ΔH and ΔV are finite displacements before buckling). The kinematic consistence between these two parts requires the equality between the vertical displacement of the flange's edge and the height variation δΔV of the yield pattern in the bracing when buckling occurs. Finally, the calculated ultimate load N_{u1} of the connection becomes a function of two indeterminate parameters : the initial value λ for the height of the yield pattern in the bracing, and the angle φ expressing the wall deformation of the bracing before buckling. Founding upon some experimental observations, we have adopted φ equal to $b_1/4$; by another way, we have calculated φ for each type of connection, assuming that the displacement of the flange's edge is equal to $b_f/100$.

CONCLUSION

The theoretical modelling turns out to be in satisfactory agreement with all the available experimental results ; besides, it provides a mechanical interpretation to the design formulae which were empirical up to that date. It also allows to develop a certain critical analysis of both the accuracy and the field of application of these formulae ; if the failure mechanisms of the different parts of a connection were well-known, this field of application could be better defined by a rational way (for example, see the figure 4).

REFERENCES

1. Afnor - Norme française NF P 22 - 255 - Assemblages soudés de profils creux ronds ou rectangulaires sur profils de type I et H. Dec. 1979.

2. Aribert, J.M., Lachal A., Mouty J. and Petit L., Failure of the welded connection between RHS bracing and H section - Welding of tubular structures - Pergamon Press 1984, pp 459 - 469.

3. Aribert, J.M., Lachal A. et Sofwan A. - Assemblages de profils creux sur profil ouvert - Etude de la ruine au niveau de la liaison soudée - Construction Métallique n° 2, 1984, pp. 5-41.

4. Aribert, J.M., Lachal A. et Sofwan A. - Modélisation par mécanismes de la ruine d'assemblages de profils creux sur profil ouvert - Construction Métallique n° 4, 1984, pp. 3 - 32.

NUMERICAL SIMULATION OF THE BEHAVIOUR UP TO COLLAPSE OF TWO WELDED UNSTIFFENED ONE-SIDE FLANGE CONNECTIONS

Miss Wajd ATAMAZ SIBAI and François FREY

IREM, Department of Civil Engineering, Swiss Federal Institute of Technology, CH-1015 Lausanne, Switzerland

ABSTRACT

Unstiffened one-side welded flange connections of hot-rolled H sections have been tested by Klein under simple bending loading; moment-rotation curves and plastic capacity of the nodes are also proposed. Here, two of Klein's nodes are computed up to collapse with a nonlinear finite element code. The numerical simulation emphasizes the importance of strain hardening; it also helps one to understand the detailed mechanical behaviour of the node, from which nonlinear spring constants can be deduced with a view to performing a beam analysis. The two nodes are also computed under pure bending action, which leads to a more severe loading condition than simultaneous bending and shear; hence it is concluded that Klein's proposals should be slightly improved.

INTRODUCTION

Within the framework of a Swiss-Austrian joint research project, Klein [1] studied unstiffened welded flange connections of hot-rolled I and H sections in an experimental way. In his thesis, he proposed simple expressions for "beam moment - beam rotation" curves ($M_b - \alpha$, see Fig. 7a and 10) and plastic limit loads of such connections.

We present here the detailed numerical study of two of these connections. Material and geometrical nonlinear effects are taken into account, although the latter is far less important than the former. We use the computer code FINELG [2] of IREM and MSM (University of Liège).

The aim of the numerical simulation is twofold :

1.- to understand deeply the mechanical behaviour of the connection up to collapse;

2.- to derive a model including nonlinear springs, which expresses the semi-rigid "moment - relative rotation" M-Φ behaviour with a view to performing a beam analysis of the connections and, further, of frames.

The selected specimens, named NR4 and NR16 in [1], are unsymmetrical one-side connections consisting of an IPE 330 beam welded to a HEB 160 column for NR4, and of a HEB 500 beam welded to a HEB 300 column for NR16 (Fig.1). The test loading is a concentrated force at the beam tip; thus the connection is submitted to both bending and shear; in practice, however, the ratio of shear is generally less important; for comparison purposes, a second load case of pure bending has been numerically performed.

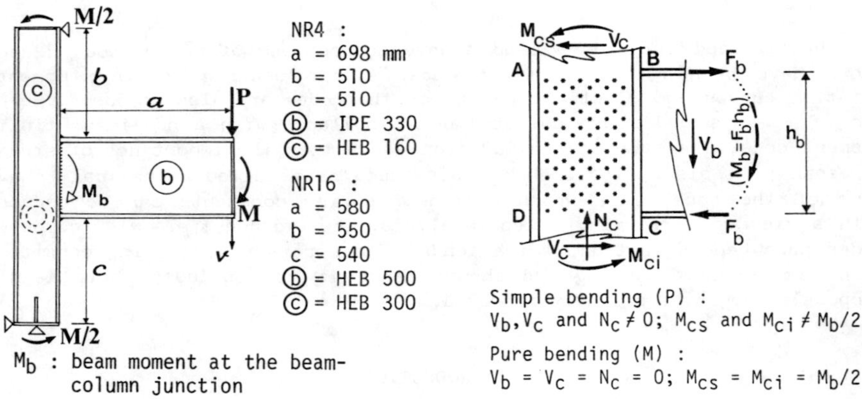

Figure 1. Test set-up (after [1]) and free-body diagram of the node. Simple bending load case : P ($\Rightarrow$ M_b = P · a); pure bending load case : M, M/2, M/2 ($\Rightarrow$ $M_b = M$).

NUMERICAL MODEL

The geometry is discretized into a 3D model of shell elements (webs and flanges) and beam elements (end plates and stiffeners). Fig. 6 gives an idea of the mesh. A geometrical imperfection is introduced in the column web at the beam compressed flange level, allowing for possible web buckling (Fig.1). The out-of-plane magnitude is taken from the rolling tolerances.

Steel is supposed to follow a piecewise linear law shown in Fig. 2; yield stress is taken from [1], strain hardening parameters and ultimate stress from [3]. The 2D elastoplastic state of stress is dealt with using the incremental flow theory and von Mises yield condition. Parabolic patterns of rolling residual normal stresses in flanges and webs are taken according to the ECCS recommendations [4]. Welding imperfections are not considered. Complete data may be found in [5].

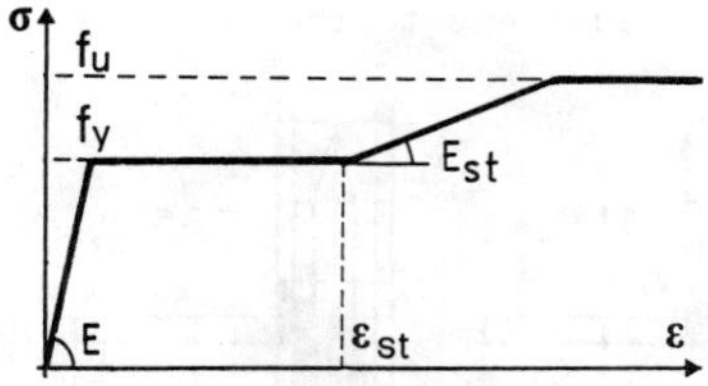

$E = 2.1 \cdot 10^5$ N/mm²
E_{st} = 4000 to 5000 N/mm²
f_y = 271 to 355 N/mm²
f_u = 395 to 485 N/mm²
ε_{st} = 1.61 to 2 %

Figure 2. Stress-strain curve (mild steel).

NUMERICAL RESULTS

Moment-deflection curves M_b-v are shown in Figs. 3 and 4 (for M_b and v, see Fig.1); simple bending curves are to be compared with the experimental ones. We conclude that :

a.- the agreement with the experimental results is satisfactory;

b.- pure bending is more severe than simple bending;

c.- the initial stiffness is larger than the experimental one;

d.- strain-hardening occurs very quickly (point B);

e.- unloading is linear elastic (Fig.4).

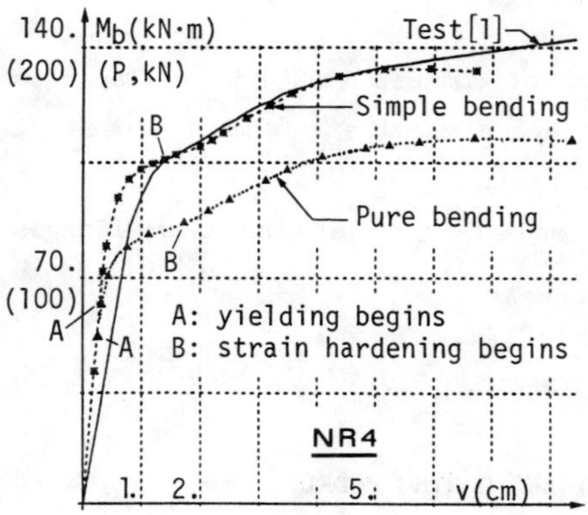

Figure 3. M_b-v curve for NR4.

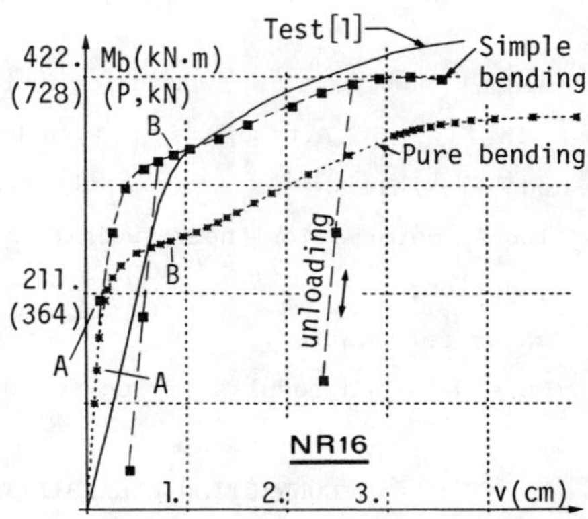

Figure 4. M_b-v curve for NR16.

Figure 5 shows the progressive spread of plasticity in the webs of connection NR4 loaded by simple bending. The flanges (not shown) are comparatively not very plastified. Hence,

f.- collapse of the connections happens mainly by total plastification of that part of the column web which is adjacent to the beam.

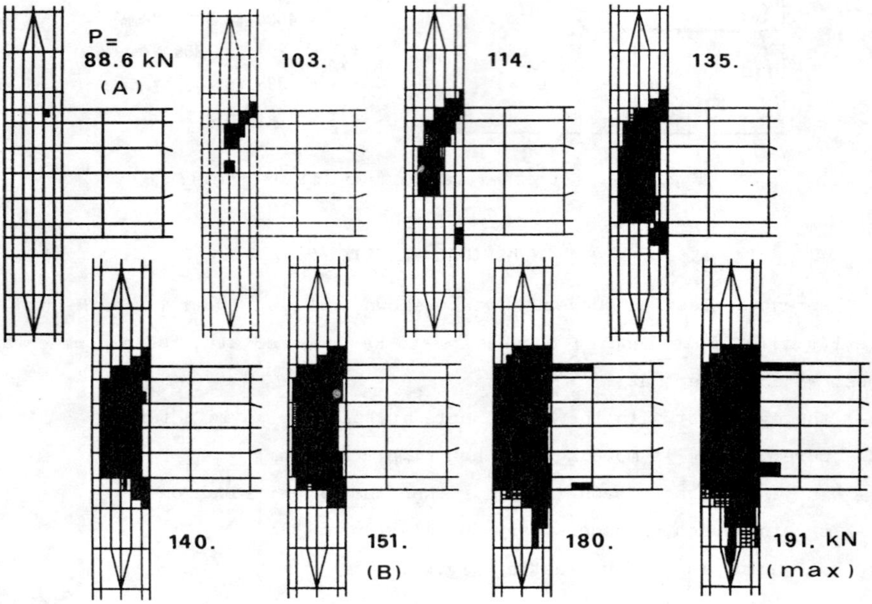

Figure 5. Web plastification of NR4 connection at various load levels (simple bending).

Global and detailed web deformation of connection NR4 at collapse is shown in Fig. 6. Although it is rather complex, the two main causes of the node flexibility [1] are clearly visible :

g.- local deformation near points B and C, i.e. at the beam flanges junction;

h.- shear deformation.

More detailed results may be found in Ref. [5].

CONNECTION FLEXIBILITY AND SPRING MODEL

If we simplify the global deformation of the ABCD web panel of Fig.6c, assuming straight edges, we obtain the quadrangle of Fig.7a; this defor-

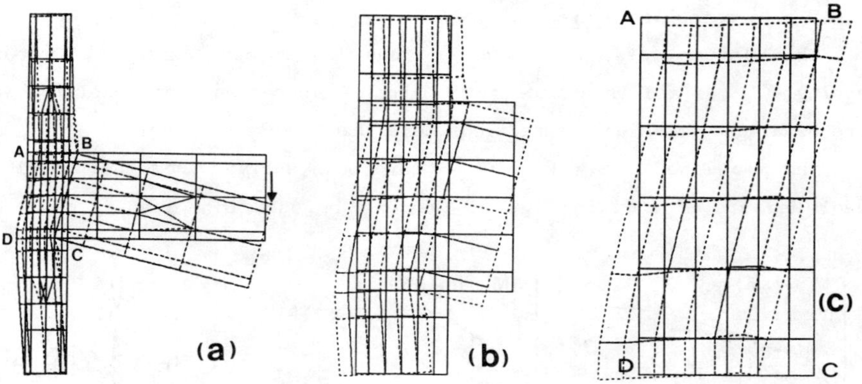

Figure 6. Web deformation of connection NR4 at maximum load (simple bending case; displacements magnified 3x); (a) full structure; (b) connection region; (c) enlarged ABCD "column-beam" web panel.

mation may be divided into two parts :

1.- the transversal effect of the beam flange forces F_b (one elongates AB, the other shortens CD) results in a relative rotation φ between the beam and the column axes; this rotation concentrates mainly along edge BC (Fig. 7b) and provides <u>a first law</u> M_b-φ;

2.- the shear effect, due to the shear force V_n in the web panel

$$V_n = F_b - V_c = (M_{cs} + M_{ci})/h_b = \Delta M_c/h_b ,$$

results in a relative rotation γ between the beam and the column axes (Fig.7c); this rotation occurs mainly along edges AB and CD and makes it

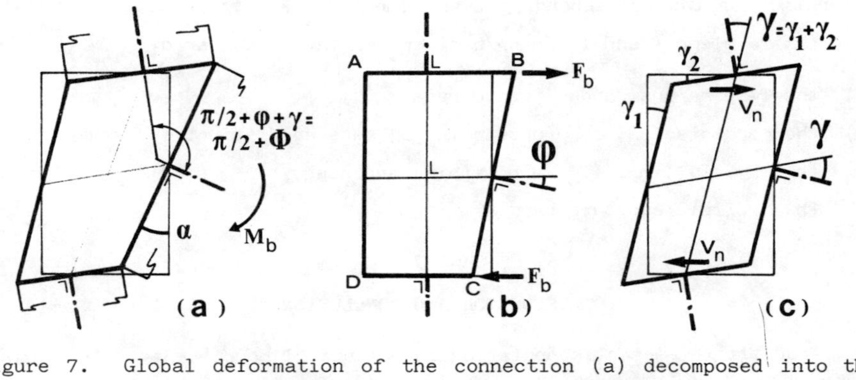

Figure 7. Global deformation of the connection (a) decomposed into the beam flange effect (b) and the shear effect (c).

possible to establish <u>a second law</u> $\Delta M_c/h_b-\gamma$.

This latter rotation is larger than the former; moreover, in pure bending, $V_c = 0$ and thus the shear force V_n increases. This explains why pure bending deforms the connections more than simple bending.

The two laws $M_b-\varphi$ and $\Delta M_c/h_b-\gamma$ have been derived from the numerical model and are plotted in Fig. 8. They can be used in a T-sping model of

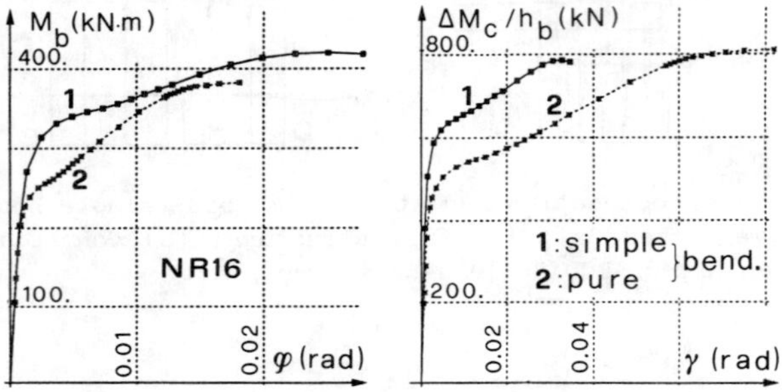

Figure 8. Relative rotation laws for connection NR16.

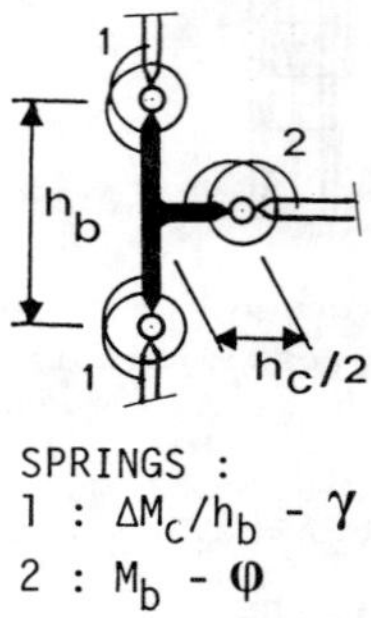

Fig.9. T-spring model for beam analysis.

the connection; this model consists of three short beam elements and springs (Fig.9). It simulates the <u>whole</u> "M-Φ" connection behaviour and can be used in a standard beam finite element numerical model.

THE T-SPRING/BEAM MODEL APPLIED TO THE TESTS

The effective rotation α of edge BC (Fig.8a) is plotted in Fig. 10 as a function of the corresponding beam moment M_b. Experimental [1], proposed [1], finite element and T-spring/beam curves may be compared :

(a) the T-sping/beam model is accurate;

(b) the tabulated values proposed in [1] make no distinction between :

(b1) yield limit (f_y) and strain hardening,

(b2) simple and pure bending.

DISCUSSION AND CONCLUSION

A 3D finite element computation is a fruitful complement to laboratory tests. A simple and accurate three node T-spring model, which applies to

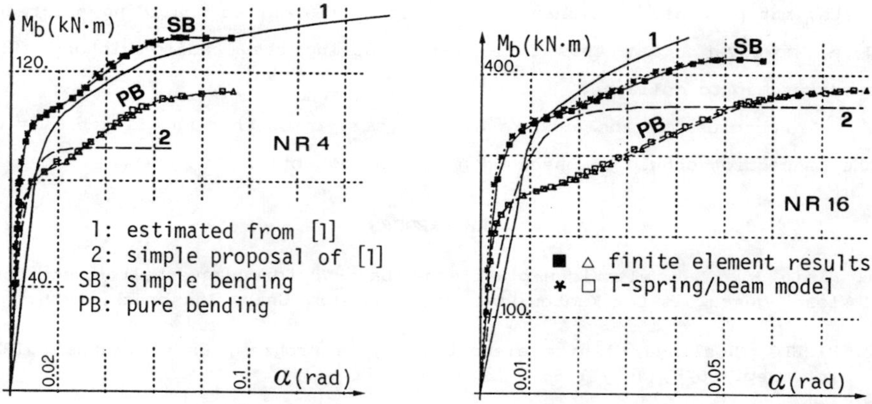

Figure 10. M_b–α curves for NR4 and NR16 connections.

beam analysis of fixed and sway frames, is derived. Any additional flexibility (due to bolted connection for instance) can easily be added to the spring laws. A two node L-spring or four node X-spring model can also be developed for other types of connections.

Reducing the one-side T connection flexibility to a single beam-to-column spring (Fig.11) appears tempting but dubious. The $\Delta M_c/h_b$–γ law can be transformed into an equivalent M_b–φ law such that beam deflections are equal in both models [5]. Hence, the whole M–Φ flexibility is concentrated into a single M_b–Φ^* spring, but the second order effects in the columns, for instance, are erroneously and unsafely reduced ($u_2 < u_1$, see Fig. 11b).

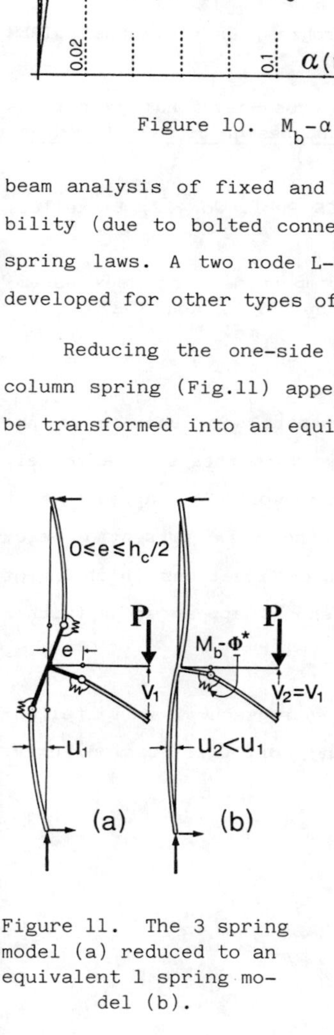

Figure 11. The 3 spring model (a) reduced to an equivalent 1 spring model (b).

Klein's proposal [1] should be refined to distinguish :

1.- simple plastic behaviour from strain hardening one,

2.- simple bending from pure bending.

Indeed, the column shear force is smaller in practice than in the laboratory tests; the ratio V_c/F_b is about 0.3 in the tests, but probably less than or equal to 0.1 in current building practice. Note, moreover, that this ratio varies during the loading history of a structure up to collapse.

In conclusion, the authors are of the opinion that the flexibility of the connection should be based on pure bending action.

REFERENCES

1. Klein H., Das elastisch-plastische Last-Verformungsverhalten steifenloser geschweisster Knoten; Dr. Dissertation, Univ. Innsbruck, 1985.

2. FINELG, Nonlinear finite element analysis program, User's manual; IREM Int. Rep. 86/6, July 1986.

3. Alpsten G.A., Variations in mechanical and cross-sectional properties of steel. ASCE-IABSE Int. Conf. on Planning and Design of Tall Buildings, Vol. Ib-9, 1972.

4. Manual on Stability of Steel Structures, ECCS Publ. No. 22, Bruxelles, 1976.

5. Atamaz Sibaï W., Simulation numérique du comportement de deux assemblages de rive soudés non raidis; IREM Int. Rep. 87/5, June 1987.

ACKNOWLEDGMENT

At the time this paper was written, the authors were unaware of the simultaneous publication of Braun's thesis [6]. This thesis is a sequel, extension and partly reformulation of Klein's work. It appears that Braun's results are in good agreement with the work presented here, especially in that range of small relative node rotations which is of practical interest. More about this subject should appear in a further paper.

6. Braun Ch., Das Momenten-Rotationsverhalten von geschweissten steifenlosen Rahmenknoten bei Querkraftbeanspruchung; Dr. Dissertation, Univ. Innsbruck, Juni 1987.

DISCUSSION
OF
LOCAL ANALYSIS PAPERS

A significant amount of attention was given to the prying force phenomenon, and distinctly different methods of computation and design code incorporation appeared to be in use in various countries. There seemed to be general agreement that prying can constitute an important additional load in tension-type bolted connections, although comments were made to the effect that prying has traditionally been over-estimated. It is clear, however, that with prying as a deformation - induced force in many types of connections, some recognition must be given to the attendant reduction in the available external load-carrying capacity of the bolt. In other words, prying reduces the bolt capacity, but only because some of it is taken to resist the deformation - induced effect. The ultimate strength of the bolt itself is not influenced.

It appears that design practices in Europe, as reflected by the recently issued Eurocode 3 - "Steel Structures" and North America (AISC Specifications for U.S.A. ; C.S.A. Standard for Canada) are similar in the way prying is treated. Thus, although there are researchers and designers that feel that prying need not be considered, the general consensus now is to incorporate the prying effect on the load side of the design equation. However, changes may be forthcoming, especially considering the importance of fatigue behaviour of tension connections, and the fact that prying in such cases will work to reduce the overall tensile strength of the bolt. Further, the Danish results cover a large range of connection types, with good correlation.

Several speakers and discussers dealt with the subjects of material strength and behaviour in general, and their application to connection performance and design in particular. Suggestions were advanced that the ultimate stress of the steel should form the basis for connection design, rather than the yield stress. However, several comments were made to the effect that this way lead to unacceptably low margins of safety against local steel cracking, due to the high degree of restraint that is likely to prevail. This is particularly accentuated for higher strength steels, where the ratio of ultimate stress to yield stress may drop well below 1,30.

Exceptions to the above obervations certainly apply in the design of bolted connections in bearing, as pointed out by numerous speakers and discussers. In these cases it is critical to bear in mind the deformation capacity of the joint, particularly in the vicinity of the last bolt in a row. Previous American allowable stress design criteria used a bearing stress of 1,5 times the ultimate value ; this has recently been reduced to 1,2, due to large hole deformations in the plate or member material at the higher stress. European criteria currently utilize a comparable bearing stress value of 3,0 times the yield stress ; this will likely be changed in the future, to reflect the use of the ultimate stress. A value of 2,0 to 2,5 may be adopted ; the former will be comparable current North American practice.

Modeling methods and solutions received significant attention, and it was clear that numerous and highly advanced finite element programs and procedures are now available for the solution of one-, two- and three-dimensional local analyses. Material and geometric properties are considered, as are the gradual spread of yielding, stability effects, and so on. It was noted that parametric investigations of this type will permit the determination of the most important factors for the behaviour and strength of the connections, as well as the separation of the individual effects. For example, several speakers pointed out that it is important and useful to determine how much of the total rotation in an end-plate connection that can be attributed to end-plate bending, how much to bolt elongation, how much to column flange bending, and so on. Finally, although some three-dimensional experimental work has been performed for beam-to-column assemblies, with reasonable correlation between tests and theory, much work needs to be done to improve the total data base.

Several contributors adressed the needs for consideration of serviceability and ultimate limit states for connections, although no general consensus nor conclusions were arrived at. It was felt that it may be premature to consider such criteria, since the responses of connections vary so widely, and the controlling factors can be widely different. For example, the question of local cracking was mentioned in several cases, as caused by weld contraction strains through lamellar tearing and other weld problems. Clearly, this focuses attention on the needs for understanding the response of the material itself under a variety of conditions, but most of all under the type of restraint that it will only experience in the connection regions of a structure. It would appear to be better to define overall modes of failure, and to identify the ones that are preferable in terms of ductility, deformation levels, and overall behaviour. This will remove some of the problems that can be encountered when purely theoretical solutions appear to predict unreasonably high local stresses, for example, especially if the latter is based on a computational model of limited sophistication.

Summarizing, numerous advanced models of behaviour are currently available for the analysis of connections, and the correlation with test results is generally good. The tools for solving most, if not all types of connection analysis problems therefore seem to be at hand.

SESSION N°2A

MATHEMATICAL MODELS

FINITE ELEMENT MODELING OF CONNECTIONS

D. Beaulieu and A. Picard
Département de Génie civil
Université Laval
Sainte-Foy, Québec, Canada
G1K 7P4

ABSTRACT

Researchers traditionally depended almost exclusively on expensive laboratory tests to understand the behavior of connections in steel structures and to derive empirical rules for design. Finite element modeling appears today as a valuable and less costly alternative (or more precisely a complement) to laboratory testing. The finite element method is no longer an esoteric and inaccessible tool in view of recent developments in the fields of numerical analysis, computer graphics and computer power. In this short paper, we intend to describe the research and development efforts done at Laval University which leaded to modeling of steel connections by the finite element method.

INTRODUCTION

The experimental approach to study the behavior of connections in steel structures will certainly remain the most popular for still some years but because of the high costs involved, researchers are increasingly looking for less costly but acceptable alternatives. The most obvious alternative is modeling by the finite element method.

Due to the highly complex nature of connections and the large number of parameters involved, numerous tests are required before an adequate set of empirical formulae is developed for the design of a specific type of connection. It appears to be more rational and more economical to develop numerical models to play with the various parameters and to check the accuracy of the numerical models against the results of an appropriate number of experimental tests. Not only are experimental tests needed to validate the models but they are also required for calibration purposes. Using finite element models therefore does not mean discarding or discrediting laboratory testing and never will.

Such a vision is shared by many researchers but it is not until lately, with the development of more powerful elements and numerical techniques, and the rapid evolution of computers and computer graphics that it is believed possible. To come up with acceptable models, however, is not a simple task and requires several years of research and development.

Since we cannot be too specific due to the limited length of the paper and the large variety of subjects to cover, we will limit ourselves, in the next sections, to a condensed presentation of the computer programs we use and to brief descriptions of some elements and techniques of resolution which are more appropriate than others to simulate the behavior of connections and some which have been developed specifically for that purpose. We will finally present some applications and point out items requiring further research and development.

STATE OF THE ART

The numerical simulation of steel connections is one of many applications resulting from developments done over the last twenty years by a group of specialists and several students at Université Laval in Québec and at Université de Technologie de Compiègne in France. A modular finite element code called MEF [1] and an interactive graphic pre post-processor called MOSAIC were created and are in continuous development. MEF contains a library of over sixty standard elements plus as many specialized elements which can be used for two and tridimensional structural and solid mechanic applications. About twenty other elements are presently under development. The standard and special elements and modules are used to solve a large variety of problems: linear elastic, nonlinear (plastic, large displacements, large strains), stability, vibration, dynamic, contact, composite materials, etc... Also, many other specialized elements have been prepared for specific applications in fluid mechanics, hydraulics, heat transfer, accoustics, form optimization, etc...

We are presently capable of adressing problems like friction, slip, contact, plate stiffening, plate stability and plasticity which are generally associated with structural steel connections. Bolting and welding can also be reasonably modeled in 2-D and 3-D applications. The types of problems we can analyze are numerous but we do not pretend to be able at the present time to offer an acceptable solution to every one of them. There is still room for improvement and more research is required, as shown later.

SPECIALIZED ELEMENTS AND MODULES

Plate and Shell Elements

The library contains a good selection of triangular and rectangular plate and shell elements which can be used for elastic, elasto-plastic and stability analyses of connections [2 - 8]. Stiffened plates in two and three dimensional applications can be properly handled.

Spring Elements

Welded and bolted connections in linear two and tridimensional applications can be grossly but properly represented by calibrated springs or bar elements [9, 10, 11]. This obviously constitutes the simplest representation.

Bolts can be represented by a combination of bar elements [10, 12, 13]. As shown in Fig. 1a, two bars are used to simulate the bolt prestress and a third is used to simulate shear. This principle was adapted to two as well as tridimensional nonlinear elasto-plastic models.

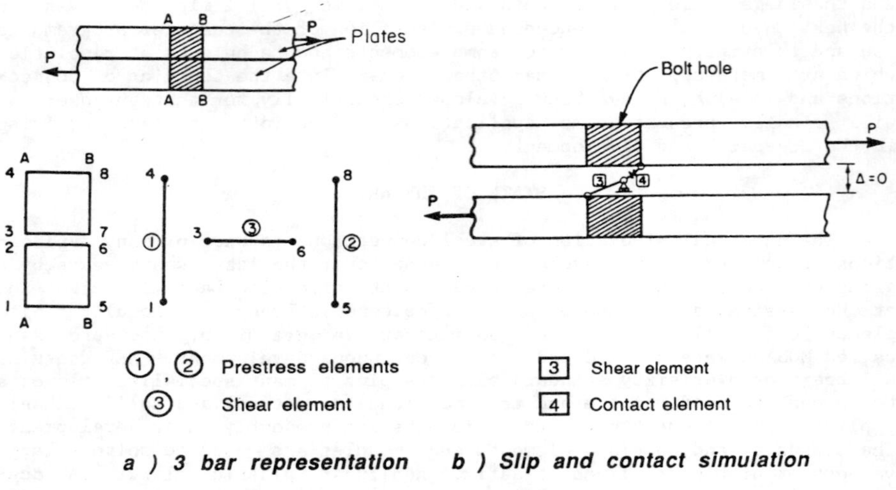

Fig. 1 -- Bolt modeling .

Contact elements
Contact between two solids can be modeled by a set of nonlinear orthogonal springs calibrated according to the nature of the contact between the surfaces [10, 14, 15]. These elements can represent separation or firm contact between solids and perfect slip or friction according to Coulomb's law.

In order to account for slip and contact in a bolted connection, a contact element is coupled with the bar element which simulates shear, as shown in Fig. 1b [10].

Weld element
Two welded plates can be roughly modeled considering they behave monolithically. A more refined model is obtained using springs, as mentioned previously, the spring constants being ultimately determined from experimental tests on the weld material [16].

For a detailed representation of the behavior of fillet welded connections, we have developed a 15 node tridimensional prismatic quadratic element which is used in combination with a 20 node tridimensional hexahedric quadratic (Brick) element to represent the connected plates [11]. These elements allow a representation of the elasto-plastic behavior of the material considering strain-hardening but were developed assuming small strains and small displacements. The angle of loading with respect to the orientation of the filled weld is automatically accounted for in the model.

The inherent mechanical properties of the weld material were obtained from recent test results [17].

Each one of these elements has been tested and calibrated against numerous published experimental and numerical results. The details can be found in the various papers and reports.

Modules

A large amount of sophisticated numerical techniques had to be developed or implemented and tested to simulate nonlinear plastic and geometric behavior as well as post-buckling behavior and, to some extent, large strains [9, 10, 11, 18]. These are contained in special modules which interact with MEF.

APPLICATIONS

At the present state of development, the MEF and MOSAIC programs and the various specialized modules and elements can be used to study a large variety of steel connections and problems. Although we are still in the early stages of our research and development program on modeling of steel connections by finite elements, we have been able to analyze a number of representative examples [10, 11]. Some of them had already been studied numerically or experimentally and were used to validate our numerical models.

Due to space limitations, it is not possible to present the results of our analyses; these will be published elsewhere. Sketches of some of the connections considered, however, are presented in Fig. 2. Problems like friction, slip, contact, plasticity, prying action and stress distribution, among others, were properly adressed. We are at the present time doing full numerical and experimental studies of standard girt-to-column connections used in light steel buildings (see Fig. 2d) and of standard column base plate connections, such as shown in Fig. 3.

FUTURE DEVELOPMENTS

Following is a list of items we have identified for future research and development:

- A major project, already under way, is to create an expert system for nonlinear resolution in which the different techniques will be used in a simple and efficient manner.

- A routine for graphical color representation of stresses in nonlinear 3-D plastic models is yet to be fully implemented.

- For the simulation of tridimensional welded connections, we shall develop an element which would consider second order terms since deformations can be quite large in fillet welds. Using proper experimental data, it shall also be possible to include residual stresses in welded connections. These two additions would certainly improve our model. Moreover, tridimensional modeling requires the use of powerful computers and the acquisition of a vectorial processor will improve our ability to perform sophisticated 3-D models.

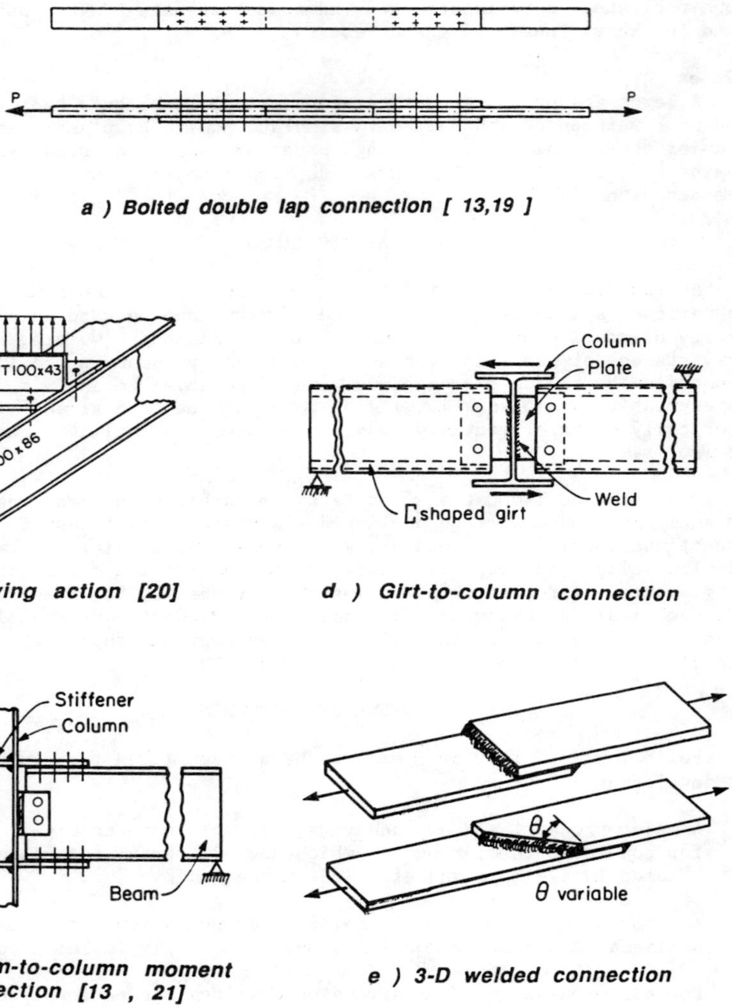

a) Bolted double lap connection [13,19]

b) Prying action [20]

d) Girt-to-column connection

c) Beam-to-column moment connection [13 , 21]

e) 3-D welded connection

Fig. 2 -- Examples of connections studied .

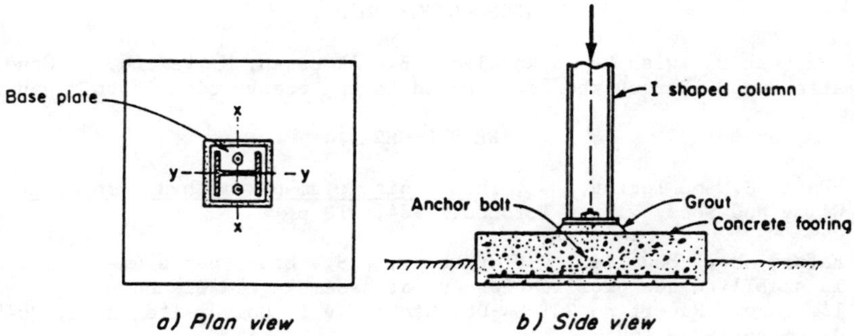

Fig. 3 -- Typical column base connection

- The bolt model shall be improved since ultimate traction and shear resistances seem to be overevaluated when compared to available experimental results. This could simply be done by an appropriate calibration of the bar elements. Besides, plastic failure criteria shall be implemented for welds, bolts and plates.

- What probably constitutes our most challenging project is the development of a global finite element computer program for frame analysis which would account for partial connection rigidity. This program would make an extensive use of interactive graphics [22] and would interact with a data base containing pertinent information on various standard connections. The program would also handle elastic, elasto-plastic, static, dynamic and stability analyses. A great deal of work is already done.

  We have developed a tridimensional beam element which accounts for geometric and material nonlinearities, residual stresses, initial deformations and post-buckling behavior [9]. The element is based on the theory of thin shells and considers the progression of plasticity through the thickness of the plates. Connections are simulated by means of 3-D springs of constant rigidity. The incremental formulation will allow us to consider nonlinear connection behavior without any great difficulty.

  The major task, as one could guess, is to fill the data base with proper information on M-Θ curves for the various types of connections. Available information would have to be gathered and compiled in an appropriate format. Missing and new data would have to come from either experimental testing or finite element modeling. This project of academic character shall keep several people busy for quite some time and is only possible within a framework of international collaboration between interested researchers.

ACKNOWLEDGEMENTS

The authors wish to acknowledge E. Akoussah, M. Fafard, G. Gendron, Y. Matte and F. Parrot who contributed to the preparation of this paper.

REFERENCES

1. Dhatt, G. and Touzot, G., The finite element method displayed. J. Wiley and Sons, N.Y. & Toronto, 1984, 510 p.

2. Fafard, M., Beaulieu, D. et Dhatt, G.S., Etude par éléments finis de la stabilité des profilés formés par assemblage d'éléments plats dans l'espace. Raport no GCT-84-06, Génie civil, Université Laval, Québec, décembre 1984, 168 p.

3. Batoz, J.L. et Ben Tahar, M., Formulation et évaluation d'un nouvel élément quadrilatéral à 12 D.L. pour la flexion des plaques minces. Université de Technologie de Compiègne, France, 1981.

4. Marcotte, L. et Dhatt, G.S., Instabilité des plaques minces semi-infinies sur fondation élastique par éléments finis. Rapport GCS-84-02, Génie civil, Université Laval, Québec, 1984.

5. Batoz, J.L. et Dhatt, G., Plaques et coques par éléments finis. Analyse linéaire et non linéaire. Institut pour la promotion des sciences de l'ingénieur, Paris, juin 1986.

6. Ben Tahar, M., Analyse élasto-plastique des plaques et coques minces par éléments finis. Thèse de doctorat, Université de Technologie de Compiègne, France, 1981.

7. Matte, Y., Développement de nouveaux éléments triangulaires pour l'analyse linéaire statique et dynamique de plaques et coques minces. Thèse de maîtrise, Université Laval, Québec, 1986.

8. Talbot, M., Comparaison de deux éléments de coques triangulaires plats utilisant une formulation lagrangienne actualisée. Thèse de maîtrise, Université Laval, Québec, 1985.

9. Akoussah, E., Beaulieu, D. et Dhatt, G.S., Etude par éléments finis de la stabilité des poteaux dans les bâtiments industriels. Rapport technique à paraître en 1987, Génie civil, Université Laval, Québec.

10. Gendron, G., Beaulieu, D. et Dhatt, G.S., Modélisation par éléments finis du comportement des assemblages boulonnés dans les charpentes d'acier. Rapport GCT-87-01, Génie civil, Université Laval, Québec, février 1987, 173 p.

11. Parrot, F., Beaulieu, D. et Dhatt, G.S., Modélisation par éléments finis du comportement des assemblages soudés dans les charpentes d'acier. Rapport GCT-87-02, Génie civil, Université Laval, Québec, mai 1987.

12. Patel, K.V. and Chen, W.F., Nonlinear analysis of steel moment connections. Journal of the structural division, ASCE, vol. 118, no 8, August, 1984, pp. 1861-1874.

13. Patel, K.V. and Chen, W.F., Analysis of a fully bolted moment connection using NONSAP. Computers and Structures, Vol. 21, no 3, 1985, pp. 505-511.

14. Lebouvier, F., Analyse non linéaire de structures planes formées de poutres et d'arcs. Thèse de docteur-ingénieur, Université de Technologie de Compiègne, Compiègne, France, 1985.

15. Mazurkiewicz, M. and Ostachowicz, W., Theory of finite element method for elastic contact problems of solid bodies. Computers and Structures, Vol. 17, no 1, 1983, pp. 51-59.

16. Richards, R.M., Rabern, D.A., Hormby, D.E., and Williams, G.C., Analytical models for steel connections. Proceedings of the W.H. Munse Symposium - Behavior of metal structures, Philadelphia, Penn., May 1983, pp. 128-155.

17. Kennedy, D.J.L. and Miazga, G.S., Behaviour of fillet welds as a function of the angle of loading. Structural Engineering Report 133, University of Alberta, Edmonton, Alta, 1986.

18. Fafard,M., Algorithmes de calcul automatique des configurations pré et post-flambement de problèmes non linéaires en calcul des structures. Thèse de doctorat à paraître en 1987, Génie civil, Université Laval, Québec.

19. Sterling, G.H., and Fisher, J.W., A 440 steel joints connected by A490 bolts. Journal of the Structural Division, American Society of Civil Engineers, Vol. 92, no St3, August 1966, pp. 101-118.

20. Fisher, J.W. and Struik, J.H.A., Guide to design criteria for bolted and riveted joints, John Wiley and Sons, N.Y., 1974.

21. Standing, K.F., Rentschler, G.P. and Chen, W.F., Tests of bolted beam-to-column moment connections. Welding Research Council Bulletin, no 218, 1976.

22. Bérubé, C., Lamontagne, R. and Picard, A., CAE System for the Design of Steel Structures. Rapport GCT-87-03, Génie civil, Université Laval, Québec, 1987.

THREE-DIMENSIONAL PHYSICAL AND MATHEMATICAL MODELING OF CONNECTIONS

A. COLSON
Laboratoire de Mécanique et Technologie
Ecole Normale Supérieure de Cachan
F94230 CACHAN
FRANCE

ABSTRACT

This paper concerns the mathematical modeling of the non linear behaviour of the connections. The interest of a 3D model rather than a 1D or 2D model is discussed regarding the problems of column stability and spatial structures. The size of the macro-element which is the physical model of the connection is defined relative to the type of frame analysis. The main parameters, i.e initial stiffness and ultimate strength, of the modeling are introduced from experimental and mechanical observations.

Different mathematical expressions, according to the loading path, are proposed, and the hypothesis of reduced isotropy gives interesting results. Finally, cyclic loading with hardening or softening is presented.

INTRODUCTION

Recent developments in the modeling of the non-linear behaviour of the steel structural connections use a 1D model, i. e. an M-ø curve. However, if we look into current industrial practice there is a lack of information regarding M-ø curve and such modeling. The required level of rigidity regarding the transmissible force at each end of a bar, for a three-dimensional description, is generally the following :

Normal force	: Rigid
Shear force/strong axis	: Rigid
Shear force/weak axis	: Rigid
Bending moment/strong axis	: Rigid, semi-rigid, pinned
Bending moment/weak axis	: Rigid, semi-rigid, pinned
Torque	: Rigid or pinned, but often neglected.

Since all the analysis methods do not require a particular model for the rigid connections, it seems that for the plane structures (normal force, shear force/strong axis, bending moment/strong axis) a 1D solution is adequate. But, in fact, due to unilateral contact, there is an obvious interaction between normal force and bending moment [1], so the level of semi-rigidity in bending depends strongly on the level of the normal force. At this stage a 2D model is necessary, more especially for the in-plane buckling conditions of the columns. Futhermore, even for the plane structures, the out-of-plane buckling conditions, i.e. for the weak axis, it is necessary to know the behaviour of the column base connection with respect to the weak axis, and including the effects of the normal force. Then a 3D model is necessary. Furthermore, such a model allows for considering torsional rigidity and bending about the weak axis for some particular spatial structures.

PHYSICAL MODELING

For the modeling and the frame analysis the overall connection is replaced by a macro-element (Figure 1), which contains all the components like bolts, welds, angles, etc. The constitutive equation will give the displacement {D} of one terminal cross section with respect to the other. Each terminal cross section is supposed to remain plane during the loading, so classical beam theory is always applicable to all of the bars that connect to the considered mode. The constitutive equation establishes a relationship between the displacement vector $\{D\} = \{u_i , \omega_j\}$ and the force vector $\{F\} = \{N_i , M_j\}$. i, j = 1, 2, 3.

At this stage such a macro-element may be geometrically integrated into any finite element program [2] that can handle non linear analysis.

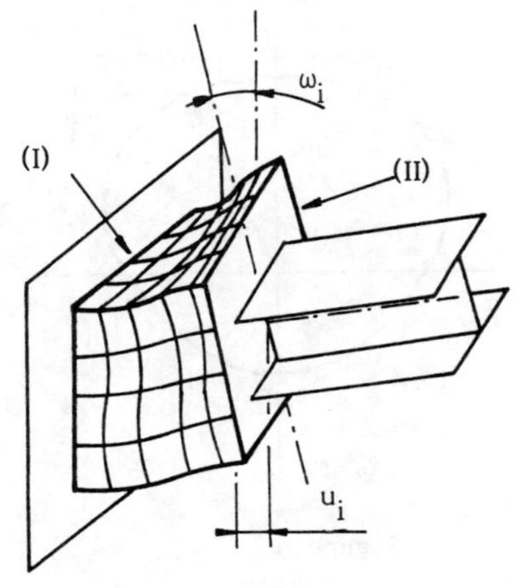

Figure 1

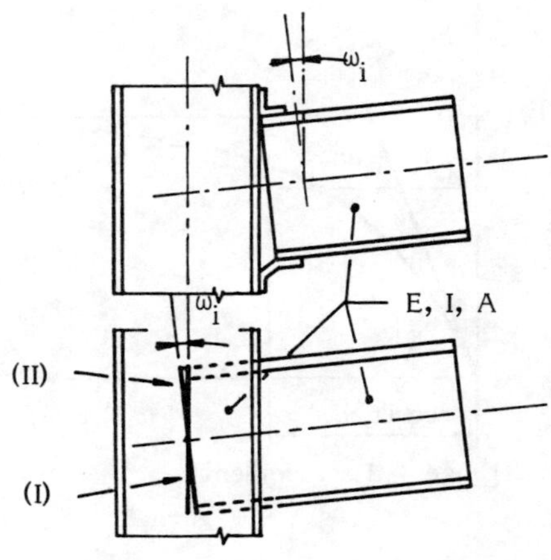

Figure 2

Size of the macro-element

Generally, regarding the St. Venant Problem, the length of this macro-element should correspond approximatively to the depth of the bar. In current applications we have chosen to reduce the size to zero (keeping the same constitutive equation between the two terminal cross sections by lengthening the bar until the neutral point (Figure 2). This is necessary, particularly for the integration into the P.E.P. program [3], which uses the plastic hinge concept rather than the progressive plastification of the bar.

The correlation with and interpretation of experimental results must be clearly defined in order to arrive at the proper comparison between theory and tests.

Main Parameters of the Modeling

Observing the general aspect of the experimental results (Figure 3), at least in the range of the currently required displacements, it is proposed to use the initial stiffness K_i and the ultimate load F_{ui} as the two main parameters. One or several other parameters must be introduced in order to describe the non linearity. In a general way, our modeling can be expressed by the following :

$$\boxed{\{D\} = f\,(\,\{F\}\,,\,[R],\,\{F_u\}\,,\,a_k)} \qquad (1)$$

with [R] : initial stiffress matrix (components K_i)

$\{F_u\}$: ultimate load vector (components F_{ui})

a_k : set of parameters inducing the non linearity.

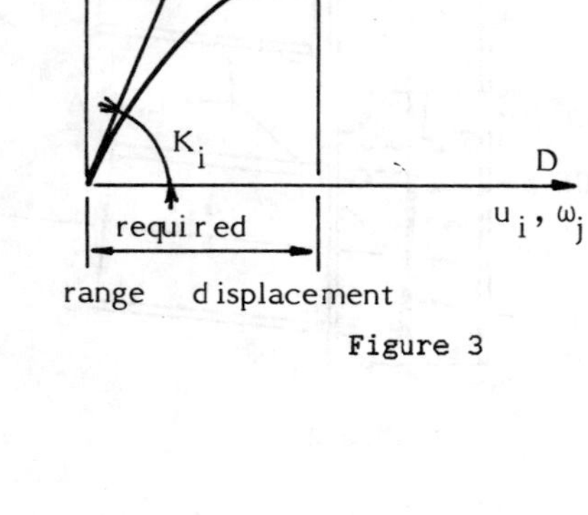

Figure 3

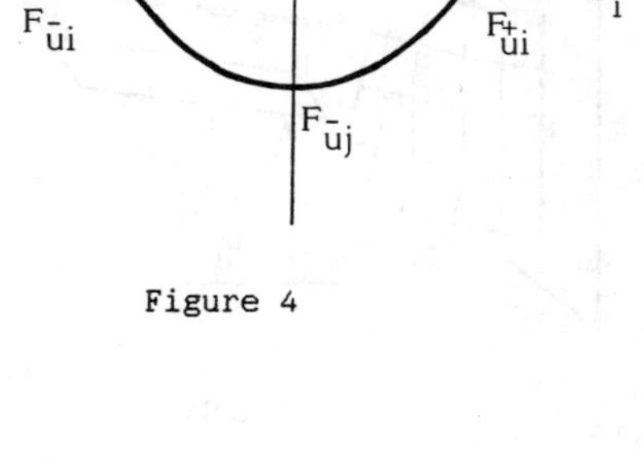

Figure 4

In view of the industrial and practical applications we have tried to minimize the number of a_k parameters (1, 2 or 3) depending on the type of loading and the required precision. After correlating with a number of test data it has been found that the initial stiffness [R] and the ultimate load $\{F_u\}$ can be obtained in a forseeable way, through a computation based on the geometric and the mechanical properties of the components of the connection. This means that for industrial practice, it is not necessary to do experiments for each kind of connection. Only the a_k parameters have to be obtained from experiments, and it is thought possible to establish a limited number of values for each family of connections. Otherwise, the partition displacement equation is proposed : $\{D\} = \{D^t\} + \{D^{an}\}$, where $\{D^t\}$ is the elastic component of the displacement and $\{D^{an}\}$ represents the inelastic part.

a. <u>Initial Stiffness</u> : [R] matrix, such as $\{D^t\} = [R]^{-1} \{F\}$

For the simple connections this value can be obtained from a local classical analysis [4] [5]. For the more complex connections we have to use the finite element method [6]. The calculation cost is limited thince it is the first step of an elastic analysis. Nevertheless, it would be interesting to developp a specific 3D F.E.M. analysis using automatic mesh generation, static condensation, substructuring techniques, etc, in order to reduce the calculation costs below those of NON SAP, ABAQUS and others.

b. <u>Ultimate load</u> (U.L.S. : Ultimate limit surface in the forces space)

Limit analysis methods have to be used to find all of the components of the ultimate load vector (Figure 4). Such methods exist for the most classical connections [5] and also for tubular connections [7]. For the all-welded connections the ultimate limit surface is generally symmetric, and can be expressed by the relationship

$$\left[\sum_{1}^{6} |f_i|^{\alpha} \right]^{1/\alpha} = 1 \text{ where } f_i = F_i / F_{ui} \qquad (2)$$

For the connections with unilateral contact (bolted, for example) the U.L.S. is unsymmetric. For example, the scheme of limit analysis (Figure 5), using the Von Mises criterion, can be used to find the U.L.S. in the compression zone due to concrete fracture [8] [9]. A classical limit analysis method applied to the anchor bolt permits the determination of the U.L.S. in the tension zone (Figure 6).

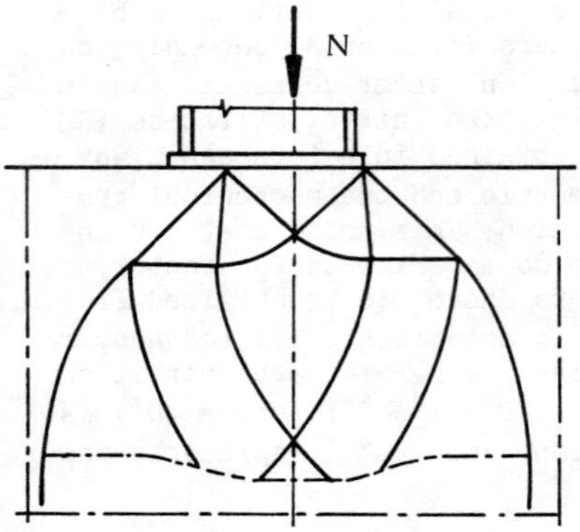

Figure 5

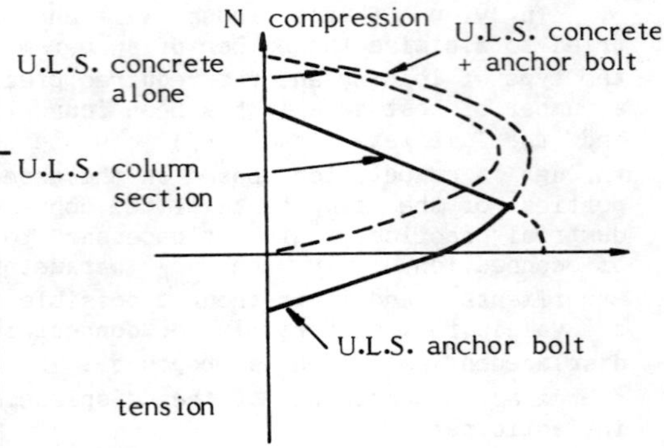

Figure 6

MATHEMATICAL EXPRESSIONS

At the moment several mathematical expressions have been developed. The experiments on the column base connections will provide the remainder, which will then be used to propose the same model for all connections in the structures (for industrial purpose). Most of them are based on the thermodynamics theory. The first model, only available for a radial monotonic loading path (Figure 7), is :

$$\{D\} = [R]^{-1} \{F\} \left(1/ (1 - \lambda^a)\right) \text{ with } \{F\} = \lambda \{F_u\} \quad (3)$$

It requires only one parameter (a) to introduce the non linearity. The term "a" assumes a value between 1.5 to 4.5, depending on the type and size of the connections (bolted, riveted, welded).

The second model [10] uses an incremental form which can incorporate cyclic and non radial loading :

$$d \{D^{an}\} = [C] [R]^{-1} g (\lambda/\delta) d \{F\} \quad (4)$$

The non radial loading path is necessary, particularly in modeling buckling of columns. In the simplified case (2D) used in the PEP program [3], the expressions are (with $n = N/N_u$ and $m = M/M_u$) :

$$du = \frac{N_u}{K_n} [dn + g. (A_n dn + B_n |dm|)] \quad (5)$$

$$d\omega = \frac{M_u}{K_m} [dm + g. (B_m dn + A_m |dm|)] \quad (6)$$

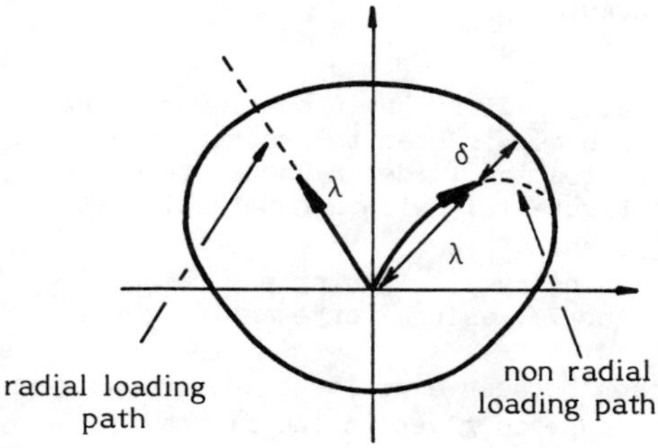

Figure 7

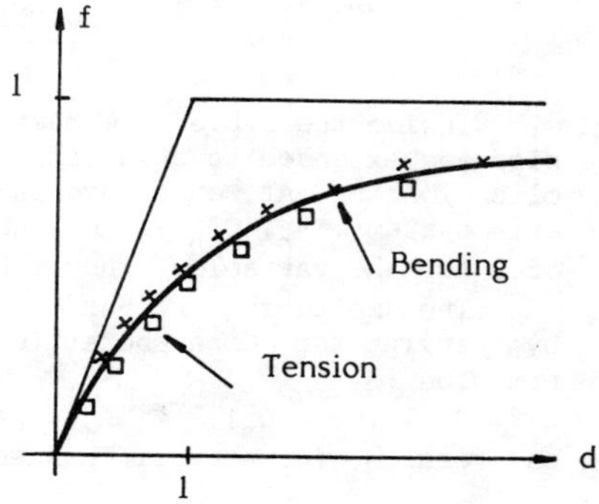

Figure 8

The third model [4] utilizes the hypothesis of reduced isotropy which is based on a normality law. This means that in the space of the reduced forces, $f_i = F_i/F_{ui}$, and in the space of the reduced displacements, $d_i = D_i/D^t_{ui}$, with $\{D^t_u\} = [R]^{-1}\{F_u\}$, the behaviour does not depend on the loading path (Figure 8). The normality law is used with respect to the first loading surface (F.L.S.), which is homothetic to the U.L.S. (Figure 9), and grows with the loading. The general mathematical form is :

$$dD_i^{an} = \frac{F^2_{ui}}{K_i}\, h\,(F)\, \overline{\frac{\partial \xi}{\partial F_i}}\, dF_i$$

Under a reasonable set of hypothesis the second and the third model give the same integrated expressions :

$$D_i = \frac{F_i}{K_i}\left[1 + C_i\,\frac{F_i}{F_{ui} - F_i}\right] \tag{7}$$

Furthermore, the third model gives $C_i = C$. This means that only one parameter is necessary to take into account the non linearity.

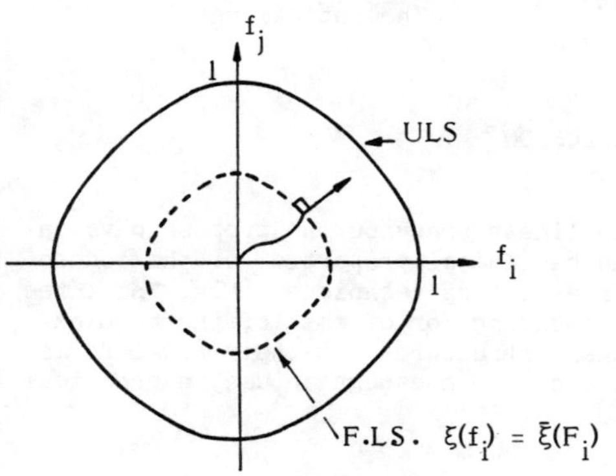

Figure 9

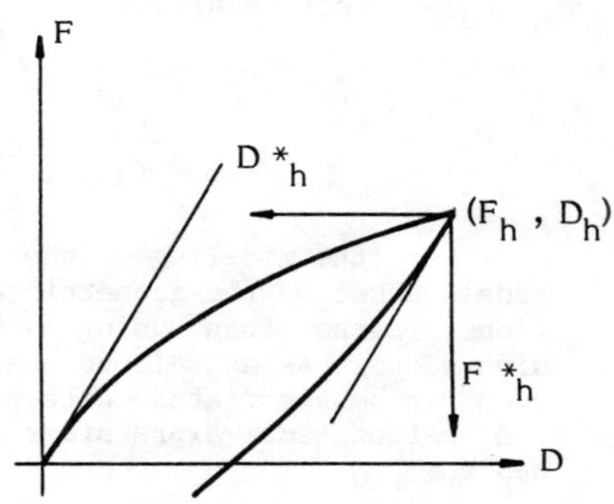

Figure 10

CYCLIC LOADING

Inside the F.L.S. the Masing rule, well known in continuum mechanics and extended to structural mechanics, is used to describe the cyclic loading. At each reversal of loading (order h) a new reference axis system (F^*_h, D^*_h) is defined (Figure 10), with a homothetic ratio of 2 for the variables, thus : $F^*_h/2$ and $D^*_h/2$.

The softening or hardening during cyclic loading may be modeled by varying the C parameter (equation 7), using, for example, the expression :

$C = C_s + (C_i - C_s)e^{-rp}$. A comparison between experiment and theoretical results, for a cyclic hardening case is given in the figure 11.

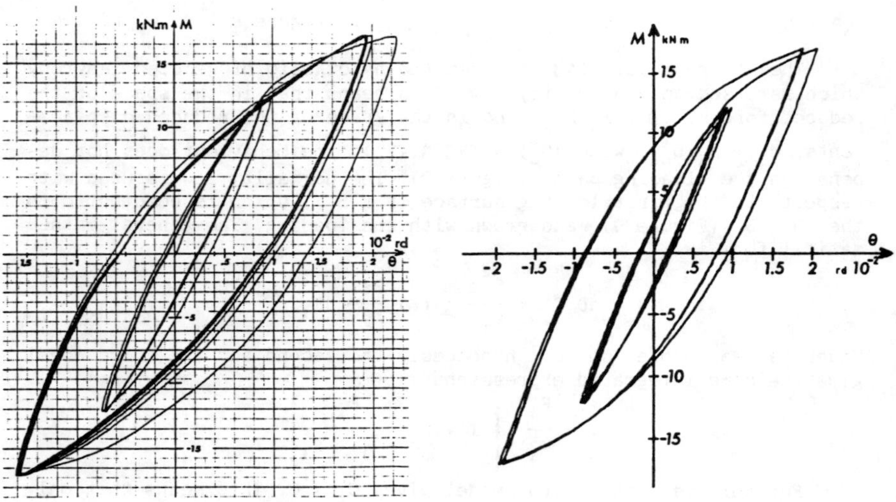

Test result   Theoretical result

Figure 11

CONCLUSION

For the modeling of the non linear behaviour we propose physical models based on the geometric and mechanical properties of the connections, rather than using a curve-fitting technique only. The three dimensional aspect allows be accounting for of most of the problems, even those associated with planar structures. In their final forms, the mathematical expressions remain simple enough to use in practical design.

ACKNOWLEDGEMENTS

The author wish to acknowledge P. Penserini who contributed to the preparation of this paper.

REFERENCES

1. Beaulieu, D., Picard, A. "Contribution des assemblages avec plaque d'assise à la stabilité des poteaux". Construction Métallique, N°2, 1985, pp3-19.

2. Poggi, C., Zandonini, R. "Behaviour and strength of steel frames with semi-rigid connections". A.S.C.E. Convention, Detroit, October 1985.

3. Galea, Y., Colson, A., Pilvin, P. "Programme d'analyse de structures planes à barres avec liaisons semi-rigides à comportement non linéaire". Construction Métallique, N°2, 1986 pp18-25.

4. Louveau, J.M. "Modélisation du comportement cyclique des assemblages de structures métalliques". Thèse de IIIème cycle. Université Paris VI 1986.

5. Yee, Y.L., Melchers, R.E. "Moment-Rotation curves for bolted connections". Journal of Structural Engineering A.S.C.E., Vol 112, N°3, 1986. pp615-635.

6. Patel, K.V., Chen, W.F. "Analysis of a fully bolted moment connection using NONSAP". Computers and Structures, Vol 21, N°3, 1985, pp505-511.

7. Mouty, J. "Calcul des charges ultimes des assemblages soudés de profils creux carrés et rectangulaires". Construction Métallique N°2, 1976, pp37-58.

8. Salençon, J. "Quelques résultats théoriques concernant la butée d'une paroi sur un coin plastique". Annales I.T.B.T.P., N°313, Janvier 1974, pp 187-194.

9. Penserini, P. "Modélisation du comportement des liaisons structures-fondations". Thèse de Doctorat. Université Paris VI. To be published.

10. Pilvin, P. "Modélisation du comportement des assemblages de structures à barres". Thèse de IIIème cycle. Université Paris VI. 1983.

MATHEMATICAL MODEL FOR SEMI-RIGID JOINTS
UNDER CYCLIC LOADS

F.M. MAZZOLANI
Istituto di Tecnica delle Costruzioni
Engineering Faculty
University of Naples (Italy)

ABSTRACT

The main features of a very sophisticated mathematical model for interpreting the actual behaviour of semi-rigid connection under alternate loading conditions are presented in this paper. After a short analytical presentation, some possible applications of this model are illustrated. It has been calibrated on the basis of experimental results and, starting from appropriate behavioural parameters ispired to recent ECCS Recommendations, it can be used to extrapolate the existing results and to interpret the semi-rigid joint performance.

INTRODUCTION

The aim of the present paper is to show some possibilities of application of the mathematical model which has been proposed in [1] for semi-rigid connections. The main data of this model are the following:

a) Among the large number of models existing in the literature, the proposed one belongs to the category of the sophisticated models.

b) The fitness with the actual behaviour of the connection is achieved by means of curve-fitting techniques which utilize up to 12 numerical coefficients.

c) It has been checked on some existing results of cyclic tests on beam-to-column connections both in steel and in reinforced concrete with a very satisfactory agreement [2].

d) It has been also used to calibrate the unified testing procedure recently proposed by the task group 1.3 "Seismic Design" of the European Convention for Constructional Steelwork (ECCS) [3], by means of appropriate behavioural parameters [4].

e) A recent research program on beam-to-column steel joints [5] provided a lot of experimental results of hysteretic relationships, which have been strictly interpreted and exploited by means of the proposed model.

THE MATHEMATICAL MODEL

The assumed moment-rotation relationship is subdivided into four branches (see fig. 1a). The increasing branch /a/ (from O to M^+) can be analytically expressed by a three terms relation

$$\varphi = \varphi_0 + \varphi_1 + \varphi_2 \tag{1}$$

where the first term φ_0 is the residual rotation of the previous cycle; the second term is the Ramberg-Osgood type rotation given by

$$\varphi_1 = \frac{M}{R(\Omega)} + \alpha\left(\frac{M}{M_2}\right)^r \tag{2}$$

the third term is the incremental rotation due to slip between M' and M", given by

$$\varphi_2 = \left[\frac{\Delta\varphi_S(\Omega)}{2} + \left[\frac{\Delta\varphi_S(\Omega)}{2} - K_1\right]\rho\left|\rho\right|^{S-1} + K_1\rho\right]\delta M \tag{3}$$

being

M_2 the moment which corresponds to a rotation twice the elastic one;

α, r, s, K_1 numerical coefficients;

$$\rho = \frac{2M - M' - M''}{M'' - M'} \quad (-1 < \rho < 1 \text{ for } M' < M < M'')\ ;$$

$$\delta M = 1 \text{ for } M' < M < M''\ ;$$

$$\delta M = 0 \text{ for } M < M' \text{ or } M > M''\ .$$

The deterioration parameters $R(\Omega)$ and $\Delta\varphi_S(\Omega)$ are given by

$$R(\Omega) = R\left[1 - \frac{\Delta R}{R}\left(\frac{\Omega}{\Omega_{max}}\right)^m\right] \tag{4}$$

$$\Delta\varphi_S(\Omega) = \Delta\varphi_{S,min} + (\Delta\varphi_{S,max} - \Delta\varphi_{S,min})\left[\frac{\Omega}{\Omega_{max}}\right]^n \tag{5}$$

depending upon the energy variation (see fig. 1b), being:

Ω_{max} the overall energy dissipated at collapse;

$\Delta\varphi_{S,min}$ ($\Delta\varphi_{S,max}$) the minimum (maximum) value of slip measured at the first cycle and at collapse, respectively;

ΔR the decrease of joint stiffness measured in the cycle before collapse;

m,n numerical coefficients.

The decreasing branch /b/ (from M^+ to O) is expressed by

$$\varphi = \frac{M - M^+}{R(M,\Omega)} + \varphi^+ \qquad (6)$$

being

$$R(M,\Omega) = R(\Omega)\left[1 - \frac{\Delta' R}{R(\Omega)}\left(\frac{M^+ - M}{M^+}\right) + \frac{\Delta'' R}{R(\Omega)}\right] \qquad (7)$$

where $\Delta'R$ and $\Delta''R$ simulate different unloading conditions (see fig. 1c). Branches /c/ and /d/ are represented by the same formulation of branches /a/ and /b/ .

THE BEHAVIOURAL PARAMETERS

The experiments on structural components under cyclic loads are carried out with particular testing procedures which usually differ each other. It doesn't allow to strictly compare the various results coming from different laboratories.

The necessity to unify such experimental procedures is unquestionable and the ECCS tasking group "Seismic Design" recently decided to work out a proposal for a recommended testing procedure for assessing the behaviour of structural steel elements under cyclic loads [3] .

The main problem was to define the fundamental parameters for the interpretation of the overall behaviour of the structural components. They have to characterize the degree of ductility, the amount of energy absorption and the deterioration of rigidity and strenght as far as the number of cycles increases.

The following behavioural parameters have been defined (see fig. 1d,e):

a) <u>partial ductility</u>: $\mu^+_{oi} = e^+_i / e^+_y$

$\mu^-_{oi} = e^-_i / e^-_y$

b) <u>full ductility</u>: $\mu^+_i = \Delta e^+_i / e^+_y$

$\mu^-_i = \Delta e^-_i / e^-_y$

c) <u>full ductility ratios</u>: $\psi^+_i = \Delta e^+_i / (e^+_i + (e^-_i - e^-_y))$

$\psi^+_i = \Delta e^-_i / (e^-_i + (e^+_i - e^+_y))$

d) <u>resistance ratios</u>: $\varepsilon^+_i = F^+_i / F^+_y$

$\varepsilon^-_i = F^-_i / F^-_i$

e) <u>rigidity ratios</u>: $\xi^+_i = tg\,\alpha^+_i / tg\,\alpha^+_y$

$\xi^-_i = tg\,\alpha^-_i / tg\,\alpha^-_y$

f) absorbed energy ratios: $$\eta_i^+ = \frac{A_i^+}{F_y^+ (e_i^+ - e_y^+ + e_i^- - e_y^-)}$$

$$\eta_i^- = \frac{A_i^-}{F_y^- (e_i^- - e_y^- + e_i^+ - e_y^+)}$$

The reference parameters F_y, e_y represent the upper bound of the quasi-elastic behaviour. They are conventionally defined by the intersection of the two tangent lines shown in fig. 1f.

INTERPRETATION OF CYCLIC RESULTS

The proposed mathematical model can be used to interpret the results of cyclic tests on structural elements [4]. The testing procedure given in [3] has been exactly simulated following the same steps. The load versus deflection hysteretic loops are automatically plotted, as it is shown in fig. 2. As the most significat parameters for comparing the cyclic behaviour, we have assumed the following:

- peak strenght (S_i), corresponding to resistance;
- secant modulus (R_i), corresponding to rigidity;
- energetic efficiency (E_i), corresponding to absorbed energy;
- ductenergy (D_i), corresponding to the product of ductility and absorbed energy.

The variations of such parameters as a function of the increasing deflection are also given in fig. 2. These curves are normalized referring to the limit elastic values (F_y, d_y). Their behaviours represent the basic elements when we compare the performance of different typologies of structural components (i.e. joints, connections,bracings,...) from the point of view of their reliability under alternate loading conditions. They can be also utilized to define the ultimate limit states which conventionally correspond to the end of the test. In fact it can occur that the structural component is submitted during the test to conditions which are out of interest from the physical point of view. Looking at these variations we can recognize that an ultimate limit state is reached when a significant reduction of the main behavioural parameters arises.
As an example, some ultimate limit states can be identifies as follow:

a) The deterioration of the material due to geometrical or mechanical effects produces a reduction of the peak load carrying capacity (S_i) during the cyclic test.
b) The excess of deformation produces a reduction of the rigidity parameter (R_i), which becomes unacceptable when leads to unstable effects on the overall structure or introduces too large residual deformations.
c) The structural component is unable to fulfil its function when the energetic efficiency parameter (E_i) is reduced below a given amount (it can happen, for instance, in seismic resistant structures).

The interpretation of the experimental cyclic result by means of these parameters and the definition of their limit values should also require the consideration of their influence on the overall behaviour of the structure which the component belongs to.

CALIBRATION OF THE MODEL

The cyclic behaviour of 14 beams-to-column specimens has been experimentally analysed in [5] . They have been designed following the current technological types communly used in rigid and semirigid connections for framed steel structures. In fig. 3 the specimens are grouped in four main categories (A,B,C,D), which cover the practical range of behavioural variability of load versus displacement relationships, as it is shown from the corresponding experimental hysteretic loops.

The proposed model has been applied to such results, giving the simulated curves plotted in the lower part of fig. 4. The upper diagrams give the ratios between simulated and experimental values for the strenght F_i and the area A_i of a half-cycle at each value of the imposed displacement v. The corresponding scatters are usually very satisfactory.

CONCLUSIONS

The above results allow to conclude that the proposed mathematical model is able to fit very closely the experimental results of beam-to-column joints. It has been satisfactory used in the interpretation, exploitation and extrapolation of testing results. It seems also a very appropriate model to be introduced in a more sofisticated analysis of the overall behaviour of semirigid framed structures.

ACKNOWLEDGMENTS

This research is supported by the Italian Ministry of Education. The research team is composed by Edoardo Cosenza, Antonello De Luca, Attilio De Martino and Ciro Faella.

REFERENCES

1. De Martino A., Faella C. and Mazzolani F.M., Simulation of beam-to-column joint behaviour under cyclic loads. Costruzioni Metalliche, n. 6 1984
2. Cosenza E., De Luca A., De Martino A., Faella C. and Mazzolani F.M., Inelastic behaviour of framed structures under cyclic loads. Euromech Colloquium 174, Palermo, October 1983
3. ECCS-CECM-EKS, Recommended Testing Procedure for Assessing the Behaviour of Structural Steel Elements under Cyclic Loads. Technical Working Group 1.3, Seismic Design, n. 45, 1986
4. De Martino A., Faella C. and Mazzolani F.M., Il comportamento ciclico dei nodi in acciaio: metodologia interpretativa. Giornate Italiane della

Costruzione in Acciaio, 10° Convegno C.T.A., Montecatini, Ottobre 1985
5. Ballio G., Calado L., De Martino A., Faella C. and Mazzolani F.M., Steel beam-to-column joints under cyclic loads: experimental and numerical approach. Eighth European Conference on Earthquake Engineering, Lisbon, september 1986

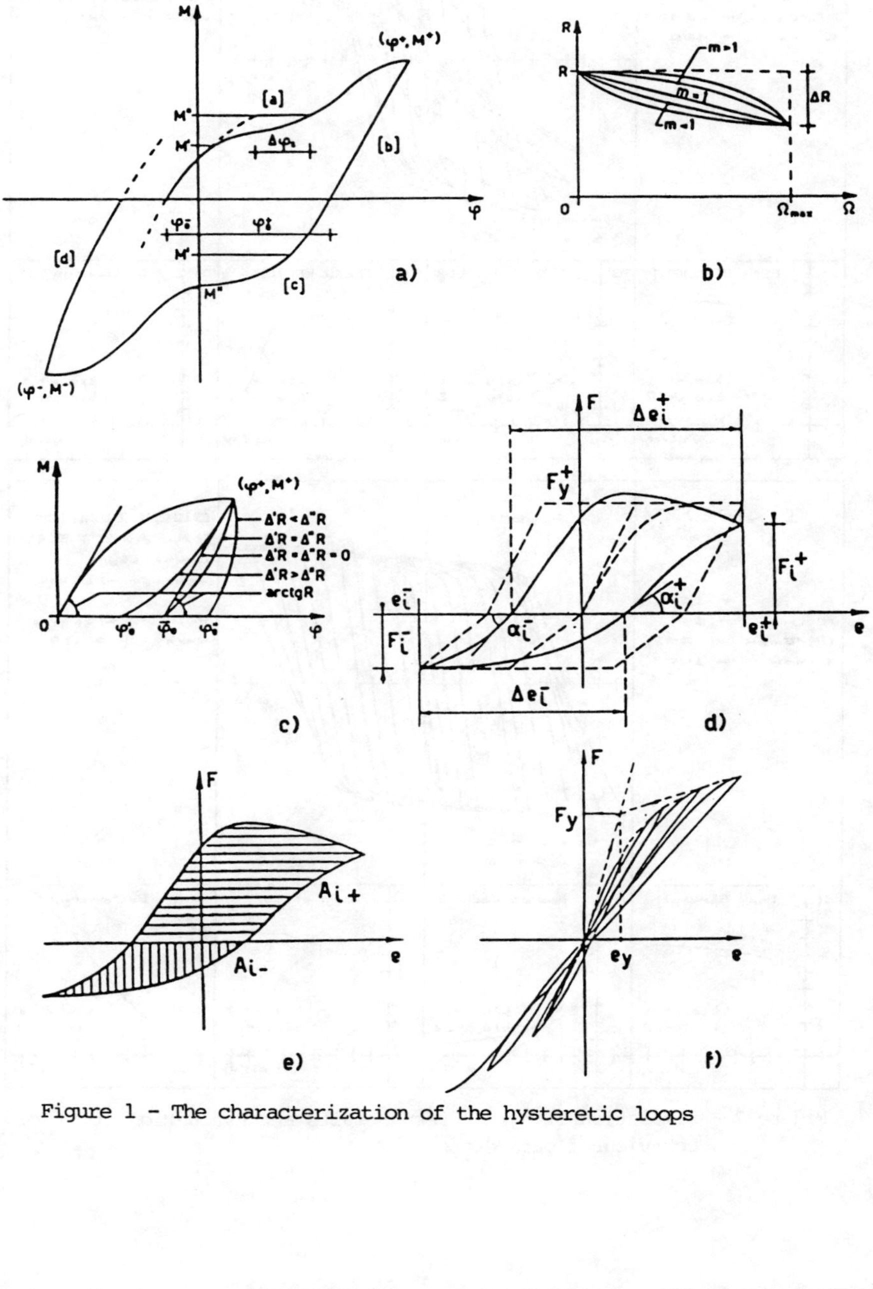

Figure 1 - The characterization of the hysteretic loops

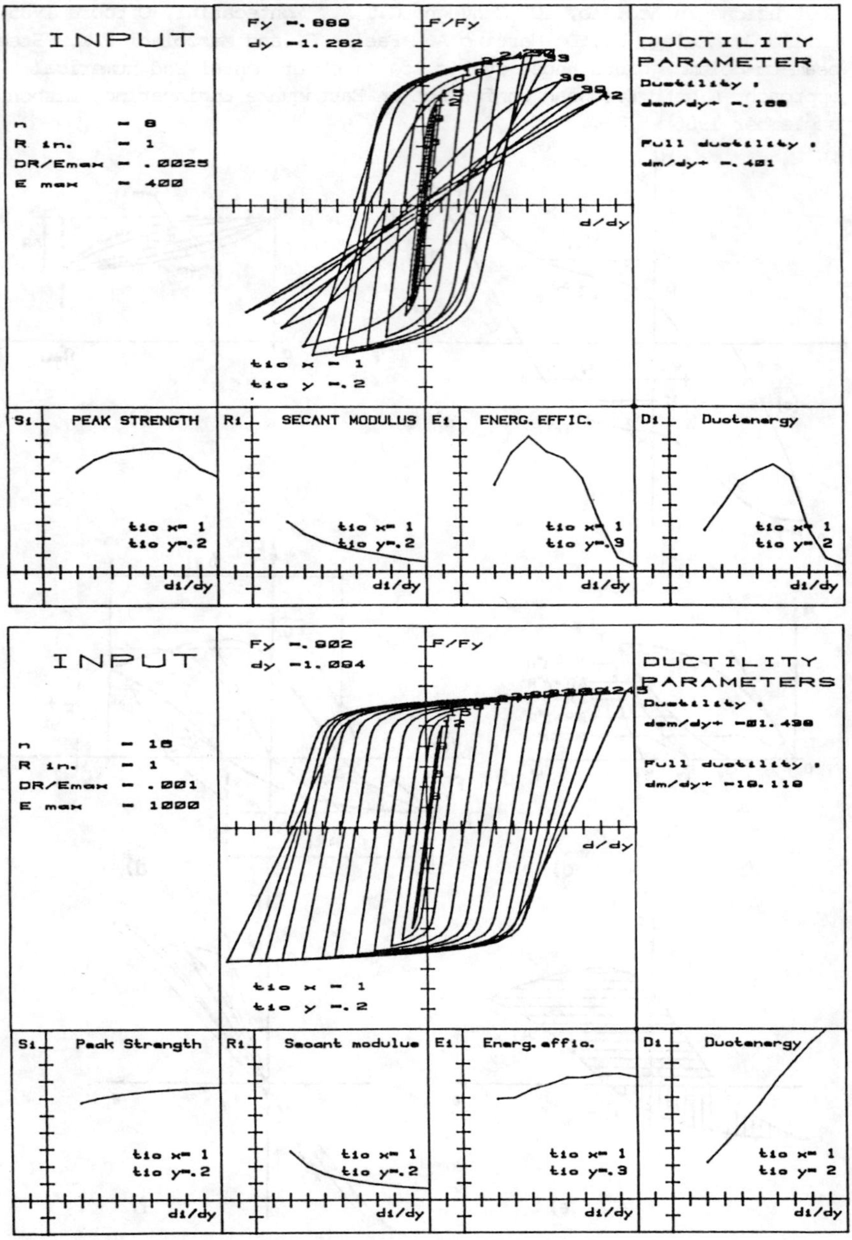

Figure 2 - Simulation of cyclic test results and evaluation of behavioural parameters

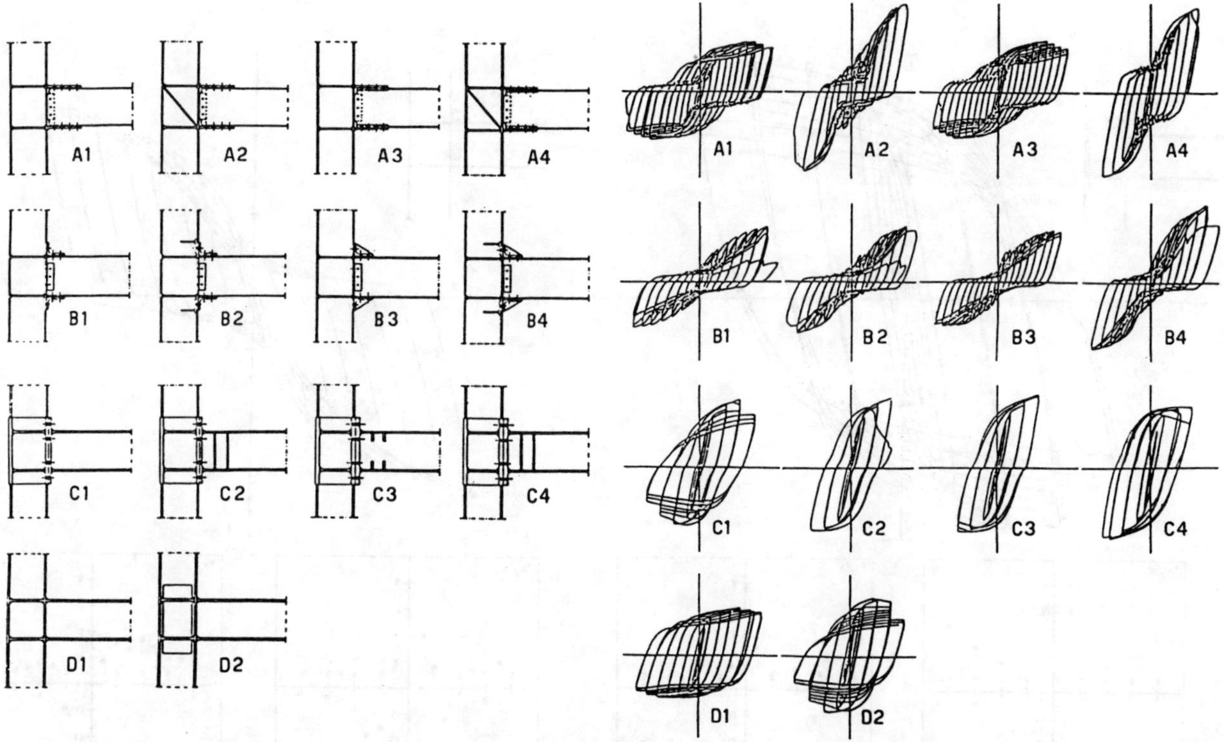

Figure 3 - Tested beam-to-column specimens and experimental hysteretic results

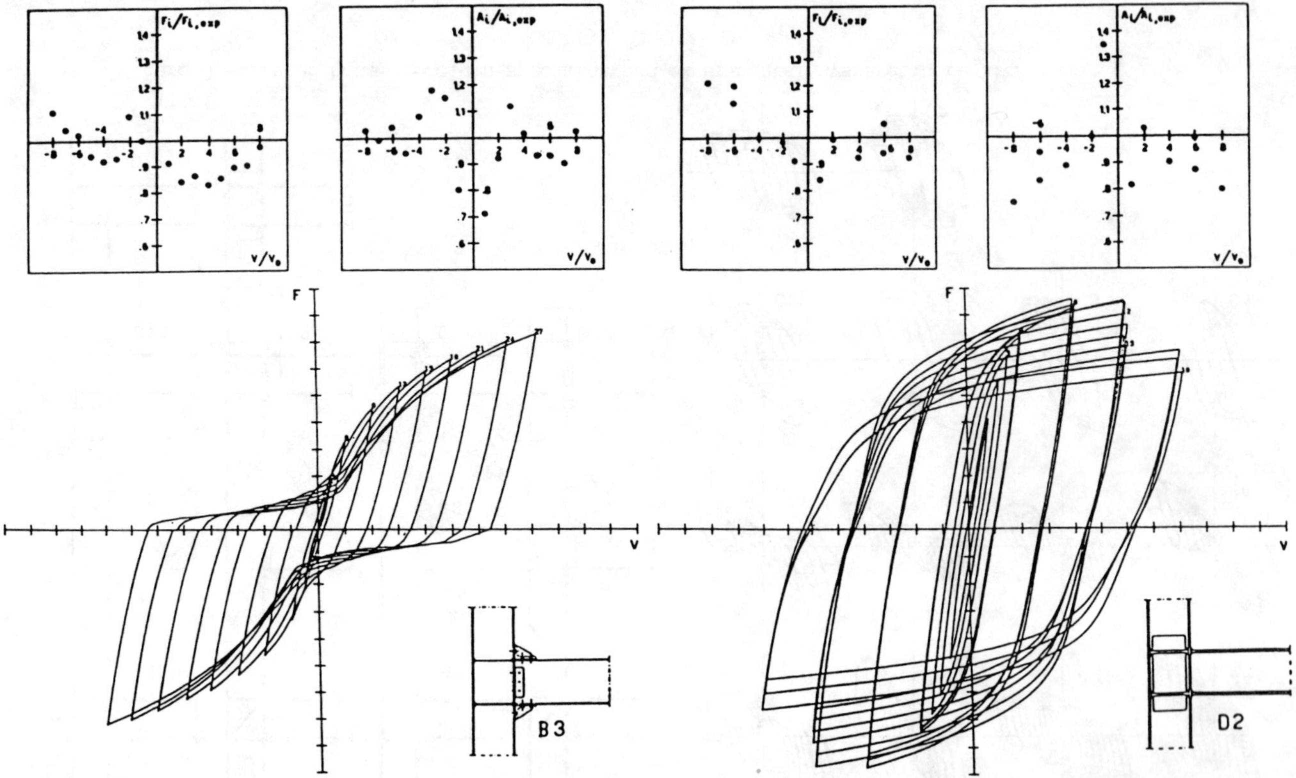

Figure 4 - Simulated hysteretic loops and comparison with experimental results for strength (F_i) and energy absorption (A_i)

Moment-Rotation Relation of Top- and Seat- Angle with Double Web-Angle Connections

by

N. Kishi[1], W.F. Chen[2], K.G. Matsuoka[3] and S.G. Nomachi[4]

ABSTRACT

In this paper, the moment-rotation relationship of top- and seat- angle with double web-angle steel beam-to-column connections is developed. In this development, the initial elastic stiffness and the ultimate moment carrying capacity of the connection are determined by using a simple analytical procedure for modeling the top-, seat-, and web-angles of the semi-rigid connections.

Using the initial connection stiffness and the ultimate moment carrying capacity so obtained, a three-parameter power model is found to be adequate for representing the moment-rotation relationship of this type of connections.

1 Associate Professor of Civil Engineering, Muroran Institute of Technology, Muroran, 050 Japan.

2 Professor and Head of Structural Engineering, School of Civil Engineering, Purdue University, West Lafayette, IN 47907 U.S.A.

3 Professor of Civil Engineering, Muroran Institute of Technology, Muroran, 050 Japan.

4 Professor of Civil Engineering, College of Industrial Technology, Nihon University, Izumi, Narashino, 275 Japan.

1. INTRODUCTION

The connection type considered here is the top- and seat-angle with double web-angle connections as shown in Fig. 1. This connection type has the inherent ductility offered by the flexural deformation of both flange and web angles in the legs attached to the column. In this paper, an analytical procedure is developed to predict the moment-rotation characteristics of this type of connections by determining its initial stiffness and ultimate moment capacity. A three-parameter power model is used here to represent the whole moment-rotation behavior of the connections. The experimental results reported by Altman et al (1982) and Azizinamini et al (1985) are used here to verify the proposed procedure.

2. FORMULATION

2.1 Initial Stiffness

Deflected configuration of the top- and seat- angle with double web-angle connections is shown in Fig. 2, based on the results of Altman et al (1982) and Azizinamini et al (1985). The beam rotated with respect to the stub column, essentially as a rigid member by pivoting about a point near the surface of the beam compression flange.

To determine the initial elastic connection stiffness R_{ki}, the connection is modeled as follows:

1. The beam and the angle legs adjacent to the beam act as rigid members.

2. Materials of the angle legs adjacent to the column flange are linearly elastic and their displacements are small.

3. The center of rotation for the connection is located at the angle leg adjacent to the compression beam flange.

4. The top angle acts as a cantilever beam in which the fixed support line is assumed to be at the fastener-nut's edge near the beam flange in leg adjacent to the column face as shown in Fig. 3.

5. The web angle acts as a cantilever beam similar to the behavior of the top angle shown in Fig. 4.

6. The resisting moment at the center of rotation is neglected.

Based on these assumptions and considering the shear deformation in the leg of the top angle, the horizontal displacement Δ_t of the heel of the top angle corresponding to the beam flange force P_t (Fig. 3) is

$$\Delta_t = \frac{P_t\, g_1^3}{3(EI_t)} \left(1 + \frac{0.78\,(t_t)^2}{(g_1)^2}\right) \qquad (1)$$

in which

EI_t = bending stiffness of the leg adjacent to the column face of top angle,

$$= E\frac{l_t (t_t)^3}{12} \quad (2)$$

$$g_1 = g_t - W/2 - t_t/2, \quad (3)$$

W = nut´s width across flats of bolt,

t_t = thickness of the top angle.

Here, the coefficient of shear deformation κ is taken as κ = 6/5 (Gere and Timoshenko, 1984).

Following the same procedure for the top angle connection, the mean value Δ_a of the horizontal displacement of the angle´s heel corresponding to total force P_a acting in one web angle (Fig. 4) is:

$$\Delta_a = \frac{P_a g_3^3}{3(EI_a)} (1 + \frac{0.78 (t_a)^2}{(g_3)^2}) \quad (4)$$

in which

EI_a = bending stiffness of the leg adjacent to the column face of web angle,

$$= E \frac{l_p (t_a)^3}{12} \quad (5)$$

$$g_3 = g_c - W/2 - t_a/2 \quad (6)$$

t_a = thickness of the web angle.

Assuming the P_a acts at the middle height of the web angle and referring to the deflected configuration shown in Fig. 2, the relationship between the horizontal displacements of each connection (Δ_t, Δ_a) and the connection rotation θ_r are:

$$\Delta_t = d_1 \theta_r \quad (7)$$

$$\Delta_a = d_3 \theta_r \quad (8)$$

in which

$$d_1 = d + t_s/2 + t_t/2 \quad (9)$$

$$d_3 = d/2 + t_s/2 \quad (10)$$

Noting that the two angles attached to beam web and column flange and taking moment about the center of rotation (Point C in Fig. 2), the moment has the value

$$M = P_t \, d_1 + 2 \, P_a \, d_3 \tag{11}$$

Substituting Eqs. (1) and (4) into Eqs. (7) and (8) respectively, and then into Eq. (12), the following relationship between M and θ_r is obtained:

$$M = \left[\frac{3(EI_t)\,(d_1)^2}{g_1\,(g_1^2 + 0.78(t_t)^2)} + \frac{6\,(EI_a)\,(d_3)^2}{g_3\,(g_3^2 + 0.78(t_a)^2)}\right]\theta_r \tag{12}$$

Thus, the initial connection stiffness R_{ki} has the value;

$$R_{ki} = \left[\frac{3\,(EI_t)\,(d_1)^2}{g_1\,(g_1^2 + 0.78(t_t)^2)} + \frac{6\,(EI_a)\,(d_3)^2}{g_3\,(g_3^2 + 0.78(t_a)^2)}\right] \tag{13}$$

2.2 Ultimate Moment

Based on the experimental results reported by Altman et al (1982) and Azizinamini et al (1985), we assume the collapse mechanisms for the top- and seat- angle connection and of the web-angle connection as shown in Figs. 5 and 6, respectively.

Since the distance between the two plastic hinges is rather short compared with each angle´s thickness, the effect of shear force on the yielding of material must therefore be considered. The estimation of the ultimate strength for each connecting part is based on the analytical procedure proposed previously by Chen and Kishi (1987) or Kishi et al (1987).

For the top- and seat- angle connecting part:

Similar to the procedure used previously by Chen and Kishi (1987), the upper plastic hinge H1 locates at a distance $t_t/2$ from the fixed point, the work equation for the mechanism shown in Fig. 5 is :

$$2\,M_{pt}\,\theta = V_{pt}\,g_2\,\theta \tag{14}$$

in which M_{pt} is the plastic moment and V_{pt} is the shear force acting at the plastic hinges in the top angle leg.

Using the Drucker´s yielding criterion (1956) for the combined bending M_{pt} and shear V_{pt}, a 4-th order equation with respect to (V_{pt}/V_{ot}^{pt}) is obtained:

$$\left(\frac{V_{pt}}{V_{ot}}\right)^4 + \frac{g_2}{t_t}\left(\frac{V_{pt}}{V_{ot}}\right) - 1 = 0 \tag{15}$$

in which

V_{ot} = plastic shear capacity of the top angle leg without coupling,

$$= \frac{\sigma_y l_t t_t}{2} \quad (16)$$

σ_y = yield stress of top angle,

g_2' = distance between the two plastic hinges,

$$= g_t - k_t - W/2 - t_t/2 \quad (17)$$

k_t = distance form the top angle's heel to the toe of the filiet as shown in Fig. 5.

The ultimate shear strength V_{pt} can be determined by solving Eq. (15). The ultimate moment at the center of rotation with no shear action is

$$M_{os} = \frac{\sigma_y l_s (t_s)^2}{4} \quad (18)$$

For the web-angle connecting part:

Based on the assumed simple mechanism as shown in Fig. 6, we consider the ultimate strength per unit length at an arbitrary section first, and then obtain the total ultimate strength by integrating them along the plastic hinge line of the web angle.

The work equation for the mechanism at an arbitrary section y in Fig. 6 with the plastic moment M_{py} and the shear force V_{py} per unit length for the one web angle is given by

$$2 M_{py} \theta = V_{py} g_y \theta \quad (19)$$

Furthermore, using the Tresca's yield criterion and the Drucker's yield criterion (1956) similar to the procedure used previously for the top- and seat- angle connection, we obtain the 4-th order equation with respect to (V_{py}/V_{oa}):

$$\left(\frac{V_{py}}{V_{oa}}\right)^4 + \frac{g_y}{t_a}\left(\frac{V_{py}}{V_{oa}}\right) - 1 = 0 \quad (20)$$

in which

V_{oa} = plastic shear capacity per unit length of the web angle leg without coupling,

$$= \frac{\sigma_y t_a}{2} \quad (21)$$

g_y = distance between two plastic hinges at an arbitrary section y,

$$= \left(\frac{g_c - k_a}{l_p}\right) y \qquad (22)$$

k_a = distance from the web angle's heel to the toe of the fillet.

The values of V_{py} can be determined by solving Eq. (20). However, the plastic shear force V_{py} becomes parabolic distribution along with angle height y which has the minimum value V_{pu} at the upper edge of the angle ($y = l_p$) and the maximum value $V_{py} = V_{oa}$ at the lower edge ($y = 0$). To simplify the analytical procedure, we assume a linear distribution for V_{py} as shown in Fig. 7. The resultant plastic shear force V_{pa} has

$$V_{pa} = \frac{(V_{pu} + V_{oa})}{2} l_p \qquad (23)$$

The above value is for one angle connected to beam web and column flange. Thus, the value for double web-angle connection is twice of Eq. (23).

Total ultimate moment capacity:

The ultimate state of the connection is shown in Fig. 8. Using the ultimate strength obtained from Eqs. (14), (15) and (18) for the top- and seat- angle part of the connection and Eq. (23), for the web-angle part, the ultimate moment capacity M_u about the center of rotation (Point C) is given by

$$M_u = M_{os} + M_{pt} + V_{pt} d_2 + 2 V_{pa} d_4 \qquad (24)$$

in which

$$d_2 = d + t_s/2 + k_t \qquad (25)$$

$$d_4 = \frac{(2 V_{pu} + V_{oa})}{3(V_{pu} + V_{oa})} l_p + l_1 + t_s/2 \qquad (26)$$

2.3 Moment-Rotation Relationship

The power model proposed by Richard (1961) and used later by Goldberg and Richard (1963) is adopted here. This model was used recently by Chen and Kishi (1987) and Kishi et al (1987) to represent the moment-rotation relationship of top- and seat-angle connections and single/double web-angle connections respectively. This power model is an effective tool for designer to execute the second-order nonlinear analysis quickly and accurately. This is because the connection stiffness can be determined directly without iterative procedure and has no negative stiffness.

Using the initial connection stiffness R_{ki} and the ultimate moment capacity M_u of the connection, the moment-rotation $(M-\theta_r)$ relationship can be represented adequately by the power model

$$M = \frac{R_{ki}\,\theta_r}{\{1 + (\theta_r/\theta_0)^n\}^{1/n}} \tag{27}$$

in which

θ_0 = reference plastic rotation

$$= \frac{M_u}{R_{ki}} \tag{28}$$

The connection stiffness R_k in Eq. (27) is

$$R_k = \frac{dM}{d\theta_r} = \frac{R_{ki}}{[1 + (\theta_r/\theta_0)^n]^{(n+1)/n}} \tag{29}$$

and the rotation θ_r is:

$$\theta_r = \frac{M}{R_{ki}\,\{1 - (M/M_u)^n\}^{1/n}} \tag{30}$$

3. EXPERIMENTAL VERIFICATIONS

To verify the power model proposed here, the tests reported by Altman et al (1982) and Azizinamini et al (1985) are used. Adding the sub-routine developed in the preceding section to our existing data based program (SCDB) by Kishi-Chen (1986, 1987), the procedures for the assessment of the $M-\theta_r$ relation can be conducted directly and automatically. The SCDB program includes the routines for analytical and numerical estimation of $M-\theta_r$ curves with the exception of those for the control of the data base for semi-rigid steel beam-to-column connections.

A comparison example for $M-\theta_r$-curves is illustrated in Fig. 9. The experimental results are compared with the analytical power model proposed here, the modified exponential model as the curve-fitting method introduced previously by Kishi-Chen (1987) and the polynomial model proposed by Azizinamini et al (1985) as Frye-Morris's model (1975). The experimental data are taken from the each level of ultimate moment capacity. Selecting a suitable value for the shape parameter n, the results obtained by the power model agree rather well with the experimental results similar to that of the polynomial and modified exponential models. It can therefore be concluded here that the proposed power model represents adequately the moment-rotation behavior of the top- and seat- angle with double web-angle connections.

4. CONCLUSIONS

In this paper, the moment-rotation relationship of top- and seat- angle with double web-angle connections is developed. The initial connection stiffness and the ultimate moment capacity of the connection are determined by using simple analytical procedures and are used as two of the three parameters in the proposed power model. The proposed power model is found in a good agreement with available results. The power model can be easily implemented in a second-order analysis.

5. REFERENCES

AISC (1986), Load and Resistance Factor Design Manual of Steel Construction, 1st Ed., AISC, November, Chicago.

Altman, Jr. W.G., Azizinamini, A., Bradburn, J.H. and Radziminski, J.B. (1982) Moment-Rotation Characteristics of Semi-Rigid Steel Beam-Column Connections, Civil Engineering Department, University of South Carolina, Columbia, South Carolina.

Azizinamini, A., Bradburn, J.H. and Radziminski, J.B., (1985), Static and Cyclic Behavior of Semi-Rigid Steel Beam-Column Connections, Civil Engineering Department, University of South Carolina, Columbia, South Carolina.

Chen, W.F. and Kishi, N. (1987), Moment-Rotation Relation of Top- and Seat- Angle Connections, CE-STR-87-4, School of Civil Engineering, Purdue University, West Lafayette, IN.

Drucker, D.C. (1956), The Effect of Shear on the Plastic Bending of Beams, Journal of Applied Mechanics, Vol. 23, No. 4, December, 509-514.

Frye, M.J. and Morris, G.A. (1975), Analysis of Flexibly Connected Steel Frames, Canadian Journal of Civil Engineering, Vol. 2, 280-291.

Gere, J.M. and Timoshenko, S.P., (1984), Mechanics of Materials, 2nd Ed., Wadsworth, Belmont, CA, 660-666.

Goldberg, J.E. and Richard, R.M. (1963), Analysis of Nonlinear Structures, Journal of the Structural Division, ASCE, Vol. 89, No. ST4, 333-351.

Kishi, N. and Chen, W.F. (1986), Data Base of Steel Beam-to-Column Connections, CE-STR-86-26, School of Civil Engineering, Purdue University, West Lafayette, IN, Two Volumes, 653 pages.

Kishi, N. and Chen, W.F. (1987), Data Base of Steel Beam-to-Column Connections, AISC Engineering Journal, Vol. 24, Chicago.

Kishi, N., Chen, W.F., Matsuoka, K.G. and Nomachi, S.G. (1987), Moment-Rotation Relation of Single/Double Web-Angle Connections.

Richard, R.M. (1961), A Study of Structural Systems Having Conservative Nonlinearity, Thesis Presented to Purdue University, West Lafayette, IN. in Partial Fulfillment of the Requirements for the Degree of Doctor of Philosophy.

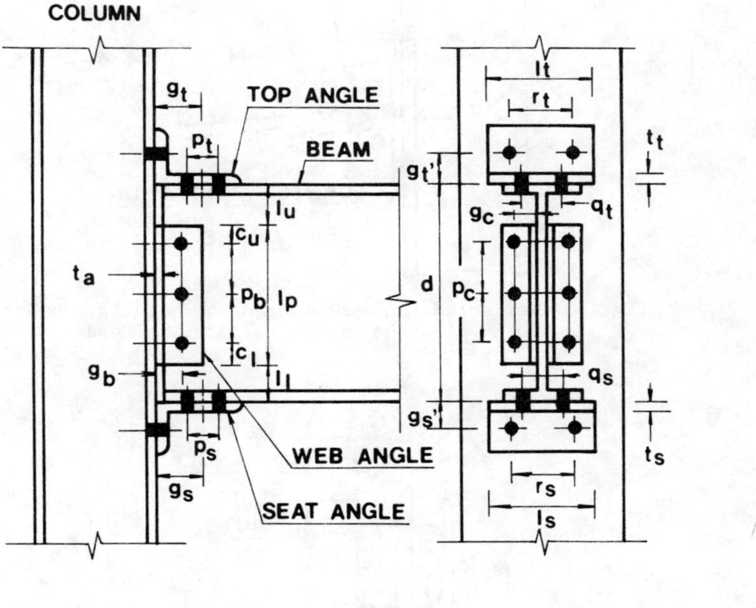

Fig. 1. Typical Top- and Seat- Angle with Double Web- Angle Connection

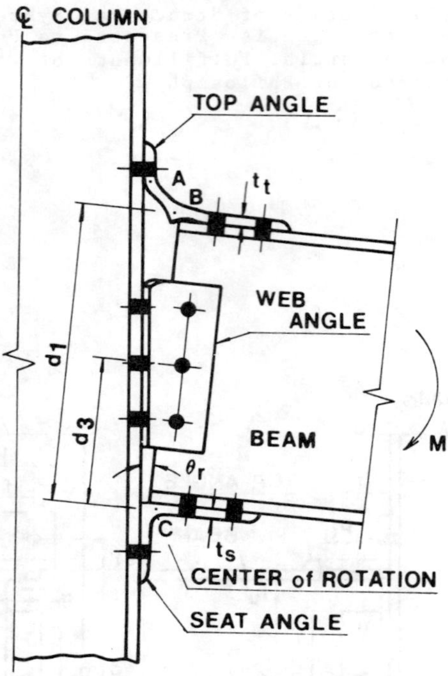

Fig. 2. Deflected Configuration of the Angles Connected to Beam and Column at the Elastic Condition

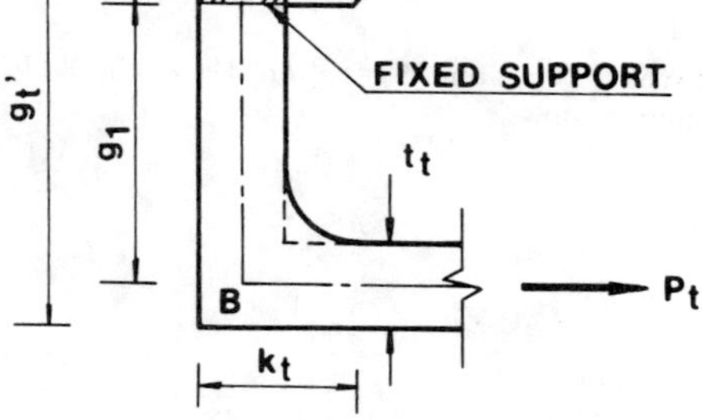

Fig. 3. Cantilever Model of the Top Angle

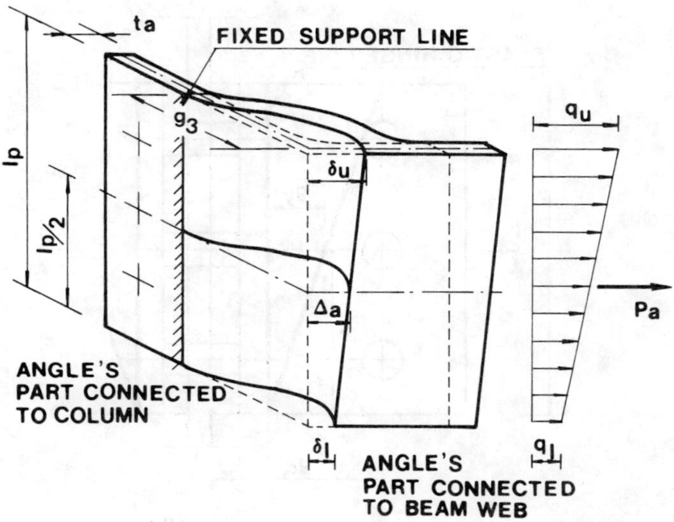

Fig. 4. Deflected Configuration for Cantilever Model of Web Angle

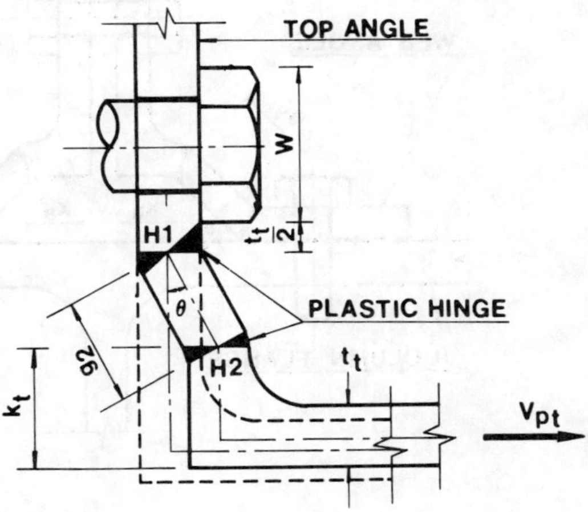

Fig. 5. Mechanism of the Top Angle at the Ultimate Condition

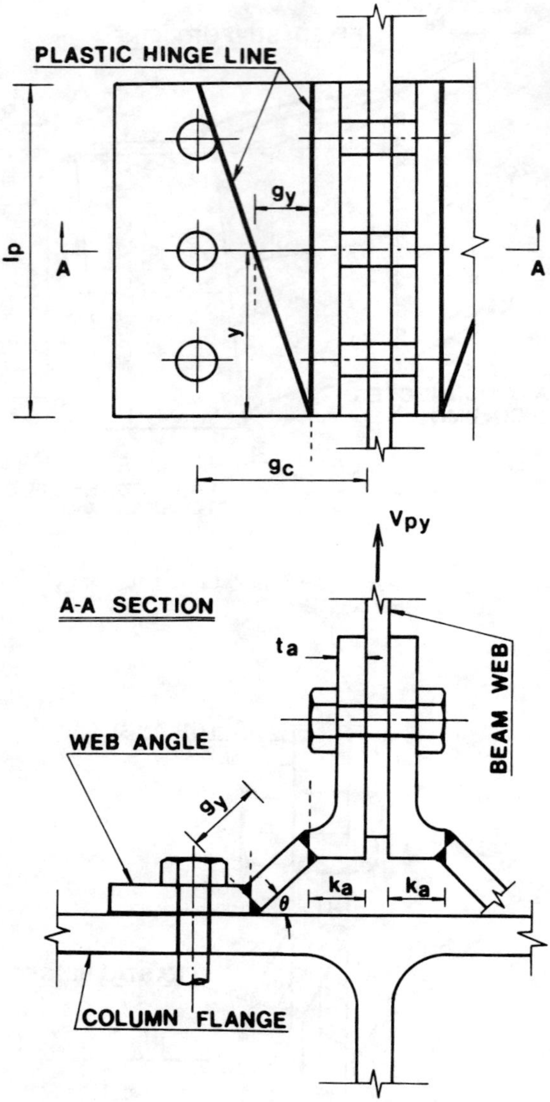

Fig. 6. Mechanism of the Web Angle at the Ultimate Condition

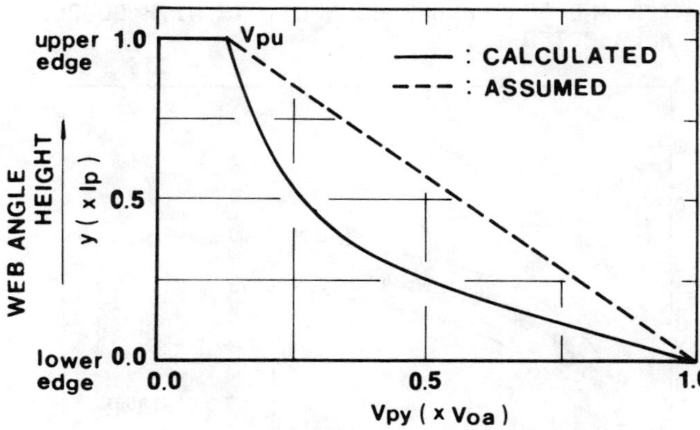

Fig. 7. Distribution of Plastic Shear Force along Web Angle Height

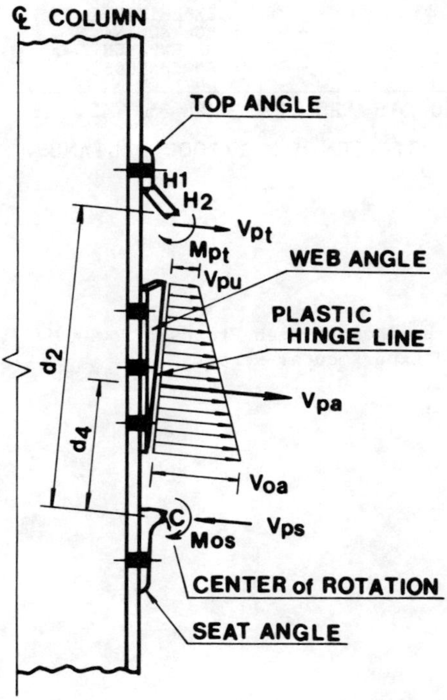

Fig. 8. Applied Forces in the Ultimate State of the Connection

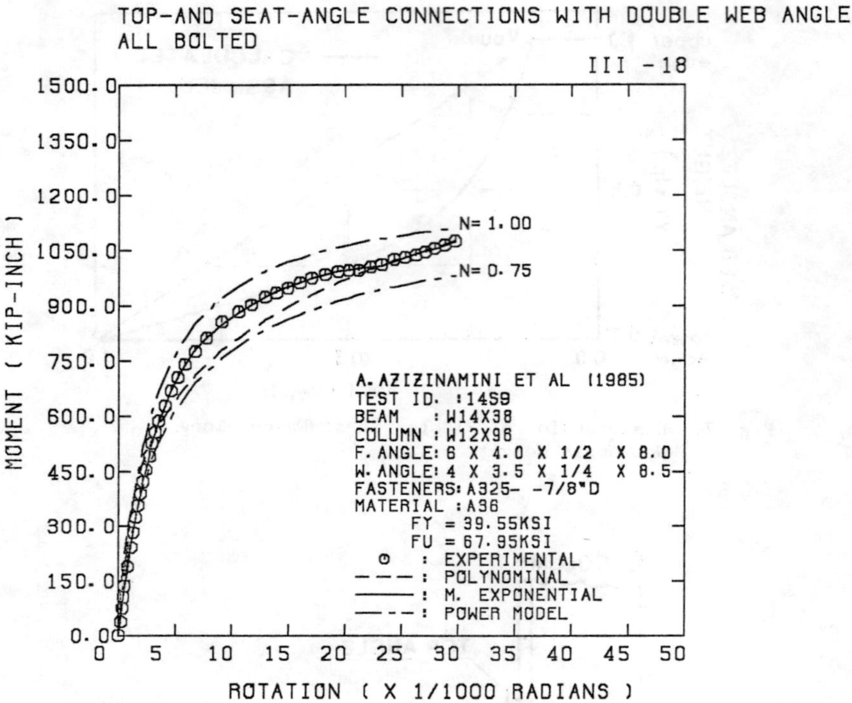

Fig. 9 Comparison Between Proposed Power Model and Experimental Results

Moment-Rotation Relation of Single/Double Web-Angle Connections

by

N. Kishi[1], W.F. Chen[2], K.G. Matsuoka[3] and S.G. Nomachi[4]

ABSTRACT

In this paper, the moment-rotation relationship of single web-angle and double web-angle beam-to-column connections is developed. In this development, the initial elastic connection stiffness is determined by using the simple bending torsion theory and the ultimate moment capacity by the simple plastic mechanism.

The complete moment-rotation relationship of the connections is represented by a three-parameter power model. The analytical model is found to be in a good agreement with available experimental results for single and double web-angle connections.

1. INTRODUCTION

Single and double web-angle steel beam-to-column connections such as those shown in Fig. 1 are widely used in practice. In this paper, an analytical procedure is developed to predict the moment-rotation characteristics of these connections by determining the initial stiffness and the ultimate moment capacity of these connections. Power model used previously by Chen and Kishi (1987) is then adopted to represent the complete moment-rotation behavior of these connections. The experimental results reported by Lipson et al (1968) on single web-angle connections, and by Bell et al (1958), Lewitt et al (1966) and Sommer (1969) on double web-angle connections are used here to verify the proposed procedure.

1 Associate Professor of Civil Engineering, Muroran Institute of Technology, Muroran, 050 Japan.

2 Professor and Head of Structural Engineering, School of Civil Engineering, Purdue University, West Lafayette, IN 47907 U.S.A.

3 Professor of Civil Engineering, Muroran Institute of Technology, Muroran, 050 Japan.

4 Professor of Civil Engineering, College of Industrial Technology, Nihon University, Izumi, Narashino, 275 Japan.

2. FORMULATION

2.1 Initial Stiffness

The general deformation pattern of the web-angle connections is shown in Fig. 2. The experimental results reported by Bell, Chesson and Munse (1958) and by Lewitt, Chesson and Munse (1966) on double web-angle connections showed the following behavior:

1. The center of rotation of the connection was near the mid-depth of the beam during the first few increments of loading.
2. The deformation and subsequent tearing of the connection angles primarily resulted from bending moment, and the effect of shear deformation on the connection behavior was relatively small.

To determine the initial elastic connection stiffness R_{ki}, we assume the web-angle to behave in the following manner:

1. The effect of shear on connection deformation is ignored.
2. The part of angle connected to column behaves linearly elastic while the part of angle connected to beam behaves as a rigid body. The connection deformation is small.
3. The part of angle fastened to column flange acts as a moderate thick plate in which the fixed support is assumed to be at the fastener-nut edge close to the beam web and the concentrated torsional moment is in equilibrium with the connection moment acting at the free edge. (Figs. 3 and 4).

Based on these assumptions and using the simple bending torsion theory, the governing differential equation with respect to angle rotation ϕ of a moderate thick plate has the form:

$$GJ \frac{d^2\phi}{dx^2} - EC_w \frac{d^4\phi}{dx^4} = - m_t \tag{1}$$

in which

J = uniform torsional constant

$$= \frac{1}{3} l_p t^3 \tag{2}$$

C_W = warping constant

$$= \frac{1}{144} l_p^3 t^3 \tag{3}$$

m_t = distributed torsional moment

E, G = Young's and shear moduli respectively.

The boundary conditions are

$$m_t = 0 \tag{4}$$

at the free edge ($x = g_1$);

$$\frac{d^2\phi}{dx^2} = 0 \tag{5}$$

$$GJ\frac{d\phi}{dx} - EC_w\frac{d^3\phi}{dx^3} = M \tag{6}$$

and at the fixed edge ($x = 0$);

$$\frac{d\phi}{dx} = 0, \qquad \phi = 0 \tag{7a,b}$$

in which

g_1 = gage distance from the fixed support line to the free edge line as shown in Fig. 4.

$= g_c - k - W/2$,

where

g_c = distance form the angle´s heel to the center of bolt holes in leg adjacent to the column face

W = the width of fastener´s nut

k = the distance from the angle´s heel to the toe of fillet.

Solving Eq. (1) together with Eqs. (4) to (7) and setting θ_r to ϕ at $x = g_1$, the relationship between the connection moment M and the rotation θ_r becomes

$$M = \frac{\kappa\cosh(\kappa g_1)}{(\kappa g_1)\cosh(\kappa g_1) - \sinh(\kappa g_1)}GJ\theta_r \tag{8}$$

in which $\kappa = \sqrt{\frac{GJ}{EC_w}}$ and putting $\nu = 0.30$,

$$\kappa = \frac{\alpha}{l_p} \tag{9}$$

where $\alpha = 4.2967$.

Introducing the new parameter β to simplify the equation, the connection moment M is given by

$$M = R_{k_i}\theta_r \tag{10}$$

where the initial connection stiffness R_{k_i} is

$$R_{ki} = G\frac{t^3}{3}\frac{\alpha\cosh(\alpha\beta)}{(\alpha\beta)\cosh(\alpha\beta) - \sinh(\alpha\beta)} \tag{11}$$

and β is defined as

$$\beta = \frac{g_1}{l_p} \tag{12}$$

Equation (11) is the initial connection stiffness of single web-angle connections. Then R_{ki} for double web-angle connections is twice the value given by Eq. (11). The initial connection stiffness R_{ki} is now expressed in terms of three parameters t, l_p and g_1.

2.2 Ultimate Bending Capacity

Based on the experimental results reported by Bell, Chesson and Munse (1958), we assume the global collapse mechanism for the web-angle connections as shown in Fig. 5. Since the distance between the two plastic hinges is of the same order of magnitude when compared with the angle thickness, we must therefore consider the effect of shear force on the yielding of the material due to bending.

The work equation for the mechanism at an arbitrary section y in Fig. 5 with the plastic moment capacity M_{py} and the shear force V_{py} is given by

$$2\, M_{py}\, \theta = V_{py}\, g_y\, \theta \tag{13}$$

The Drucker´s yield criterion (1956) for the combined bending moment M_{py} and shear force V_{py}, has the form

$$\left(\frac{M_{py}}{M_0}\right) + \left(\frac{V_{py}}{V_0}\right)^4 = 1 \tag{14}$$

in which M_0 and V_0 are respectively the pure plastic bending moment capacity and shear force capacity per unit length of web angle.

Using the Tresca´s yielding criterion, we have

$$M_0 = \frac{\sigma_y\, t^2}{4} \tag{15}$$

$$V_0 = \frac{\sigma_y\, t}{2} \tag{16}$$

in which σ_y is the yield stress of web angles.

Substituting, Eqs. (13), (15) and (16) into Eq. (14) and rearranging the 4-th order equation with respect to (V_{py}/V_0) is obtained as

$$\left(\frac{V_{py}}{V_0}\right)^4 + \frac{g_y}{t}\left(\frac{V_{py}}{V_0}\right) - 1 = 0 \tag{17}$$

The value of V_{py} can be determined by solving Eq. (17). The plastic shear force V_{py} has a parabolic distribution along the angle height y. It has the minimum value V_{pu} at the upper edge of the angle ($y = l_p$) and the maximum value $V_{py} = V_0$ at the lower edge ($y = 0$). To simplify the analytical procedure, the variation of V_{py} is assumed to have a linear distribution as shown in Fig. 6. The resultant plastic shear force F is

$$F = \frac{V_{pu} + V_0}{2} l_p \tag{18}$$

Taking the moment at the lower edge of the angle, the ultimate moment capacity M_u of the connection is obtained as

$$M_u = \frac{2V_{pu} + V_0}{6} l_p^2 \tag{19}$$

Equation (19) is for single web-angle connections. Thus, for double web-angle connections, the value M_u is twice that of Eq. (19).

2.3 The Moment-Rotation Relationship

The power model, originally proposed by Richard (1961) and later applied by Goldberg and Richard (1963), was used recently by Chen and Kishi (1987) to represent the complete moment-rotation relationship of top and seat angle connections. This power model is found to be an effective tool for designer to execute a second-order nonlinear analysis quickly and accurately. This is because the connection stiffness can be determined directly from the expression without recourse to an iterative procedure. It has no negative stiffness in the power model. We therefore adopt this model for the present web-angle connections.

Using the initial connection stiffness R_{ki} and the ultimate moment capacity M_u of the connections, the moment-rotation (M-θ_r) relationship can be represented adequately by the power model in the form

$$M = \frac{R_{ki}\,\theta_r}{\{1 + (\theta_r/\theta_0)^n\}^{1/n}} \tag{20}$$

in which

θ_0 = a reference plastic rotation

$$= \frac{M_u}{R_{ki}} \tag{21}$$

n = shape parameter

The connection stiffness R_k in Eq. (20) is

$$R_k = \frac{dM}{d\theta_r} = \frac{R_{ki}}{[1 + (\theta_r/\theta_0)^n]^{(n+1)/n}} \tag{22}$$

and the rotation θ_r is

$$\theta_r = \frac{M}{R_{ki}\ \{1 - (M/M_u)^n\}^{1/n}} \tag{22}$$

3. EXPERIMENTAL VERIFICATIONS

To verify the power model proposed here, the tests reported by Lipson et al (1968) for single web-angle connections, and by Bell, Chesson and Munse (1958), Sommer (1969) for the double web-angle connections are used. Adding the subroutine of the proposed power model to the data base program (SCDB) developed previously by Kishi and Chen (1986) at Purdue University, the procedures for experimental verifications of the present model can be conducted in a direct and automatic manner.

Figures 7 and 8 show the comparison of test results with predicted initial connection stiffness R_{ki}, in which test results are based on the Kishi-Chen's modified exponential $M\text{-}\theta_r$ curve-fitting expression. The analytical results agree in general with the experimental values.

The comparison of $M\text{-}\theta_r$ curves with tests is illustrated by Figs. 9 and 10. Figure 9 shows the results for a single-web connection with t = 0.25" and l_p = 5.5". Figure 10 shows the comparison for a double web-angle connection as reported by Lewitt et al (1966). In these figures, the experimental results are compared with analytical values based on the Kishi-Chen's modified exponential model (1987), polynominal model proposed by Frye-Morris (1978) and the power model presented here. The analytical results using the power model agree quite well with the experimental results. It can therefore be concluded here that the proposed power model represents adequately the moment-rotation behavior of web-angle connections.

4. CONCLUSIONS

In this paper, analytical moment-rotation relationships of single and double web-angle connections are developed. The initial connection stiffness and the ultimate moment capacity of these connections are first determined by simple analytical procedures. The three-parameter power model is then used to represent the complete moment-rotation relationship for these connections. The proposed model is found to be in a good agreement with available experimental results.

5. REFERENCES

Bell, W.G., Chesson, E., Jr., and Munse, W.H., (1958), Static Tests of Standard Riveted and Bolted Beam-To-Column Connections, A Progress Report of an Investigation Conducted by the University of Illinois Engineering Experiment Station.

Chen, W.F. and Kishi, N. (1987), Moment-Rotation Relation of Top- and Seat-Angle Connection, CE-STR-87-4, School of Civil Engineering, Purdue University, West Lafayette, IN 47907.

Drucker, D.C. (1956), The Effect of Shear on the Plastic Bending of Beams, Journal of Applied Mechanics, December, 509-514.

Frye, M.J. and Morris, G.A. (1975), Analysis of Flexibily Connected Steel Frames, Canadian Journal of Civil Engineering, Vol. 2, 280-291.

Goldberg, J.E. and Richard, R.M. (1963), Analysis of Nonlinear Structures, Journal of Structural Division, ASCE, Vol. 89, No. ST4, 333-351.

Kishi, N. and Chen, W.F. (1986), Data Base of Steel Beam-to-Column Connections, CE-STR-86-26, School of Civil Engineering, Purdue University, West Lafayette, IN, 47907, Two Volumes, 653 pages.

Kishi, N. and Chen, W.F. (1987), Data Base of Steel Beam-To-Column Connections, AISC Engineering Journal, Vol. 24, Chicago.

Lewitt, C.W., Chesson, E. Jr. and Munse, W.H. (1966), Restraint Characteristics of Flexible Riveted and Bolted Beam-to-Column Connections, Department of Civil Engineering, University of Illinois.

Lipson, S.L. (1968), Single-Angle and Single-Plate Beam Framing Connections, Canadian Structural Engineering Conference, Toronto, Ontario, 141-162.

Sommer, W.H. (1969), Behavior of Welded Header Plate Connections, Master Thesis, University of Toronto, Toronto, Ontario.

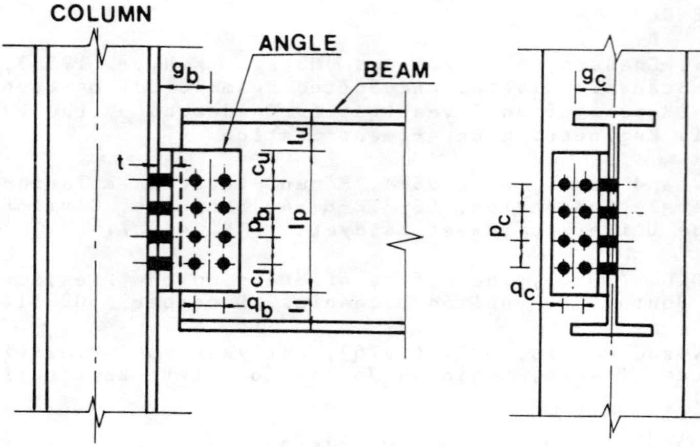

(a) Typical single web-angle connection

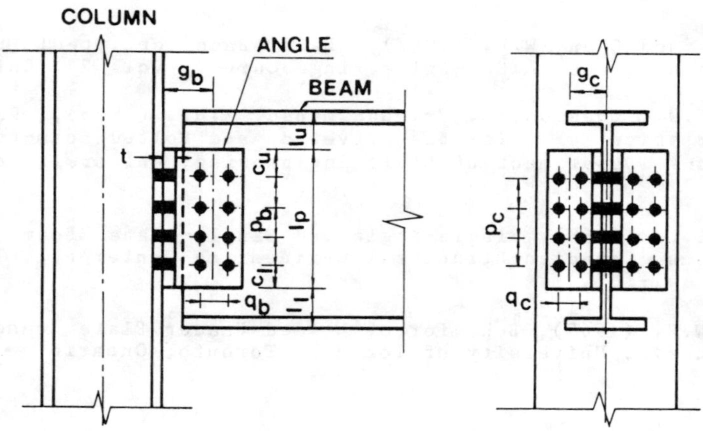

(b) Typical double web-angle connection

Fig. 1. Typical web-angle connections

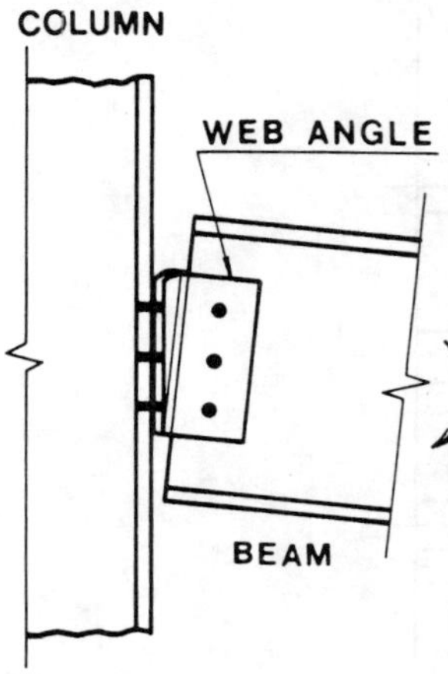

Fig. 2. General deformation pattern of the angle of web-angle connections

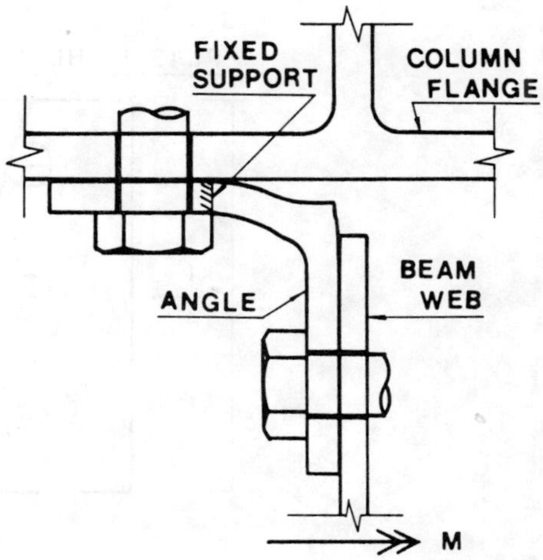

Fig. 3. Deformation of the angle during the initial loading

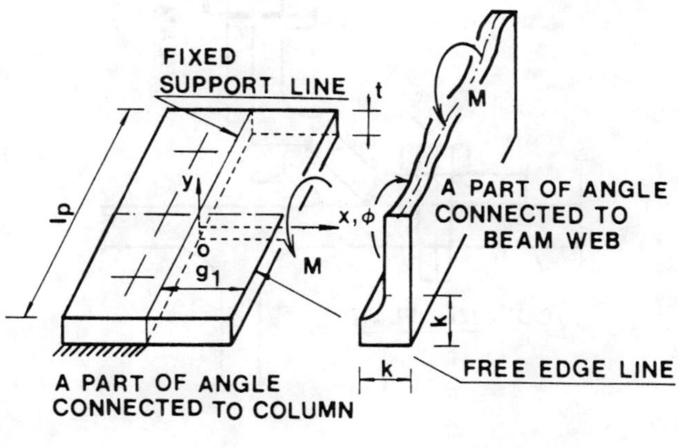

Fig. 4. The moderate thick plate modeling of web-angle

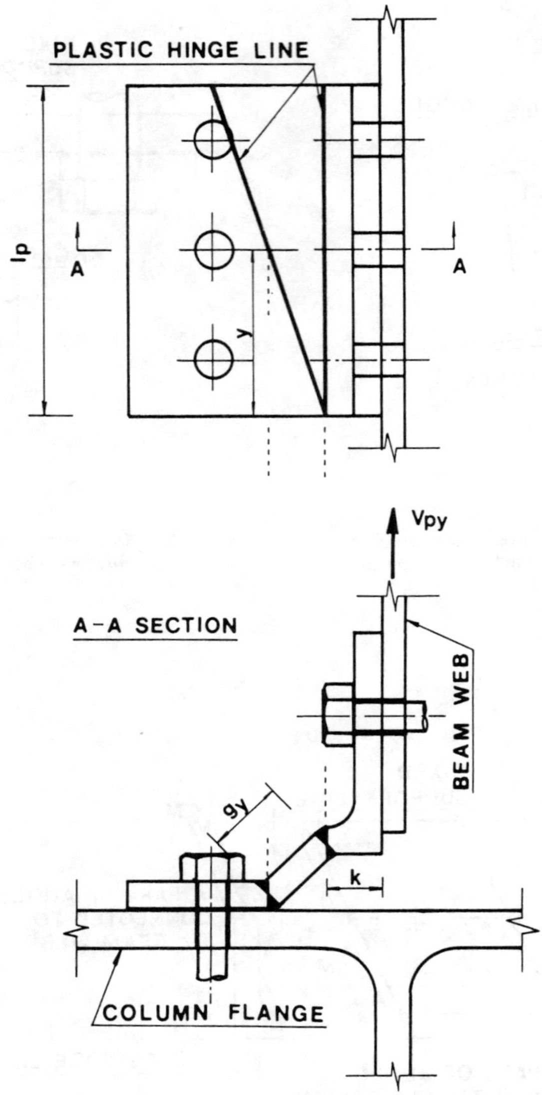

Fig. 5. Mechanism of the part of the angle connected to the column flange at the ultimate condition

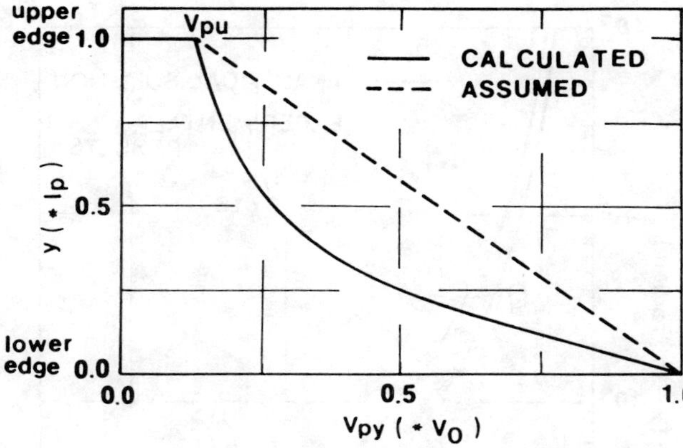

Fig. 6. The distribution of plastic shear force

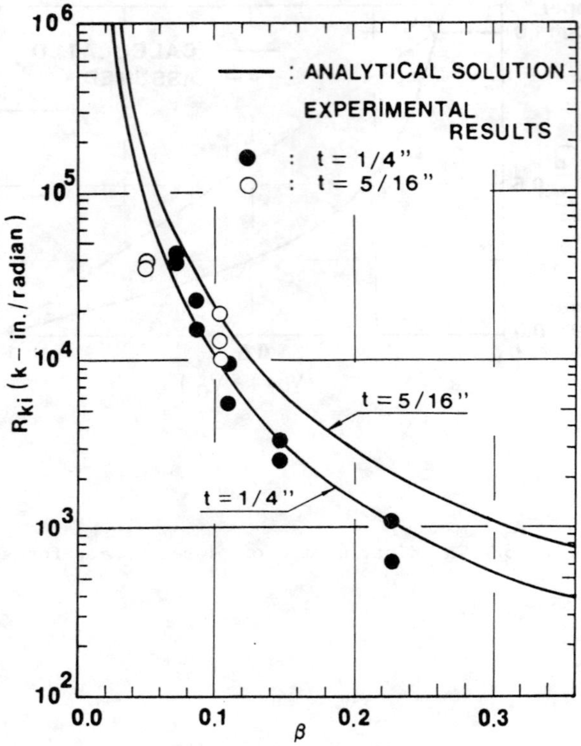

Fig. 7. Initial connection stiffness R_{ki} for single web-angle connections

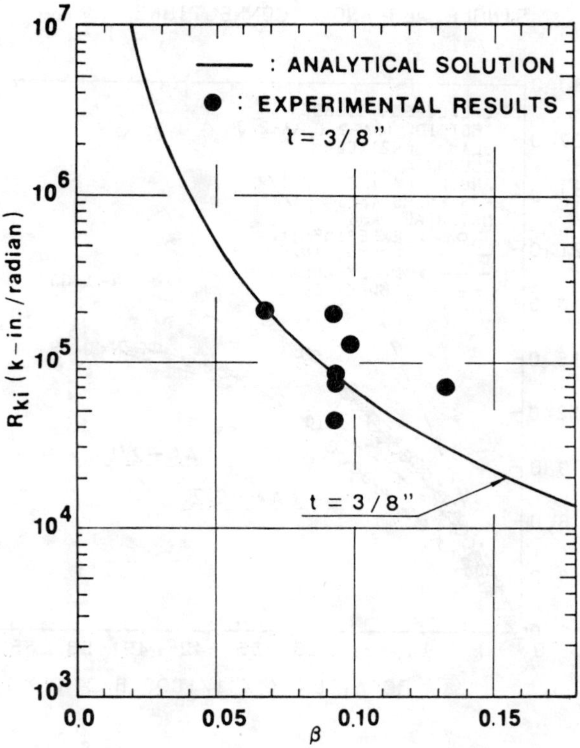

Fig. 8. Initial connection stiffness R_{ki} for double web-angle connections

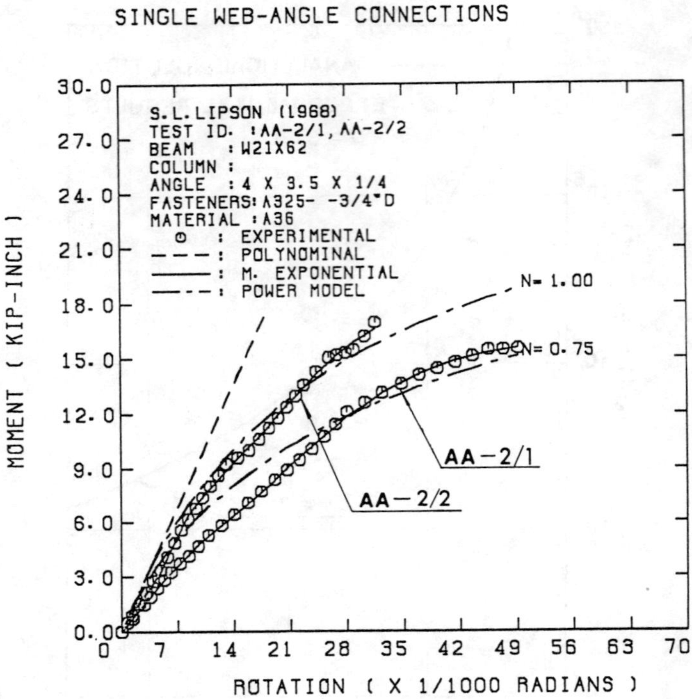

Fig. 9. Comparison of single web-angle connections with t=0.25" and lp=5.5"

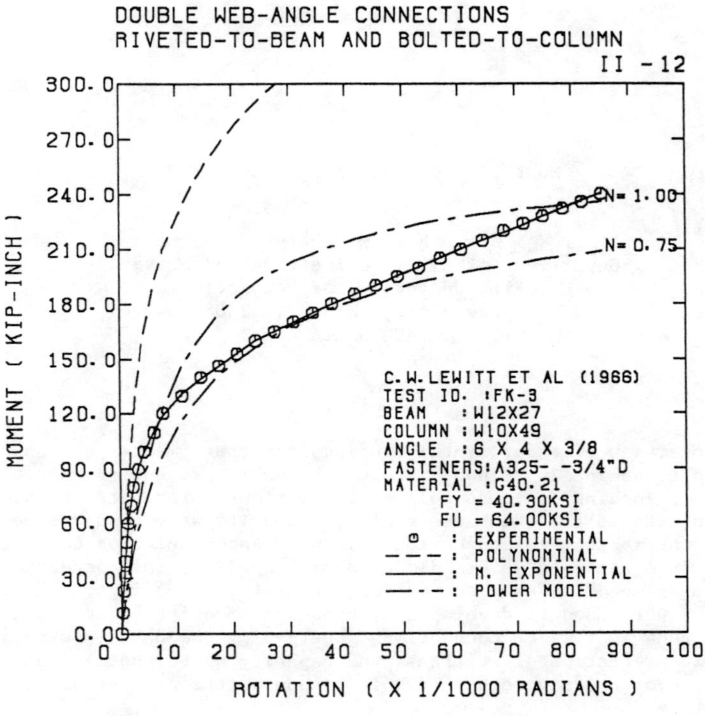

Fig. 10 Comparison of a double web-angle connection

MODELLING OF COLUMN-BASE BEHAVIOUR

R.E. Melchers
Department of Civil Engineering and Surveying
The University of Newcastle
Newcastle, N.S.W., 2308
AUSTRALIA

ABSTRACT

An overview of experimental findings for the moment-rotation behaviour of two-bolt, nominally "pinned", column-bases is presented. Monotonic and alternative loading cases as well as the various parameters of interest are described. Baseplate thickness and bolt capacity were found to be the main factors contributing to connection stiffness and connection capacity. After the baseplate has been deformed due to high loading, the connection tends to behave "pinned" as assumed for subsequent lower loadings. Physical parameters were used to develop a mathematical model to describe the moment-rotation behaviour. For connections governed by the baseplate response, reasonable prediction of stiffness and capacity up to the maximum was obtained. For cases governed by bolt response, the prediction was less satisfactory.

INTRODUCTION

Rules exist in many design codes and in standard texts for the proportioning of column-bases. Typically the parameters considered to be of importance are (i) the thickness of the baseplate; (ii) its plan dimensions; (iii) the weld between column section and baseplate, and (iv) the diameter and strength of the bolts. Generally, the nature of the grout layer between baseplate and concrete foundation is not considered in design, although the strength and extent of the concrete foundation itself is considered.

Interestingly, experimental results on which such recommendations are based are not extensive. For realistic column-bases subject to concentric loading, Bijlaard [1] gave a series of experimental results for connection strength, showing that the strength of the grout layer was particularly important. It was argued that in combination with rules derived from beam column connections, the strength of eccentrically loaded column-bases could also be calculated. However, no experimental verification appears to have been carried out.

Much earlier, Salmon et al [2] had considered eccentrically loaded and nominally "fixed" column-bases. They obtained a maximum strength for moment capacity as limited by the ultimate strength of the bolts, the baseplate in bending and the concrete in compression. They used a model based on the behaviour of reinforced concrete sections. The extensive yield-line patterns given in [1] were not yet in evidence, nor was there consideration of the layer. No experiments appear to have been carried out.

More recently, Picard and Beaulieu [3] gave a rather similar result based on very similar reasoning. However, they also reported tests on 15 connections in which column rotations relative to base rotations were recorded. As might be expected, the nominally "fixed" connections displayed considerable flexibility. This was related to the elastic stiffness of the connections through socalled "fixidity factors". It was found that high axial loads increased the stiffness for small deformations, as also might be expected. Regarding strength prediction, it was found that this was overestimated (sometimes grossly) relative to experimentally observed capacity.

The stiffness properties of connections in steel frames are of importance for frame deflection prediction, and for stability calculations [4, 5]. It is therefore somewhat surprising that for beam-column connections, for which quite extensive experimental and theoretical work has been done [6], there is only limited information about load-deformation behaviour useful for frame analysis [7, 8]. The aim of the present research activity is to try to provide such information also for column-bases. Most effort has thus far been reserved for the column-baseplate-concrete foundation interface region; preliminary studies are in progress for the concrete-soil interface region.

SUMMARY OF TESTS ON "PINNED" BASE CONNECTIONS

The results obtained from full-scale, industry representative column-base connections are described elsewhere in detail [9]. The bases were nominally "pinned", that is, only two holding-down bolts were used, placed symmetrically about the column-web and along the column minor axis. Every attempt was made to replicate good field practice in such matters as bolt tightening, packing placement and grouting. The test program consisted of

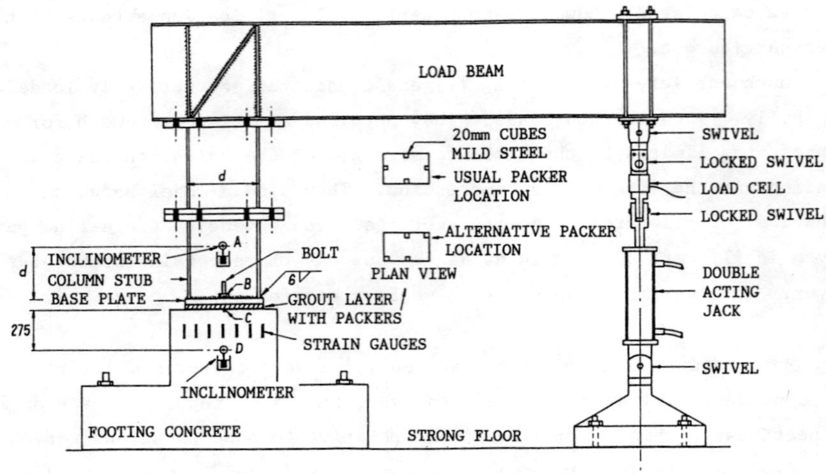

Figure 1. Schematic of Test Rig

26 tests, including 3 different column sizes (460 UB; 310 UC; 310 UB), baseplate thicknesses of 12, 16, 20, 25 and 30 mm and bolt sizes of M20 and M25. In all cases a 6 mm fillet weld was used to connect the column to the baseplate.

To simulate the shims used to align the base during erection of the frame, four 20 mm steel cubes were placed under the base, one each at the mid-points of each baseplate edge.

The loading was applied to the connection through a column stub and lever-arm system as shown in Figure 1. This allowed the ratio moment/axial load to be varied between tests.

It was found that the moment-rotation behaviour of the column-bases depended largely on baseplate thickness. For relatively thin baseplates, behaviour tended to be initially elastic and then became partly plastic. Eventually, strain hardening of the baseplate deformation mechanism governed behaviour. This is illustrated in Figure 2. The baseplates deformed either in the yield-line pattern of Figure 3b or 3c; the pattern of Figure 3a, while theoretically feasible, was not observed in the present test series.

For thicker baseplates, overall behaviour tended to be governed more by bolt deformation characteristics (see Figure 4); in some cases the bolts fractured.

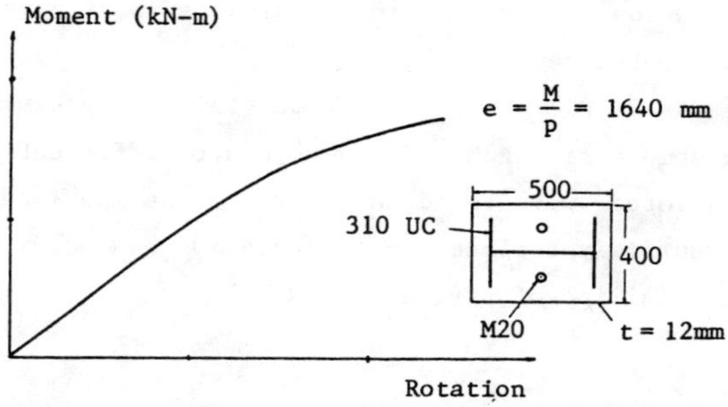

Figure 2. Typical Plate-Dominated Response

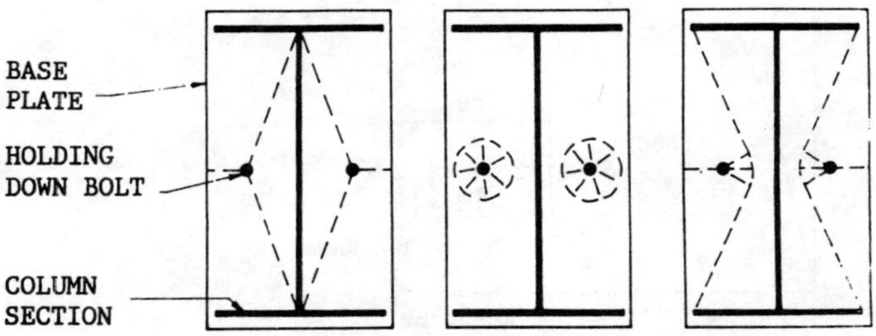

Figure 3. Baseplate Yield-Line Mechanisms

In summary, the following observations were made [9]:

1. Increasing baseplate thickness markedly increased connection stiffness;
2. Baseplate plan size appeared to have only a marginal effect on stiffness;
3. Increased bolt size led to greater stiffness, and, for sufficiently thick baseplates, to higher maximum capacity;
4. Increased column size for given plate size increased connection stiffness;
5. Increased loading eccentricity led to slight increases in bending stiffness and moment capacities;

6. The location and type of the packing (shims) appeared to be of no importance;
7. There was little evidence of grout failure, presumably because axial compression zone forces were not sufficiently large;
8. Prying forces were found to occur only at low loads; at high loads sufficient plate-grout separation had occurred to remove the possibility of prying action.

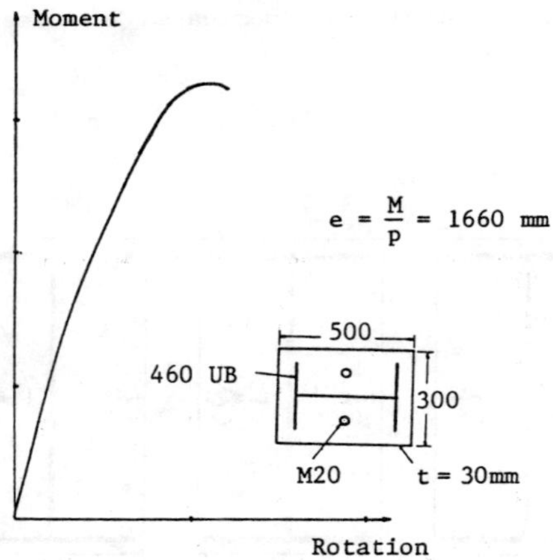

Figure 4. Typical Bolt-Dominated Response

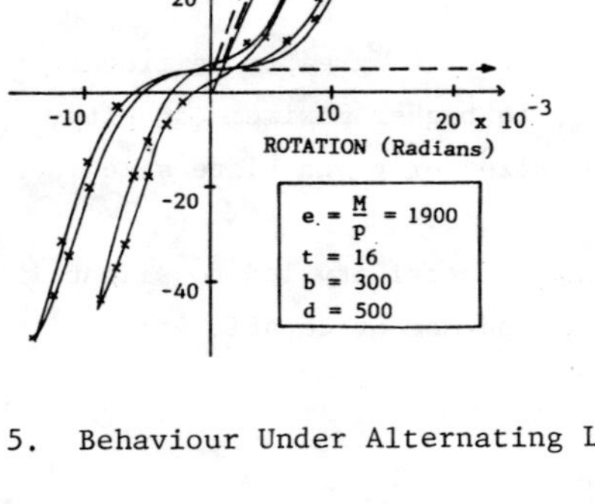

Figure 5. Behaviour Under Alternating Load

One specimen was subjected to alternate loading, see Figure 5. It will be seen that there is evidence of energy dissipation, and also, at higher loads, of permanent baseplate deformation such that reversal from positive rotation to negative rotation can occur under (close to) zero applied moment. This indicates that for bases with sufficiently thin baseplates the connection will indeed behave as a "pin", provided sufficiently high previous loading has taken place to deform the baseplate.

MATHEMATICAL MODELLING

Because it is not feasible to test all possible design combinations for column-base connections, it is appropriate to develop theoretical relationships between applied moment M, axial force P and connection rotation θ. A previous attempt [8] to do so for beam-column connections indicated that a relationship between moment capacity and connection rotation of the form:

$$M = M_p\{1 - \exp[-(K_i - K_p + c\,\theta)\frac{\theta}{M_p}] + K_p\,\theta \qquad (1)$$

might be appropriate. Here M_p is the plastic moment capacity; $K_i = dM/d\theta$ as $\theta \rightarrow 0$ is the initial (elastic) stiffness of the connection; $K_p = dM/d\theta$ as $\theta \rightarrow \infty$ is the strain hardening stiffness of the connection and c is a (dimensional) "constant" found by fitting to the data. It governs the rate of transition from linear elastic behaviour to strain hardening.

K_i may be found from a consideration of the elastic deformation of all components of the connection, in a manner similar to that described in [8]. In the present case, elastic deformation of the baseplate due to the bolt forces as well as that due to the outstand of the baseplate on the compression side was modelled using a finite difference program.

K_p was estimated by substituting the strain hardening modulus E_{st} for E in the calculations for K_i relevant for the baseplate. This method of approach is a well-known technique. Deformation of the bolts was not considered in this way because bolts exhibit very little, if any, strain hardening. Furthermore, strain hardening was found to be associated with the plate dominated failure modes (see Figure 3).

M_p was obtained in the usual way from ideal-plastic theory. Yield stresses obtained from samples rather than nominal values were used.

The results obtained appear to reasonably correspond with the experimentally obtained values, at least for cases dominated by plate behaviour. For such cases, $c \simeq (0.1 - 0.2) \times 10^6$ kNm. Rather less satisfactory is the correspondence between theory and experiment for intermediate and for bolt-dominated cases. Full details have been given elsewhere [10] but Figure 6 indicates some typical results. It is clear that the value of the "constant" c appears to depend on the relative contributions of the baseplate and the bolts to the connection strength.

PRACTICAL IMPLICATIONS

Nominally "pinned" base connections of the type considered herein clearly do possess moment capacity. This remains the case provided the connection is not deformed beyond elastic limits, which would appear quite likely to be the case for much of the life of a typical structure. At greater deformation, baseplate deformation may develop, depending on the thickness of the plate. Thin plates will deform at lower loads and hence will more easily allow the connection to conform to its assumed behaviour mode. This is illustrated by the cyclic test result of Figure 5.

It is possible, with a sufficiently thick baseplate and high capacity bolts to attain the yield point in the column compression flange, indicating that in some cases, quite high moment capacity and high stiffness is possible. The implications for deflection (and also stability) analysis are obvious. What is not so clear is the effect this might have on existing foundations, and whether this matters.

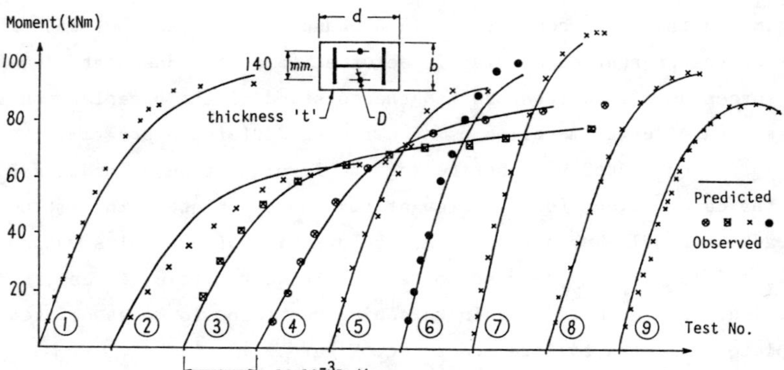

Figure 6. Correspondence Between Test Results and Predicted M - θ Relationship for Tests 1 - 9

CONCLUSIONS

An overview has been given of test results on nominally "pinned" base connections and a preliminary attempt at their mathematical modelling. It was shown that such connections always have strength and stiffness in bending, and, in cases of thick baseplates and adequate bolts, these factors can be significant. However, deformation of the baseplate at higher loads will render the "pinned" design assumption true for subsequent relatively low connection rotation. The present work did not address the possible influence of the behaviour of the soil-foundation interaction, a matter of considerable importance.

ACKNOWLEDGEMENTS

The author is indebted to Mr. K.K. Hon, Graduate Student, who performed the tests and carried out many of the calculations.

REFERENCES

1. Bijlaard, F.S.K., Rekenregels voor het ontwerpen van kolomvoetplaten en experimentele verificatie, Report No. BI-81-51 TNO-IBBC, Delft, 1982.

2. Salmon, C.G., Schenker, L. and Johnston, B.G., Moment-rotation characteristics of column anchorages, J. Struct. Div., A.S.C.E., 1955, 81, April.

3. Picard, A. and Beaulieu, D., Behaviour of a simple column-base connection, Can. J. of Civil Engrg., 1984, 12, pp. 125-136.

4. Galambos, T.V., Influence of partial base fixity on frame stability, J. Struct. Div., A.S.C.E., 1960, 86, ST5, pp. 85-108.

5. Colson, A., Galea, Y. and Pilvin, P., Structures à barres et modélisation du comportement des assemblages, Final Report, 12th Congress I.A.B.S.E., Vancouver, 1984, pp. 661-667.

6. Jones, S.W., Kirby, P.A. and Nethercot, D.A., The analysis of frames with semi-rigid connections, A State of the Art Report, Department of Civil and Structural Engineering, University of Sheffield, U.K., 1979.

7. Frye, M.J. and Morris, G.A., Analysis of flexibly connected steel frames, Can. J. Civil Engrg., 1975, 2, pp. 280-291.

8. Yee, Y.L. and Melchers, R.E., Moment-rotation curves for bolted connections, J. Struct. Engrg., 1986, 112, pp. 615-635.

9. Hon, K.K. and Melchers, R.E., Experimental behaviour of steel column bases, J. Const. Steel Res., (submitted).

10. Hon, K.K. and Melchers, R.E., Moment-rotation curves for "pinned" column bases, (submitted for publication).

ON THE NONLINEAR BEHAVIOR OF JOINTS IN STEEL FRAMES

Ferdinand Tschemmernegg
o.Univ.Prof.Dipl.-Ing. Dr. techn.
Head of Institute of Steel and Timber Construction
University of Innsbruck

ABSTRACT

The paper introduces a method to determine analytically the nonlinear moment-rotation-relationship for joints in structural steel frames. This method was developed on the basis of extensive series performed at the Institute of Steel and Timber Construction, University Innsbruck, Austria. The field of application includes welded joints as well as bolted joints.

INTRODUCTION

At the design of structural steel frames problems are mainly encountered during the design of joints, as the elastic-plastic behavior of the joints is obviously not sufficiently clarified. Furthermore, the influence of joint deflection on the overall behavior of the frame cannot be neglected, especially in the design of unbraced frames.

This paper reports on the results of tests performed on joints and on a method established on the basis of these test results which makes it possible to analytically determine the moment-rotation-relationship for structural steel joints.

MATERIALS AND METHODS

Fig. 1 shows a typical welded joint (fig 1 a) as well as a typical bolted joint (fig 1 b).

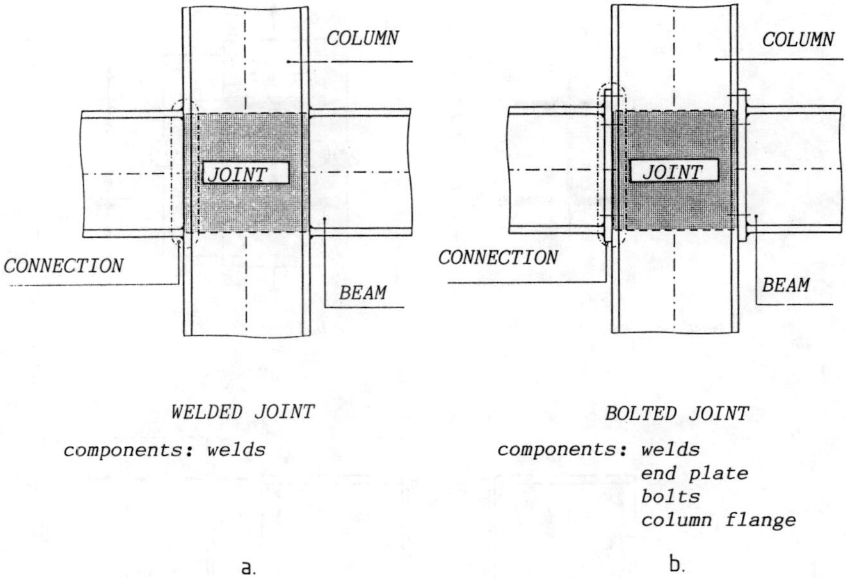

Figure 1

Basically there has to be made a distinction between the flexibility of the joint itself and the flexibility of the connection between joint and the adjacent beam. This distinction together with a macroscopic view of the joint is decisive for the following approach. By this way, it is possible to investigate the behavior of the very complex mechanism taking place inside the joint by separating the influence of joint and connection. In order to analyze the behavior of joint and connections, several different test series were undertaken, as summarized in fig. 2.

Test series 1 served to isolate the forces introduced by the beam flange into the joint and to study this effect separately.
Test series 2, performed on symmetrically loaded joints, allowed for studying the additional effect of beam web on the load introduction into the joint.
Test series 3 - representing an unsymmetrically loaded joint - was performed to study the combined effect of load introduction and shear acting on the joint.
Test series 4 - basically done in the same configuration as series 3 - was undertaken to deliver information on the additional influence on joint behavior of axial forces acting in the column.

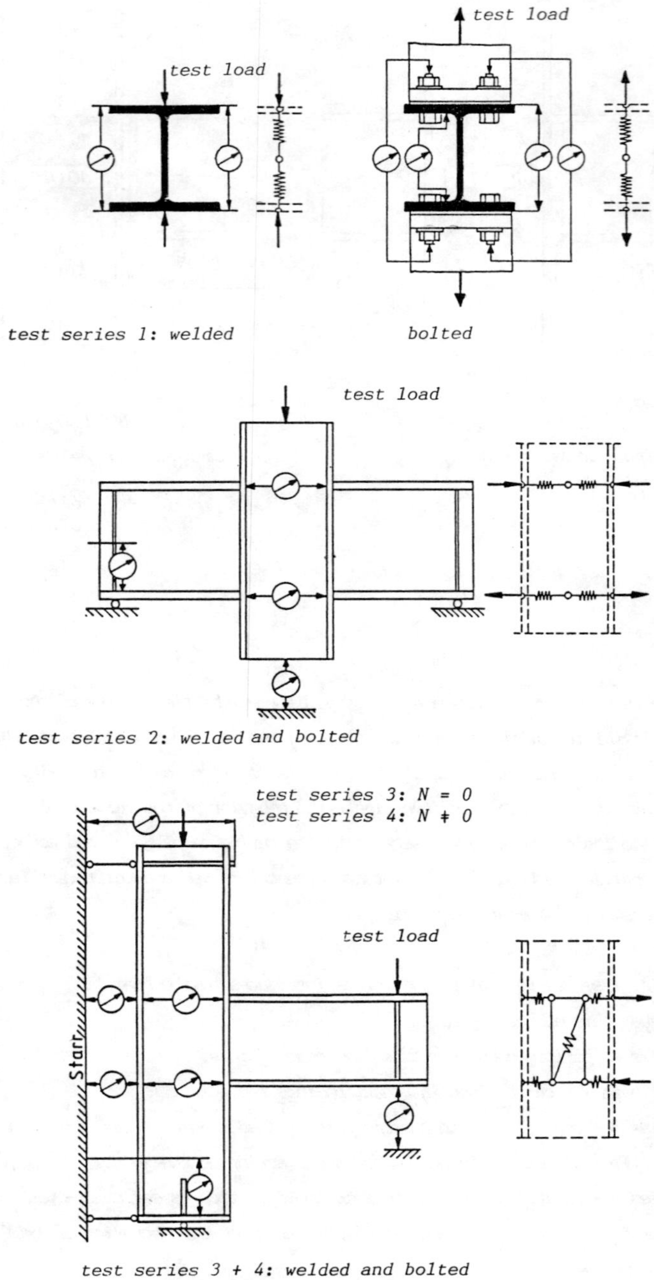
test load
test load
test series 1: welded
bolted
test load
test series 2: welded and bolted
test series 3: N = 0
test series 4: N ≠ 0
test load
Starr
test series 3 + 4: welded and bolted

Figure 2

Refering to measuring methods several publications [1], [2], [3] were issued. In fig. 2 the test specimen and the measuring equipment are drawn schematically.

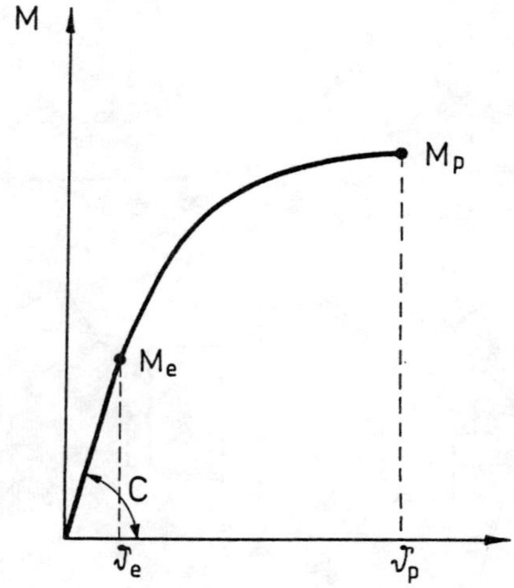

Figure 3

The target of the research programm was to analyze separately the elastic-plastic moment-rotation-curves for load introduction and shear. As represented in fig. 3 the moment-rotation-curve can be determined by the following 4 characteristic values;

- *elastic limit moment*
- *plastic limit moment*
- *elastic limit of rotation*
- *plastic limit of rotation*

RESULTS

Basing on the measurements and a systematic evaluation, the joint was modelled in a step-by-step method with the result that the elastic-plastic behavior of the joint can be described in a pertinent way.

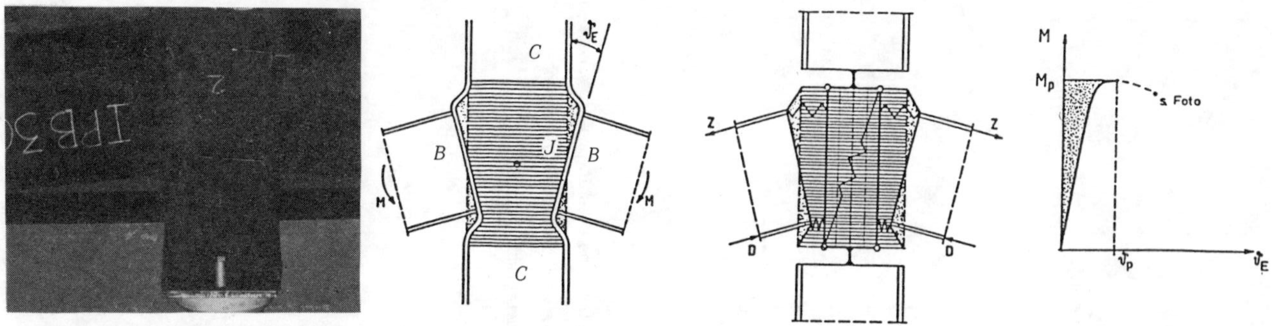

a. symmetrical joint - deformation due to load introduction (ϑ_E)

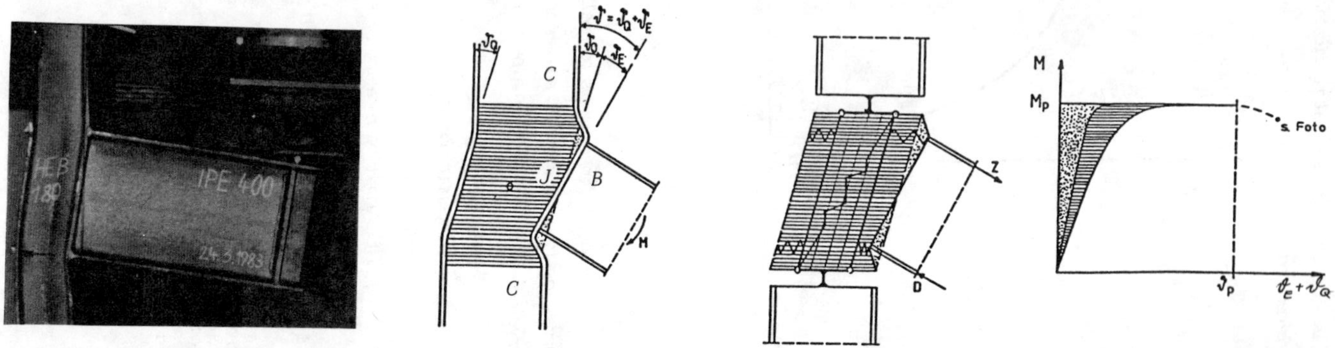

b. unsymmetrical joint - deformation due to load introduction (ϑ_E) and shear (ϑ_Q)

Figure 4

Fig. 4 a, b shows a symmetrical and unsymmetrical joint at the plastic limit.

By this way the general spring model given in fig. 5 was developed incorporating the following springs:

- *shear spring,*
- *load introduction springs,*
- *connection springs.*

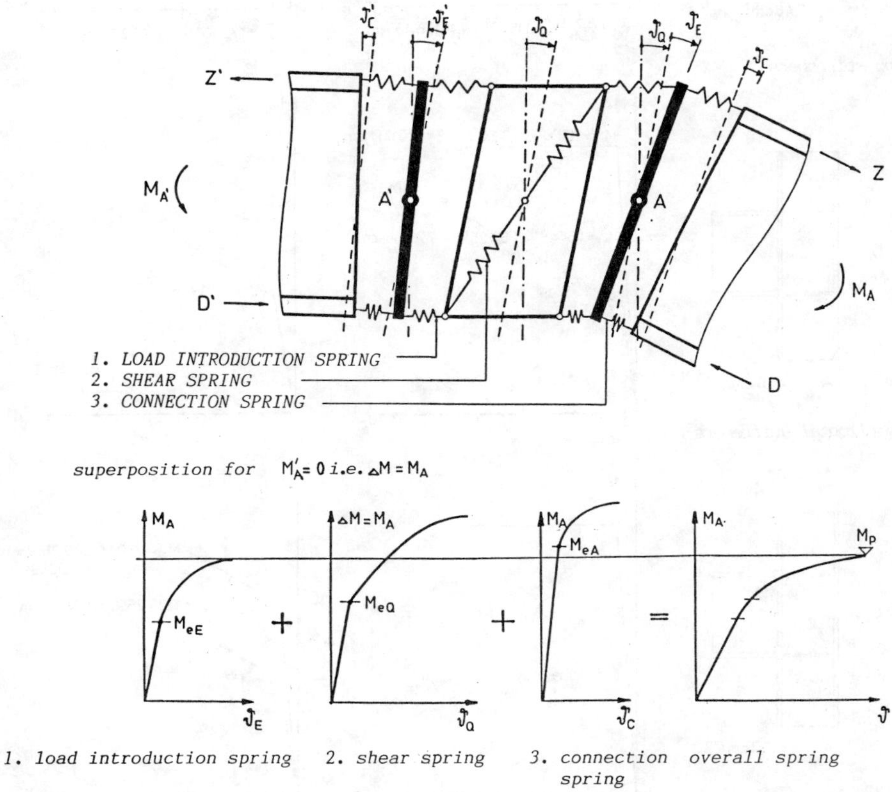

Figure 5

These springs are assembled together in series.

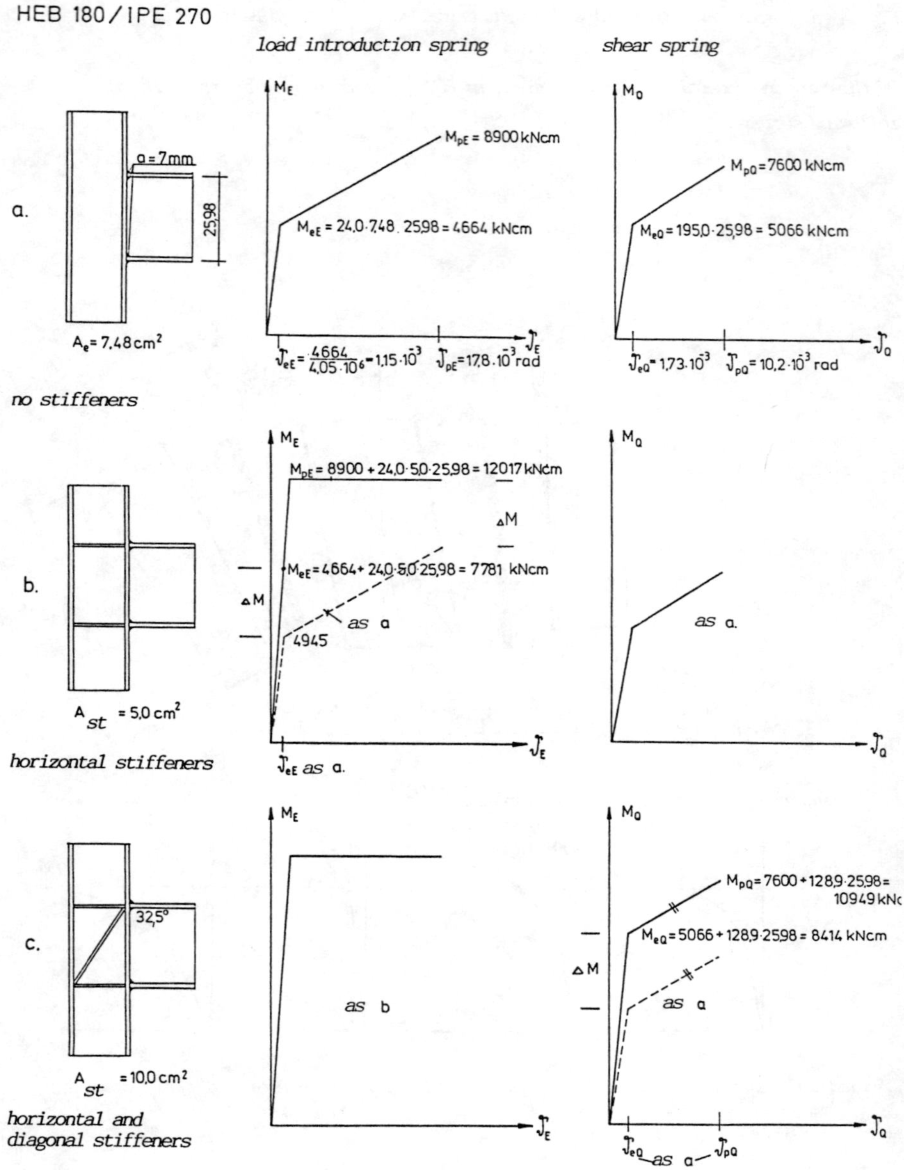
HEB 180/IPE 270
load introduction spring
shear spring
a.
a = 7mm
25,98
A_e = 7,48 cm²
no stiffeners
M_{pE} = 8900 kNcm
M_{eE} = 24,0·7,48 . 25,98 = 4664 kNcm
$\vartheta_{eE} = \frac{4664}{4{,}05 \cdot 10^6} = 1{,}15 \cdot 10^{-3}$
$\vartheta_{pE} = 17{,}8 \cdot 10^{-3}$ rad
M_{pQ} = 7600 kNcm
M_{eQ} = 195,0·25,98 = 5066 kNcm
$\vartheta_{eQ} = 1{,}73 \cdot 10^{-3}$
$\vartheta_{pQ} = 10{,}2 \cdot 10^{-3}$ rad
b.
A_{st} = 5,0 cm²
horizontal stiffeners
M_{pE} = 8900 + 24,0·5,0·25,98 = 12017 kNcm
ΔM
M_{eE} = 4664 + 24,0·5,0·25,98 = 7781 kNcm
as a
4945
ϑ_{eE} as a.
as a.
c.
32,5°
A_{st} = 10,0 cm²
horizontal and diagonal stiffeners
as b
M_{pQ} = 7600 + 128,9·25,98 = 10949 kNc
M_{eQ} = 5066 + 128,9·25,98 = 8414 kNcm
as a
ϑ_{eQ} as a ϑ_{pQ}

Figure 6

Load Introduction and connection spring are acting in series on the applied moments M_A and M_A' respectively, while the shear spring only acts on the difference of applied moments $\Delta M = M_A - M_A'$ (fig 5).

The case shown in fig. 4 ist the welded connection; in this case the springs for the connection can be assumed infinitely rigid. For other connection types moment-rotation behavior could without problems be represented by an elastic-plastic spring.

The overall bearing capacity of the joint is determined by the capacity of the weakest spring.

The results of the tests performed made it possible to supply a method for calculating analytically the characteristic values of the springs for any combination of beams or columns out of the range of European profiles. The moment-rotation-relationship can be evaluated separately for load introduction + shear. The result of such a calculation is shown in fig. 6.
The method for calculation is applicable to unstiffened joints as well as stiffened joints.

D I S C U S S I O N

The moment-rotation curves being determined, calculation of frames can be performed now in a simple way with consideration of elastic-plastic behavior of the joints. This method of frame design is described in detail in [4].
One of the conclusions is that the flexibility of joints is of great influence especially in the design of unbraced frames.

R E F E R E N C E S

1. *KLEIN,H.: Das elastisch-plastische Last-Verformungsverhalten M- steifenloser, geschweißter Knoten für die Berechnung von Stahlrahmen mit HEB-Stützen. Diss.Univ. Innsbruck, 1985.*
2. *HUMER,Ch.: Das Momenten-Rotationsverhalten von steifenlosen Rahmenknoten mit Kopfplattenanschlüssen. Diss.Univ. Innsbruck, 1987.*
3. *BRAUN,Ch.: Das Momenten-Rotationsverhalten von steifenlos ausgeführten Rahmenknoten bei Schubbeanspruchung im überkritischen Bereich. Diss.Univ. Innsbruck, 1987.*
4. *SZS: Rahmentragwerke in Stahl unterer besonderer Berücksichtigung der steifenlosen Bauweise. Structural Steel Association Switzerland, to be published in 1987.*

SEMI-RIGID CONNECTIONS IN FRAMES, TRUSSES AND GRIDS

Dr.-Ing. Ch. Stutzki, Dr.-Ing. J. Lopetegui
and
Prof. Dr.-Ing. G. Sedlacek

Lehrstuhl fuer Stahlbau, RWTH Aachen, W.-Germany

SUMMARY

Several models for the analysis of non-linear connections are presented. The models are suitable to describe the nonlinear behaviour of connections according to the problems to be solved. These models are embedded in a general computer program for the analysis of space frames with geometrical and physical nonlinearity. Three examples are presented in which the connection behaviour is of great influence on the global structure.

BEAM TO COLUMN CONNECTIONS IN BRACED FRAMES

The characteristic property of a connection in a frame is its moment-rotation behaviour. For some commonly used connections the M-ϕ-curves are given in fig.1 in a qualitative way. Not only the linear flexibility, but the whole range of the M-ϕ-curve is of importance, especially, if an ultimate limit analysis for the frame is performed.

Two influences are neglected in the commonly used static analysis:
1) The size of the connection. Usually the adjacent members are connected to an idealized node which is the point of intersection of the axes of the members.
2) The flexible, nonlinear M-ϕ-characteristic.

With a simple model both influences are taken into account, (fig.2). The truss members 1 and 2 in fig.2b are assumed to be rigid, whereas member 3 works as a control member with an axial- force/elongation- behaviour that is controlled by a given curve (fig.2c). If the axial force in member 3 is N then the theoretical moment in joint is $M = N \cdot c$, and the rotation of the joint is $\phi = \Delta l/c$.
The controlling curve for member 3 is defined as a polygon.

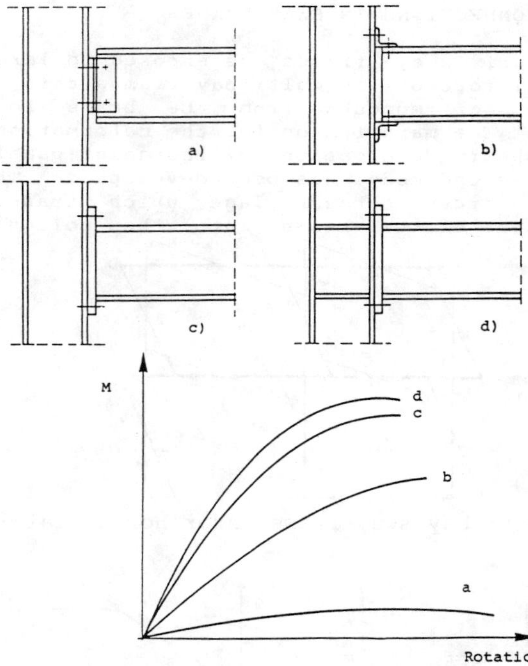

fig.1 : Typical Connections and their moment- rotation-behaviour

The values of the polygon are given as input-data to the program , and may be derived from previous calculations for the joint behaviour or from tests. The model can be inserted into every plane or spatial framework.

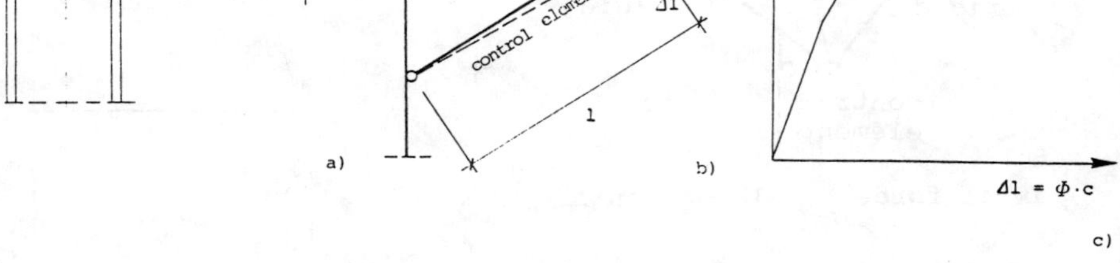

fig.2 : truss model for a beam to column- connection

BEAM-TO-COLUMN CONNECTIONS IN SWAY FRAMES

The connections of sway frames are exposed to large moments due to horizontal forces. In multi-bay frames this leads to antimetric action of moments from the beams to the inner columns (fig.3). The main reason for the deformation of the joint is the shear deformation of the joint panel (fig.4). For this case a second model has been developed. This model consists of a truss subassemblage which transfers the differential moment of the beams into the column (fig.5).

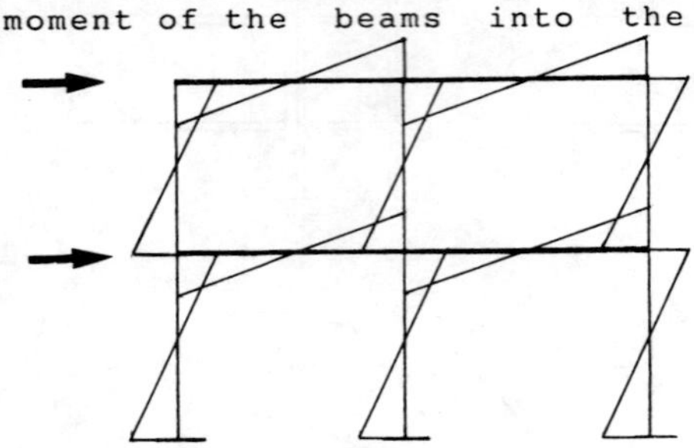

fig. 3 : multi-bay sway frame under horizontal load

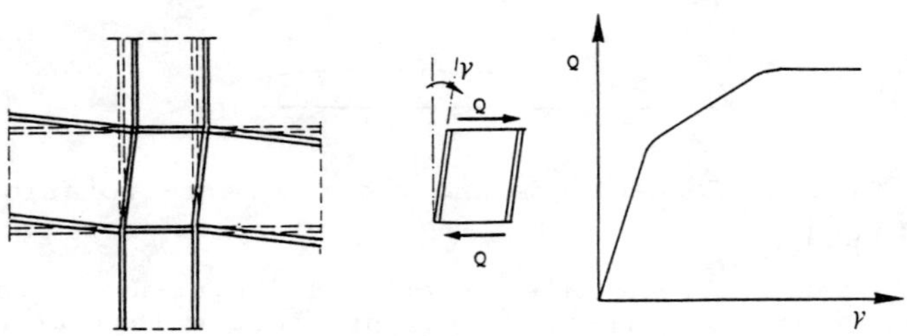

fig.4 : joint deformation due to shear forces

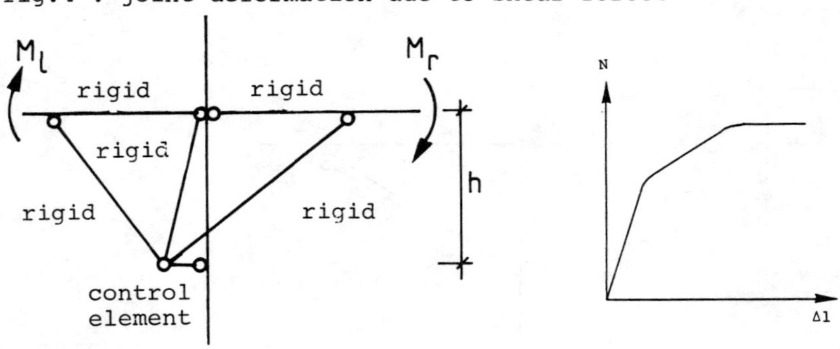

axial force N, elongation Δl

fig.5 : truss model for a shear connection

Also in this model the axial- force/elongation- behaviour of the control element is controlled by a diagram. The diagram is described as a polygon, and its values may be taken from test results or are derived from a previous calculation for the shear panel. The differential moment in the joint is $M = h \cdot N$ where N is the axial force in the control element, and the angle of the shear deformation is $\gamma = \Delta l/h$.
The models for the symmetrical loading and for antimetrical loading can easily be combined, (fig.6) with two control elements for symmetric moment, according to fig.2, and one control elements for the differential moment according to fig.5.

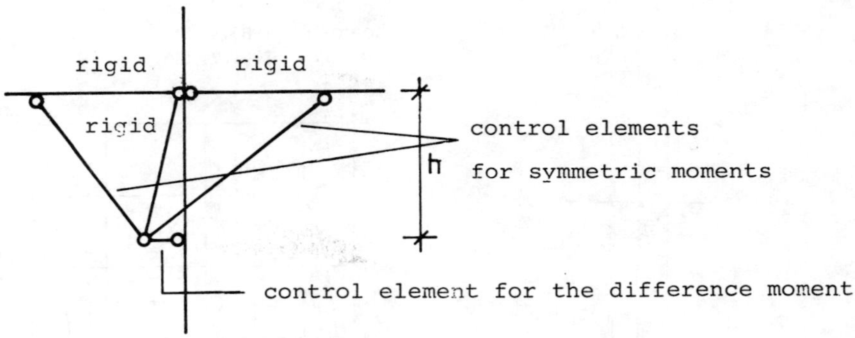

fig.6 : combined model for symmetric and antimetric moments

EXAMPLE 1

Many calculations and comparisons with other methods and test results have proved that the model is able to simulate the joint behaviour and its influence on the total structure with good accuracy. It is embedded in a computer program for space frames with geometrical and physical nonlinear beam elements. The analysis is performed incrementally with an equilibrium iteration in each step. Only the initial stiffness matrix is used, according to the iteration method of orthogonal load-deformation-states /2/. It is not necessary to form a tangent stiffness, neither for the beam elements nor for the joint stiffness. As an example for the application a small study is presented on the influence of different connection types on the ultimate limit state of a plane frame (fig.7).

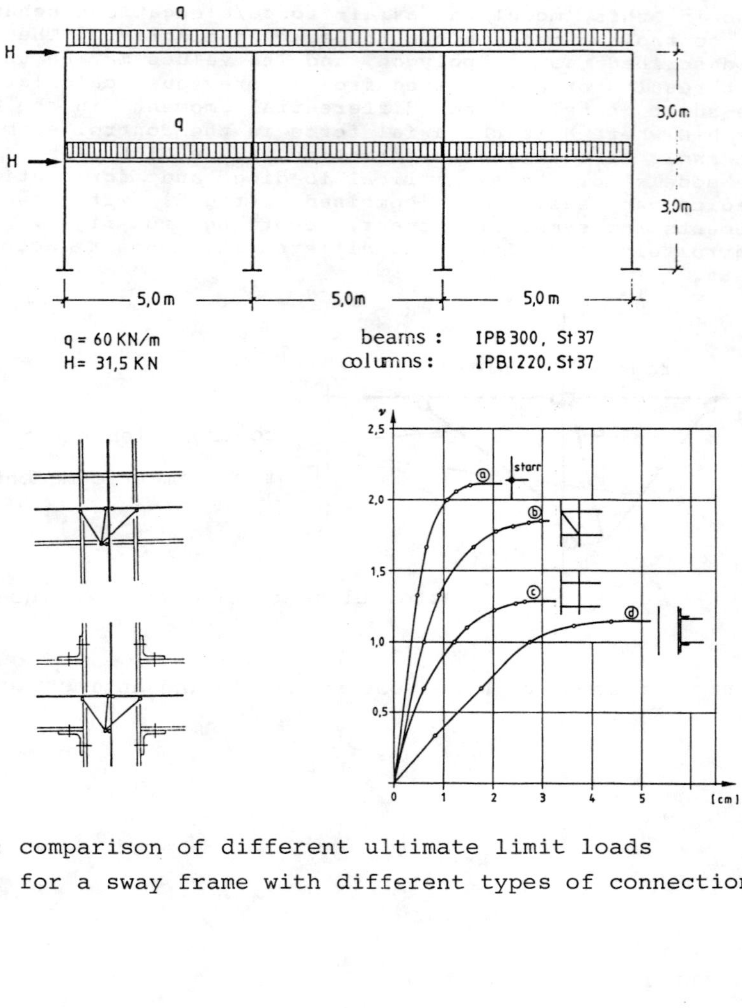

fig.7 : comparison of different ultimate limit loads
for a sway frame with different types of connections

BOLTED CONNECTIONS IN TRUSS MEMBERS OF TRANSMISSION TOWERS

Investigating the consequences of irregular foundation settlement of transmission towers, one ends up with axial forces in some truss diagonals which are much higher than the failure load of these diagonals. This is the result, if a linear analysis is performed with one of the commonly used computer programmes. However failure has not been observed, even if settlements of two or some more centimeters occur. The 'intelligence' of the real structure and some second order

effects provide some additional safety which is not taken into account in a linear static analysis.
A truss model, taking into account bolt slipping (fig.8), has been used to investigate the influence of this connection behaviour on the distribution of internal forces, if in addition to the regular loading, foundation settlement occurs. The behaviour of the truss element is linear, until the value of friction R_H is reached, then slipping may take place, until the possible sliding distance, due to tolerances between bolt and hole, is reached. From then onwards, linear behaviour is assumed again. A complete tower (fig.9 shows the lower part of it) has been analysed, using the described model, with varying values for friction and sliding distance.

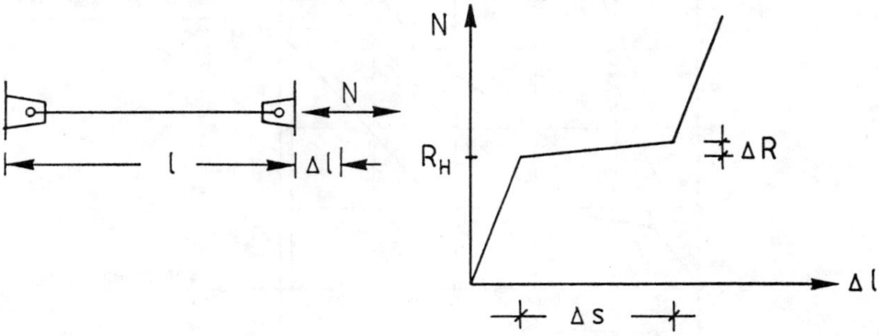

fig.8 : truss model with bolt slipping at the joints

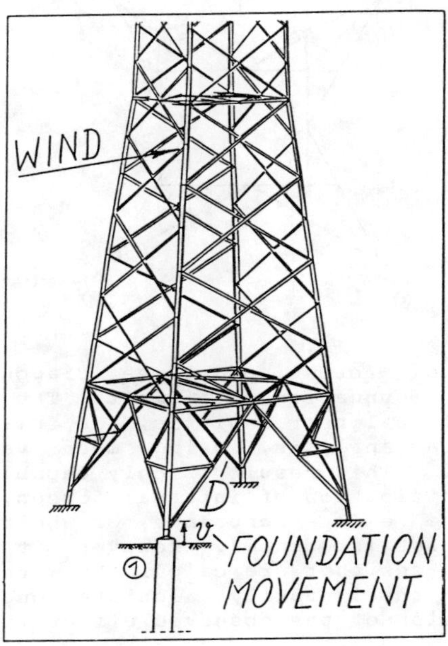

fig. 9

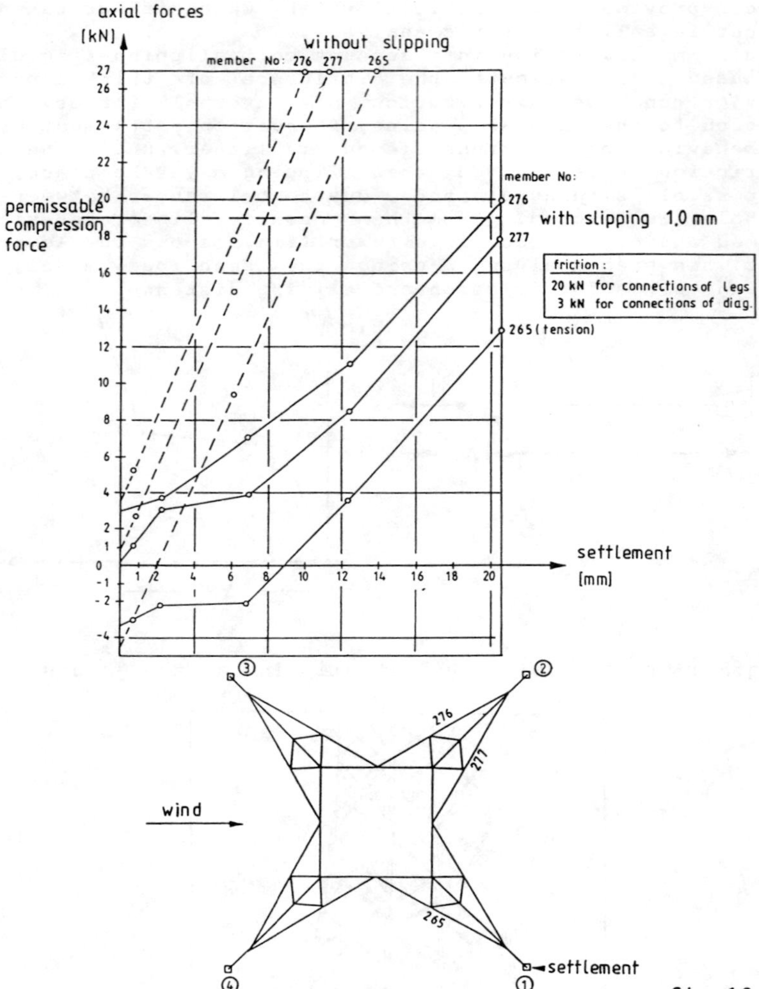

fig.10

In fig.10 the axial forces of three diagonal members are plotted against the foundation settlement. The dashed lines are the result of a linear analysis, the full lines are the result of a nonlinear analysis, using mean values of bolt slipping of 1 mm. This result is only capable of showing a tendency in the distribution of internal forces, not the real forces, because the exact occurance of bolt slipping and friction in each joint is unknown. However it can be seen clearly that the forces that are calculated with bolt slipping are much lower than the linearily calculated ones, which gives a good interpretation of the observed effect (that no failure occurs).

COUPLINGS IN A PONTOON BRIDGE

Pontoon bridges consist of floatable, quasi-rigid boxes of steel that can be connected to each other quickly and without much effort. This means a kind of construction that is simple to handle even in rough sea conditions. For these reasons the connection mechanism has large tolerances in order to achieve a quick assemblage. The static analysis of this type of structure is complicated contrary to the simplicity of its construction, because the deformation behaviour of the couplings determines the global behaviour of the bridge during the crossing of a vehicle.
The connection behaviour is represented in fig.11. The moment-rotation behaviour is bilinear: for small relative rotations of two pontoons the connection behaves like a hinge, and for large rotations the connection is rigid. A simple model has been developed to simulate this type of coupling (fig.12). The shear force can be transmitted by two rigid members, connected by a hinge. The bending moment is transmitted by a control member which behaves according to the diagram in fig.11b, depending on the relative rotation .

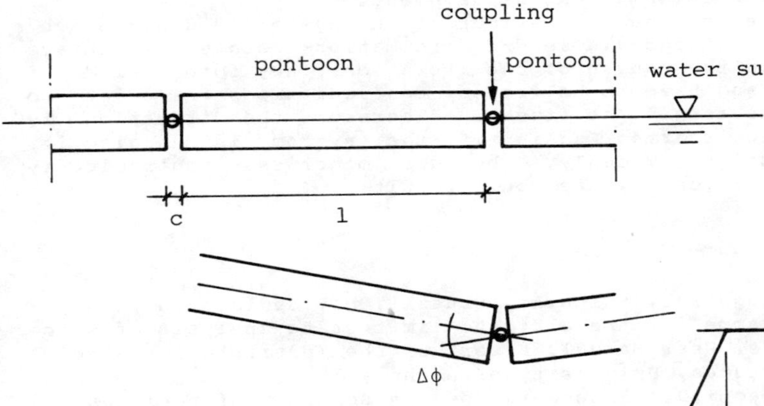

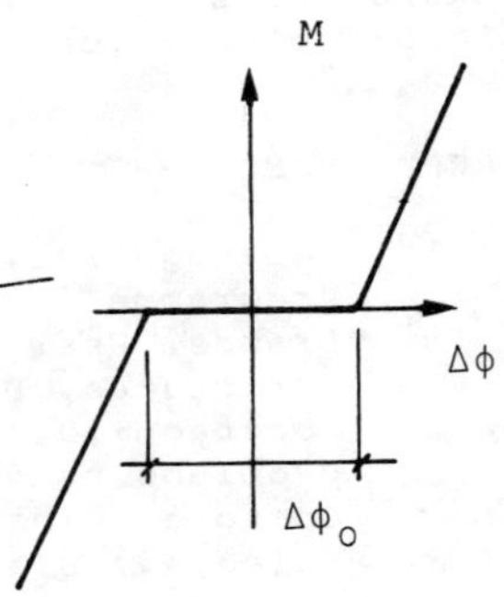

fig.11 : coupling of pontoons

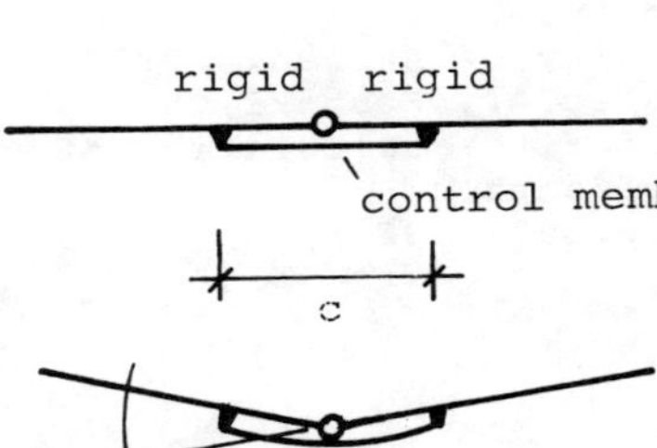

fig.12 : model for the coupling

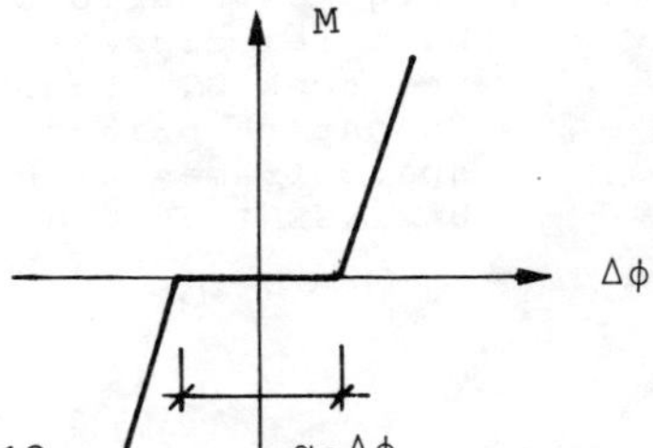

fig.13 : parameter controlled gap

The equilibrium iteration for the complete structure is very difficult, especially for low loading, because any foresight of the deformations may lead to completely wrong results if they are extrapolated from a position with hinges to a final position where the connections are rigid. In order to overcome this difficulty in the analysis, the incrementation is performed not in the load, but in the tolerance $\Delta\phi_0 \cdot \alpha$ of the couplings, with a stepwise increase of the factor α (fig.13) from 0 to 1. With this kind of procedure calculations with very large gaps are possible.

NONLINEAR PROCEDURE

The actual behaviour of a nonlinear system is not described by a tangent stiffness matrix but by so called load- deformation states /4/. A load-deformation-state consists of a deformation increment v_i and of the difference of the reaction forces before and after the application of the incrementl deformation v_i, that is: $R(v_0 + v_i) - R(v_0)$ (v_0 is the total deformation of a previous step). The reaction forces can easily be determined applying the previously described constitutional laws with the deformation as argument.

The complete nonlinear behaviour of the system is described by a set of orthogonal load- deformation- states which are generated in the course of the iteration. They form a kind of a subspace and have to be free of contradictions in order to simulate the tangent stiffness. Therefore the iteration is stopped and started again if the system is changing its characteristics suddenly, because otherwise contradictory load- deformation- states would result.

REFERENCES

/ 1/ Lopetegui,J., Saleh,A., Stutzki,Ch., Sedlacek,G., Programme for the ultimate limit state analysis of space frames. User Manual. Report of the Institute for Steel Structures, University of Aachen, 1987.

/ 2/ Lopetegui,J., Procedure for the solution of nonlinear problems using load - deflection - states. Doctoral dissertation, University of Aachen, 1983

/ 3/ Saleh,A., Ultimate limit state analysis of frames with large deflections and plasticity. Doctoral dissertation, University of Aachen, 1982

/ 4/ Sedlacek,G., Lopetegui,J., Saleh,A., Stutzki,Ch., Computer- orientated methods for the calculation of space frames with nonlinear effects. BAUINGENIEUR, Vol.60, 1985, pp.297- 305

FLEXIBILITY FORMULAE AND MODELLING OF JOINT BEHAVIOUR IN GIRDERS MADE OF RECTANGULAR HOLLOW SECTIONS

A. Czechowski, J. Kordjak, J. Bródka
Metal Structures Research
and Design Centre "Mostostal"
Krucza 16/22, 00-926 Warsaw, Poland

ABSTRACT

This paper deals with the flexibility problem of stepped (unequal width) joints in RHS girders. First of all modes of flexibility for different joints are defined with relevant elastic formulae being established. Then it is shown how those linear characteristics can be extended into non-linear range of joint behaviour. Some comparison with test results is also given. Having known (calculated) joint flexibility one can take it into account in the structural analysis by using so-called micro-bar model of a joint.

INTRODUCTION

In last 15 years a large amount of research, stimulated and co-ordinated mainly by CIDECT and IIW, has been done in the domain of RHS joints [1]. Also in Poland, in "MOSTOSTAL" Centre, an extensive research programme has been carried out, including tests on isolated joints [2 - 4], as well as on complete trusses [6] and, recently, Vierendeel girders.

Since the joint strength itself is not the only aspect of the structural behaviour, some part of investigations has been concerned with the problem of joint flexibility and adequate calculation model. It should be mentioned here, that relatively small number of papers exist in which this particular problem is undertaken, although its implications are practically important and in general must not be ignored in design.

The subject of further considerations are unreinforced truss and moment joints (Fig. 1) consisting of RHS members directly welded together, in which bracing - to - chord (or beam - to - column) width ratio is equal or less than 0.8. Such

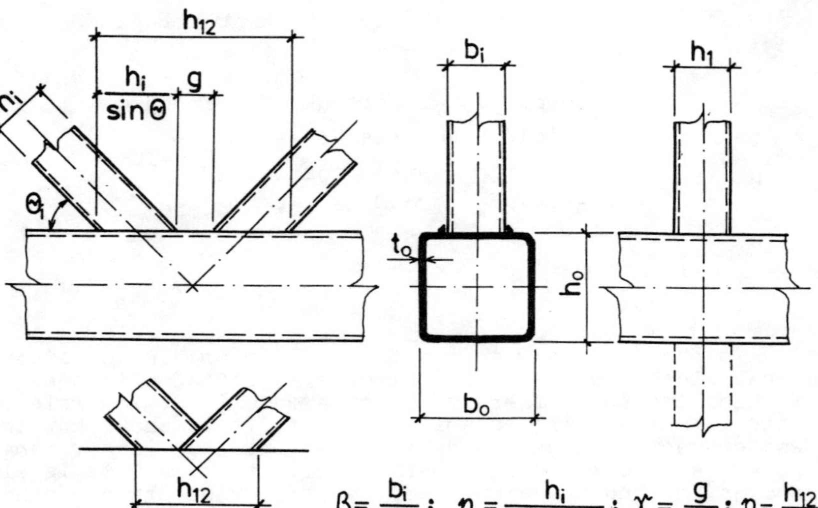

$$\beta = \frac{b_i}{b_o};\ \eta = \frac{h_i}{b_o \sin\Theta_i};\ \gamma = \frac{g}{b_o};\ \eta_{12} = \frac{h_{12}}{b_o}$$

Figure 1. RHS joints - geometrical parameters

joints have to be treated as flexible ones, otherwise than rigid or semi - rigid connections in conventional steel/work systems.

ELASTIC FLEXIBILITY FORMULAE

In order to make the term "flexibility" more precisely comprehended, regarding actual deformations of joints, three basic modes of flexibility are specified (see Fig. 2):

A - translational (typical for axially loaded T and X - joints),

B - antitranslational; push - pull mechanism (typical for truss joints with relatively big gap between bracings),

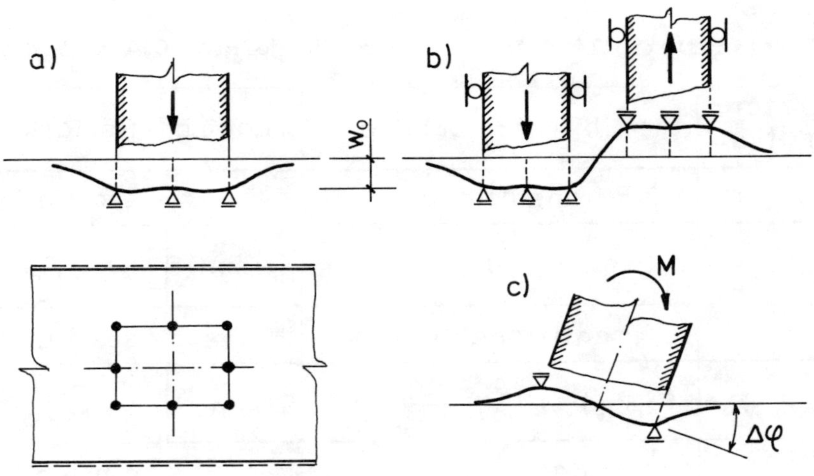

Figure 2. Modes of flexibility

C - rotational (typical for moment T and X - joints, as well as for overlap truss joints).

The above mentioned flexibility cases have then been solved analytically. The analytical approach accepted for this purpose can be characterized, as follows:

a) Chord face is considered first as to be a simply supported strip plate, loaded by one or two rigid hollow punches (bracing members), acting translationally or rotationally respectively;

b) Actual continous plate - punch interaction is replaced by a discrete one, existing in a definite number of points (supports) along the punch perimeter (see Fig. 2a);

c) Using Green's function for a strip plate deflection problem a system of linear equations is established, where unknowns are reactions in points of contact. This procedure is to be repeated for each new set of contact points coordinates;

d) Solution of the above mentioned equations enables a characteristic deflection due to a unit load to be determined;

e) Since the chord face is not simply supported, but is

TABLE 1

Elastic flexibility formulae for RHS joints ($0{,}4 \leq \beta \leq 0{,}8$)

Defor-mation	Flexibility - formula*	Loading	Remarks
	Φ_A - translational		see Fig.2a)
$w_o =$	$f_i(\beta,\eta_o)\cdot\omega_A(\beta)\,\frac{b_o^2}{Et_o^3}$	$\times N\sin\Theta$	$\omega_A = \frac{1}{2}(0{,}91-\sqrt{\beta})$
	Φ_B - antitranslational		see Fig. 2b)
$w_o =$	$f_i(\beta,\eta)\cdot\omega_B(\beta,\gamma)\,\frac{b_o^2}{Et_o^3}$	$\times N\sin\Theta$	$\omega_B = \omega_A\cdot\gamma/\gamma_o \leqslant \omega_A$ $\gamma_o = \sqrt{1-\beta}$
	Φ_C - rotational		see Fig. 2c)
	$\Phi_{C(\varphi)} =$		
$\Delta\varphi = \frac{2w_o}{\eta\cdot b_o} =$	$f_i(\beta,\eta_o)\cdot\omega_C(\beta,\eta)\cdot\frac{2}{\eta\cdot Et_o^3}$	$\times M$	$\omega_C = 2\omega_A/[1-(k-1)\cdot\beta]$ $k = \eta/\beta \geqslant 1$
	$\Phi_{C(N)} =$		
$w_N = \Delta\varphi\frac{z_o}{2} =$	$f_i(\beta,\eta_o)\cdot\omega_C(\beta,\eta)\cdot\frac{z_o^2}{\eta_{12}Et_o^3}$	$\times N\sin\Theta$	
	Φ_{BC} - translational-rotational		see Fig.5
$w_N = w_{NB} + w_{NC} =$	$(\Phi_{NC} + \Phi_{NB})$	$\times N\sin\Theta$	
	$\Phi_{NB} = \frac{\eta_{12}}{\eta_{12}-\gamma}\left(\Phi_B - \frac{2\gamma}{\eta_{12}+\gamma}\Phi_{C(N)}\right)$		for $\gamma = 0$
	$\Phi_{NC} = \frac{1}{\eta_{12}-\gamma}\left(\eta_{12}\Phi_{C(N)} - \frac{\eta_{12}+\gamma}{2}\Phi_B\right)$		$\Phi_B \equiv 0$; $\Phi_{NC} = \Phi_{C(N)}$

*/ $f_i = f_1 = 0{,}25(\sqrt{\eta_o}+\beta)/(1+\beta)$ - for one side joints

$f_i = f_2 = f_1(1+0{,}04/\beta)$ - for two side joints

E - Young's modulus

t_o, b_o - chord wall thickness and width

restrained at the edges by adjacent walls, a frame analogy has been used to take this effect into account by introducing a correction factor.

After having obtained the solutions for different joint parameters approximate flexibility functions of those dimensionless parameters were established. Final elastic flexibility formulae for truss and moment joints are presented in Table 1. There is also included a translational - rotational flexibility typical for gap truss joints. This formula consisting of two flexibility components is based on the assumption, that inner deflection is proportional to antitranslational flexibility, whereas the outer one is proportional to rotational flexibility of a K - joint treated as a T - moment joint ($\eta \rightarrow \eta_{12}$). Thus both flexibility components can be consequently determined.

NON - LINEAR BEHAVIOUR

This problem is closely associated with elastic and ultimate strength, thus with so - called plastic reserve, which being a difference between those two limits, is a specific measure of joint flexibility. In this sense rigid connections exhibit relatively low plastic reserve, whereas joints of low stiffness show a remarkable post - yield behaviour. Many different geometrical, material and technological factors make the problem of joint behaviour including strength very complex and extremely difficult to be solved analytically, although some attempts using finite element method are known [5]. Nevertheless even such an apparently accurate approach need to be experimentaly verified.

In this paper a semi - empirical approach is proposed. Elastic (linear) load - deformation relationship can be extended into non - linear range of behaviour provided that the ultimate strength and the "yield - line" load capacity of a joint are known. Non - linear secant stiffness formula is given in Fig. 3 and compared with experimentally observed T - joint behaviour in Vierendeel girder. A fairly good correlation is obtained for exponent n = 2. For other joints exponent n may be

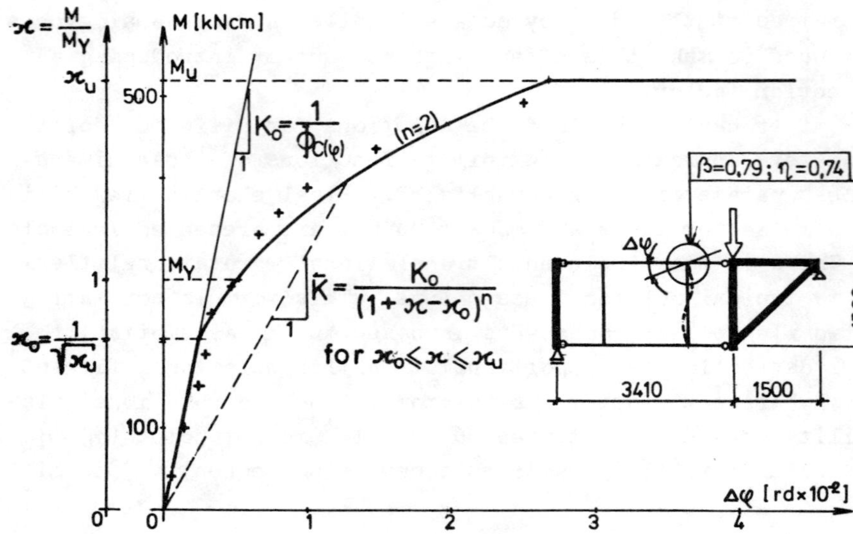

Figure 3. M - Δφ relationship for a moment joint. Non - linear stiffness formula compared with experimantal data

different.

MICRO - BAR MODEL OF A JOINT

After having determined joint flexibility the problem is how to take it into account in statical analysis of a whole structure. For this purpose a micro - bar model of a joint has been devised. The idea of this model is illustrated using a general example of a gap truss joint with translational - rotational flexibility. Respective components of flexibility are taken into account by adequate reducing cross - section area of bracing members (see Fig. 4) and by introducing a system of micro - bars which is "responsible" for rotation only (see Fig. 5).

In contradistinction to conventional systems with pin - ended or rigidly connected members, that one involving micro - bar models of flexible joints may be called "second order sy-

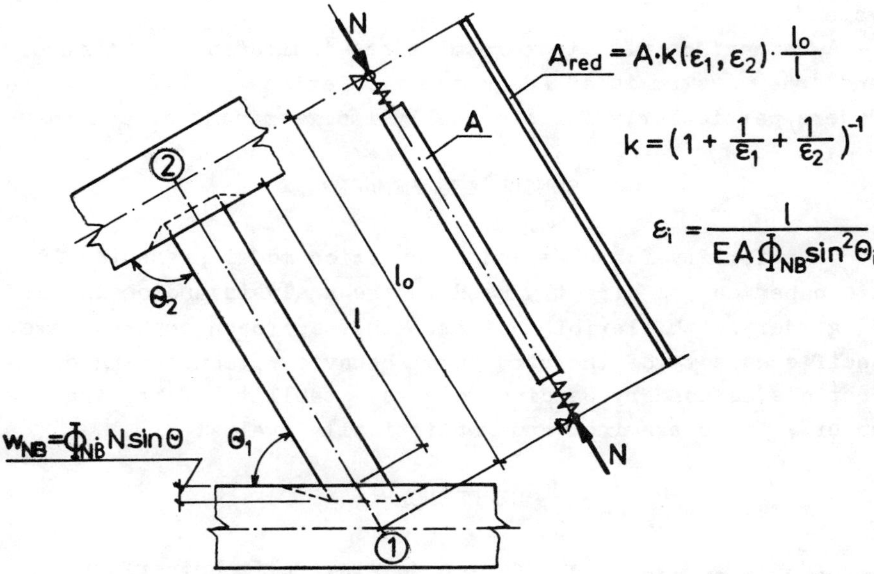

Figure 4. Equivalent cross - section area of a bracing member reduced due to translational flexibility of joints

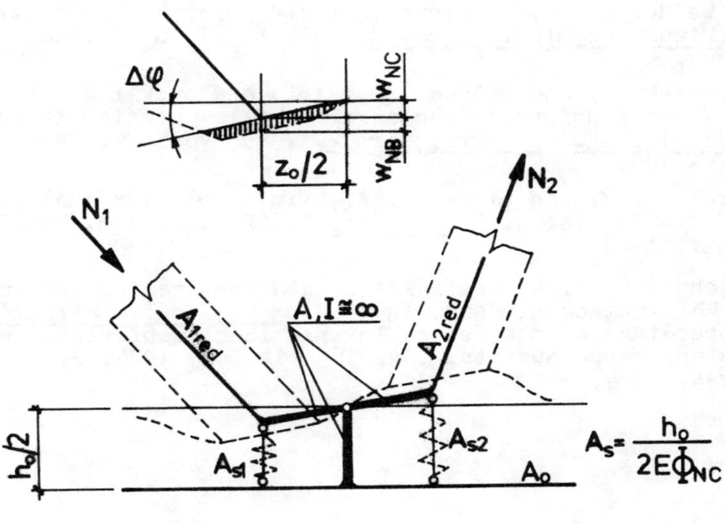

Figure 5. Micro - bar model of a gap truss joint with translational - rotational flexibility

system".

A comparison with test results has indicated [6], that second order system is an adequate calculation model for lattice girders, particularly for statically indeterminant structures.

CONCLUDING REMARKS

Flexibility formulae and calculation model presented in this paper can be directly used in the analysis and design of RHS girders with flexible joints. This approach enables some specific aspects of the structural behaviour, concerning deflections, secondary bending moments, stability of bracing members, to be examined and realistically evaluated.

REFERENCES

1. Wardenier, J., Hollow section joints, Delft University Press, Delft, 1982.
2. Czechowski, A. and Bródka, J., Étude de la résistance statique des assemblages soudés en croix de profilés rectangulaires, Construction Metallique, n° 3, 1977.
3. Życiński, J., Czechowski, A. and Bródka, J., Static strength of welded K - type joints in rectangular hollow sections, Archiwum Inżynierii Lądowej, vol. XXXII, n° 3, 1986 (in polish).
4. Szlendak, J. and Bródka, J., Yield and buckling strength of T, Y and X joints in rectangular hollow section trusses, Proc. Instn Civ. Engrs, Part 2, 1985, vol. 79, Mar., pp. 167 - 180.
5. Korol, R. M. and Mirza, F. A., Finite element analysis of RHS T - joints, J. Struc. Div., ASCE, vol. 108, No ST 9, Sept., 1982.
6. Czechowski, A., Investigation into the statics and strength of RHS lattice girders, In Welding of Tubular Structures, Proceedings of the Second International Conference held in Boston, Massachusetts, USA, 16 - 17 July 1984, Pergamon Press, 1984.

DISCUSSION
OF
MATHEMATICAL MODELING PAPERS

The primary thrust of the papers of this session was that complete (= global) connection modeling for behaviour and strength may be the preferable method, in the long run. Although it presents numerous difficulties, as pointed out by several participants, especially in the numerical treatment of elasto-plastic phenomena that also may involve strain-hardening, it will clearly make if easier to apply semi-rigid concepts to frame analyses if the complete moment-rotation curves for the connections are specified.

Several discussers addressed the problems that are associated with contact problems of the kind that commonly occur in column footing and end-plate connections. Whether special finite element types were used for this (for example, with zero stiffness after contact is lost, but regaining it if cyclic loads force a re-contact), or non-linearly elastic springs were utilized, the end result appeared to be good correlation between tests and theory. These techniques were also applied to reflect prying action for bolts, as well as friction and the loss of same for slip-resistant connections.

Although it was agreed that it would be preferable to utilize overall connection response characteristics in the frame response analysis, it would still be necessary to determine the primary response parameters, in the fashion as indicated by the local analysis session. Specifically, initial stiffness and ultimate strength data must be determined very carefully.

The effects of cyclic loading were discussed by a number of contributors. The problems are particularly complex if full load reversal occurs, both in terms of the needs of the modeling scheme as well as those of the eventual application to frame analysis itself.

Several questions were raised regarding the evaluation of column footings and their performance. Thus, structural details such as base plate to concrete foundation separation were considered, as were anchor bolts, grout and the like. However, there appeared to be general agreement that this is one area of connection research that is in need of significant additional studies.

Many of the discussers emphasized the need for concentrating on connections that are commonly used in industry. For example, North American practice for many types of connections now involves the use of slotted holes, to facilitate the erection work. It is clear that such will have a significant impact on the moment-rotation characteristics, at least following the first load reversal or after bolt slip has taken place.

Overall, the many global connection models that have been developed appear to hold excellent promise for more general applications to connection analysis. The need for being able to assess the importance of the various physical details was stressed by all presenters as well as discussers.

SESSION N°2B

CLASSIFICATION

STRUCTURAL PROPERTIES OF CONNECTIONS IN STEEL FRAMES

J.W.B. Stark
Eindhoven University of Technology,
Den Dolech 2, 5612 AZ Eindhoven,
The Netherlands

TNO Institute for Building Materials and Structures
P.O. Box 49, 2600 AA Delft
The Netherlands

and

F.S.K. Bijlaard
TNO Institute for Building Materials and Structures
P.O. Box 49, 2600 AA Delft
The Netherlands

ABSTRACT

Modern design codes allow the design of structures to be based on the actual load deformation characteristic of connections. In this paper a classification of connections is proposed. Essential is that the classification depends on the design method of the connection e.g. elastic method versus plastic method. It is necessary that designers are able to determine the M-Ø characteristics of connections. Brief information is given on available connection test data. These data are essential to develop analytical models to be used in design calculations. Two possible approaches for the determination of such models are discussed.

INTRODUCTION

In structural steelwork the connections between members play an important role. From an economic point of view the costs for design and fabrication form a considerable part of the total costs. From a structural point of view the properties of the connections fundamentally influence the response of the structure to actions. In the past, when 'working stress' design was normally used, the connection design was based on rather simple though not necessarily economical assumptions.

The connections were assumed to behave either as hinges (simple construction) or as infinitely stiff (rigid construction). The forces on the connections then followed from an elastic analysis of the structure.

The parts of the connections such as end plates and angles, welds and bolts, could subsequently be dimensioned. Even now this design procedure seems to be used in the majority of cases. However, the introduction of limit state design, including practical rules for plastic design, requires a more realistic treatment of the connections. When using these methods, the designer is confronted directly with the fact that for a better insight into topics such as the stability of columns and frames and for a minimum cost design of members and connections, understanding of the behaviour of connections is essential.

Another factor is that modern computer programs, now available to the majority of designers, allow a more sophisticated treatment of connections without an appreciable increase in calculation costs. Finally, also the use of automatic NC drilling and sawing equipment in the fabricators shop influences the cost relationship between various forms of connections, leading to a need to minimise the number of welded stiffeners.

Modern design codes e.g. Draft Eurocode 3, allow the design of the structure to be based on the actual load-deformation characteristics of the connections. It is of course essential that designers are able to determine those characteristics, preferably in the form of a mathematical model. The available information on this point will be reviewed in an IABSE-report that is under preparation now [1]. It is drafted in liaison with ECCS-TWG 8.2 and should be considered as complementary to [2] dealing with analysis and design of steel frames.

This paper, to be presented at the "CACHAN-WORKSHOP' is an introduction to [1] and is aimed to collect the latest information.

DEFINITION OF PROPERTIES OF JOINTS

A steel frame, whether considered as a plane or spatial structural system, essentially consists of linear members joined together by connections (Fig. 1).

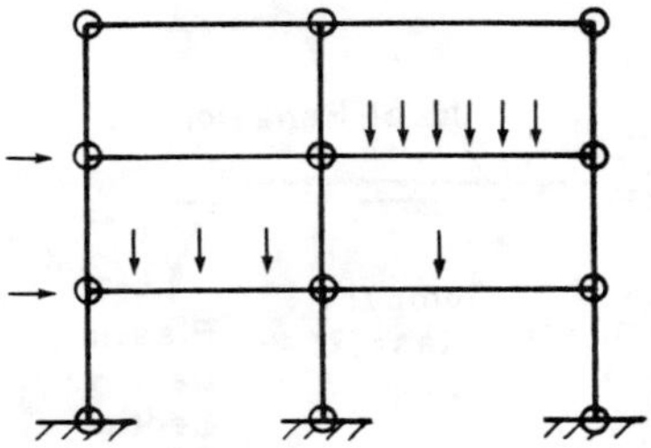

Figure 1. Combination of linear members and connections in a plane structural system.

The response of the system to actions (loads) is in general influenced by the structural properties of both members and connections. The relevant properties of these elements are <u>strength</u>, <u>stiffness</u> and <u>deformation</u> capacity (ductility). Assuming for the present that bending is dominant, all three parameters can be presented in a moment-rotation (curvature) diagram of the type illustrated qualitatively in Fig. 2.

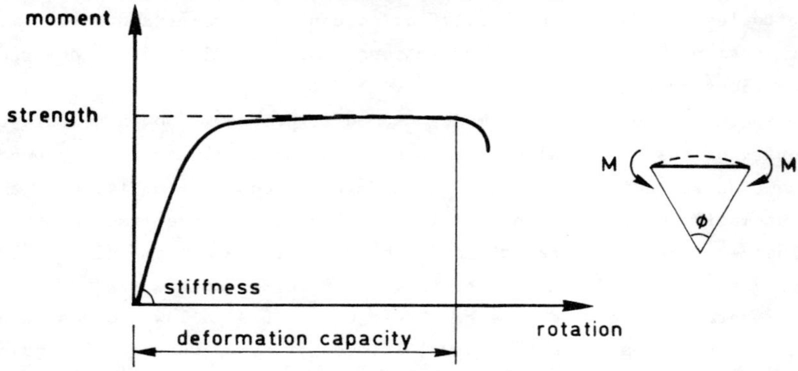

Figure 2. Presentation of structural parameter in a M-Ø curve.

The properties of a member with parts in compression are dependent upon section geometry and the yield stress of the steel mainly due to the effect of local buckling.

In most codes the different types of cross-section are classified dependent on their respective moment-rotation properties. This procedure is used because no practical methods are available to determine the actual M-Ø relation for each individual member which can be used in a simplified design.

Depending on the class of the cross-section (form of M-Ø relation: see Fig. 3) different analysis models may be used.

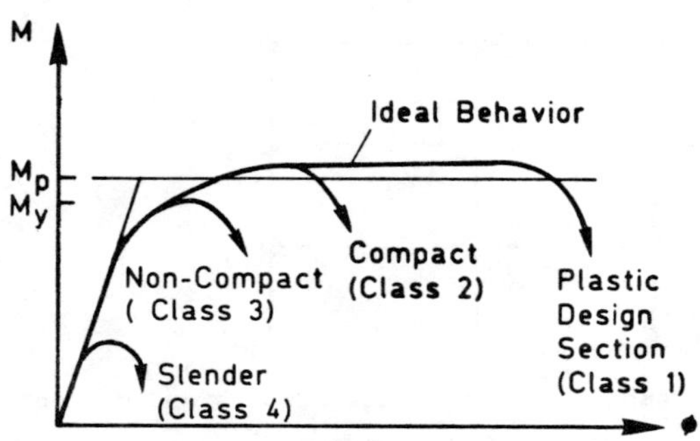

Figure 3. Load-deflection relationships.

As an example the possibilities for ultimate limit state verification according to the Draft Eurocode 3 are given in Table 1.

Class of member	Capacity of cross-section	Analysis of system
Class 1	plastic	plastic
Class 2	plastic	elastic
Class 3	elastic	elastic
Class 4	reduced stress or effective section	elastic

TABLE 1

The structural properties of connections can also be presented in a M-Ø diagram. In Fig. 4 a set of M-Ø curves for connections with different types of behaviour is shown.

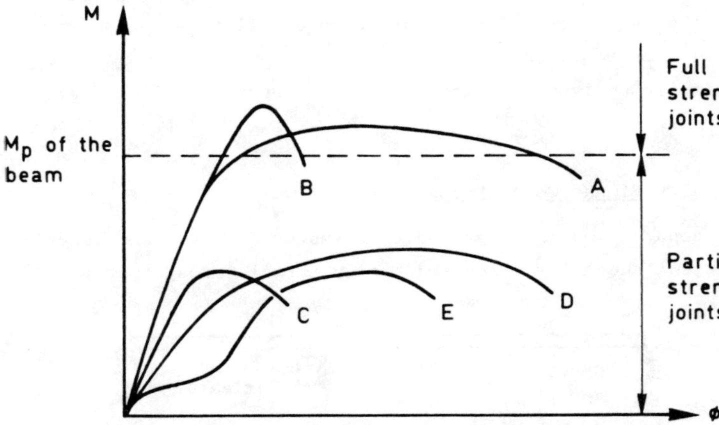

Figure 4. Moment-rotation curves of beam-to-column connections.

For the use of plastic design the connections can be classified in the following categories:

<u>Nominally Pinned Connections</u>

This type of connections is designed to transfer shear and normal force only. The rotation capacity of the hinge should be sufficient to enable all the plastic hinges necessary for the collapse mechanism to develop.

Full Strength Connections (connections A and B in Fig. 4)

The moment capacity is greater than that of the member. A plastic hinge will not be formed in the connection but in the member adjacent to the connection. In theory no rotation capacity is required for the connection. If the connection has little deformation capacity (connection B) an extra reserve of strength should be required to account for possible overstrength effects in the member.

Partial Strength Connections (connections C, D and E in Fig. 4))

The moment capacity is less than that of the member. A plastic hinge will be formed in the connection, so sufficient rotation capacity is required. For this reason connection C is unsuitable. Curve E is typical for a bolted bearing type connection, showing slip due to the clearance of the holes. For a detailed treatment of the requirements, reference is made to [3]. This explains how, especially when using partial strength connections in plastic design, information on all three parameters of the M-Ø relation is required.

In elastic design traditionally two categories of connections were considered:

Nominally Pinned Connections (Fig. 5)

The connections are assumed to transfer only the end reaction of the beam (vertical shear force and eventually normal force) to the column.

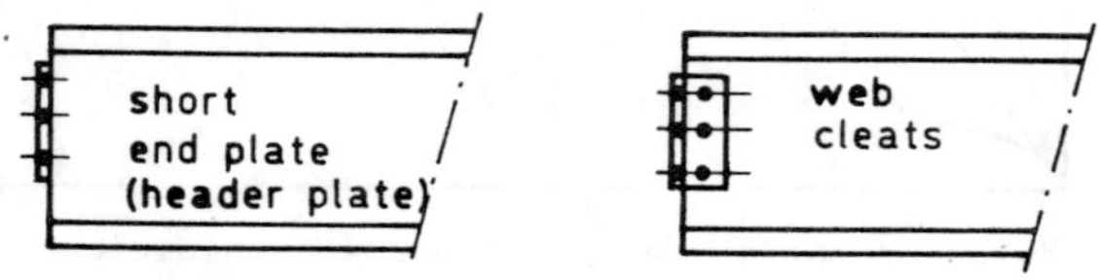

Figure 5. Nominally pinned connection.

They should be capable of accepting the resulting rotation without developing significant moments, which might adversely affect the stability of the column. In many countries it is common practice to design structures on a simply supported basis and then to provide connections which are in effect semi-rigid. A typical example is the flush end plate detail shown in Fig. 6. This may be unsafe due to insufficient rotation capacity of the connection.

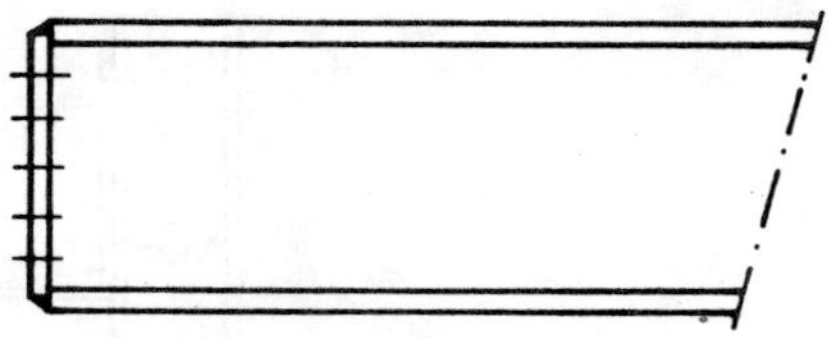

Figure 6. Flush end plate connection.

Rigid connections (Fig. 7)

Rigid connections are used to transfer moments as well as end reactions. Design assumes joint deformation to be sufficiently small (stiffness large) that any influence on the moment distribution and the structure's deformation may be neglected.

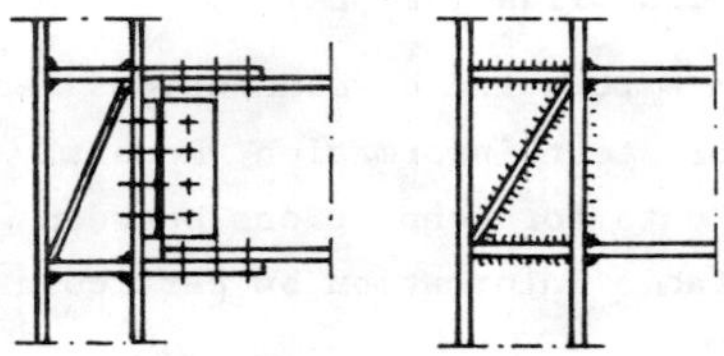

Figure 7. Rigid connections

Practical realisation of the assumptions of rigid construction often leads to relatively expensive connection details. Thus in cases where stiffness is not required for stability reasons (e.g. in braced frames), the use of nominally pinned connections is usual. However, practical forms of this type will transfer some moments and contribute to the structural stiffness. On the other hand, rigid connections can be simplified and made less expensive by the omission of stiffeners and other parts and accepting some flexibility. To fill the gap between pinned and rigid connections, a third category is defined and accepted in most modern codes.

Semi Rigid (semi flexible) Connections (Fig. 8)

These connections are designed to provide a predictable degree of interaction between members based on actual or standardized design M-Ø characteristics of the joints.

Figure 8. Semi rigid (semi flexible) connections.

In conclusion it seems to be evident that there is a need to present information the on M-Ø relations of joints in a form which can be readily used by designers.

CONNECTION TEST DATA

Design models for the M-Ø relation of connections can only be based on a systematic evaluation of test information. Recently prepared survey's [4, 5, 6] of available test data for connections between beams and columns permit the existence of suitable information on particular connection types to be checked.

Of much importance is the Steel Connection Data Bank (SCDB) at Purdue University [7]. Digitised versions of over 300 experimentally obtained M-Ø curves covering 7 different connection types, are stored in the computer. The tests in this collection are mainly from sources published in the USA, Canada and the UK. On the European continent also much test information is available, e.g. [8, 9, 10, 11]. These tests are now collected in the framework of the Eurocode-work and will be stored in a data base at the University of Aachen. Combination and exchange of these two data bases is recommended.

ANALYTICAL METHODS FOR THE DESCRIPTION OF THE M-Ø RELATION

Inclusion of connection flexibility into the analysis of the behaviour of members or complete frames requires that M-Ø relations for the connections are available to the designer in a suitable mathematical format.

Two essentially different approaches are possible:

i . An analytical model is developed to represent the M-Ø relation found by tests on a specific connection as closely as possible. Strictly speaking for use of this method test results should be available for every connection to be used in the actual structure.

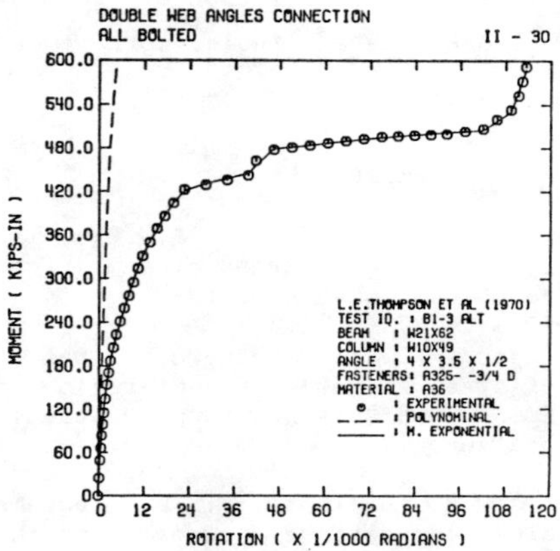

Figure 9. Analytical model to represent the M-Ø relation [7]

ii. An approximate method based on linearization of the M-Ø relation. This approach offers the possibility to develop semi-emperical models for the calculation of the relevant parameters of the schematized M-Ø relation. An example of such a linearisation is given in Fig. 10.

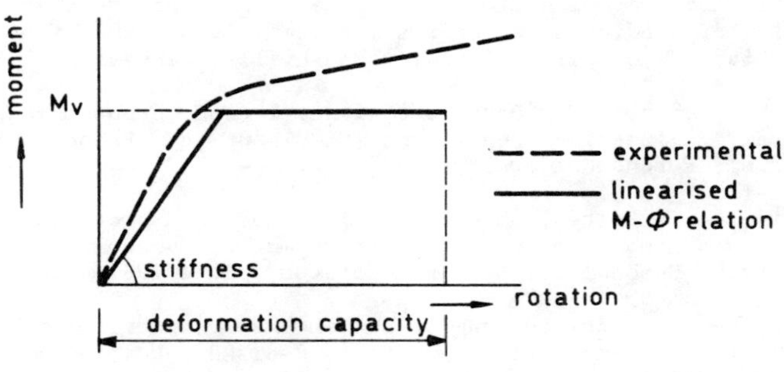

Figure 10. Linearised M-Ø relation

Formulae for the calculation of the moment capacity $\hat{M}_v$ are already included in Draft Eurocode 3. Formulae for the determination of the stiffness are also available for a number of connection types and will be included in an appendix of the revised EC3. For braced frames and floor beams requirements for the rotation capacity can easily be determined. For those applications a classification method as used for members can be applied. For unbraced frames the required rotation capacity can not simply be determined in a general way and needs to be calculated in each individual case. So for use in unbraced frames formulae for the calculation of the minimum available rotation capacity should also become available. Work is underway to develop these formulae.

REFERENCES

1. Bijlaard, F.S.K., Nethercot, D.A., Stark, J.W.B., Tschemmernegg, F., Zoetemeijer, P., Structural properties of semi-rigid joints in steel frames. IABSE paper to be published Dec. 1987.

2. Anderson, D., Bijlaard, F.S.K., Nethercot, D.A., Zandonini, R., Analysis and design of steel frames with semi-rigid connections. IABSE Survey. Prepared by ECCS-TWG 8.2, Task Group on Stability of semi-rigid frames. To be published June 1987.

3. Witteveen, J., Stark, J.W.B., Bijlaard, F.S.K., Zoetemeijer, P., Welded and bolted beam-to-column connections. Journal of the Struct. Div., Proceedings of the ASCE, Vol. 108, no St2, Febr. 1982.

4. Nethercot, D.A., Utilisation of experimentally obtained connection data in assessing the performance of steel frames. Connection Flexibility and Steel Frames, ASCE, ed W.F. Chen, 1985, pp 13-37.

5. Nethercot, D.A., Steel beam to column connections - A review of test data and their application to the evaluation of joint behaviour on the performance of steel frames. CIRIA report, RP 338, 1985.

6. Goverdham, A.V., A collection of experimental moment-rotation curves evaluation of predicting equations for semi-rigid connections. Doctoral dissertation, Vanderbilt University, Nashville, Tennessee, 1984.

7. Kishi, N., Chen, W.F., Data base of steel beam-to-column connections, Structural Engineering report no CE-STR-86-26, School of Civil Engineering, Purdue University, 1986.

8. Zoetemeijer, P., Bolted connections with flush end plates and haunched beams. Tests and limite state design methods. Report 6-81-17, Delft University of Technology, Stevin Laboratory, June 1981.

9. Zoetemeijer, P., Bolted beam-to-column knee connections with haunched beams. Tests and computations. Report 8-81-23, Delft University of Technology, Stevin Laboratory, December 1981.

10. Zoetemeijer, P., Proposal for the standardization of extended end plate connection based on test results. Tests and analysis. Report 6-83-23, Delft University of Technology, Stevin Laboratory, 1983.

11. Zandonini, R., Zanon, P., Semi-rigid and flexible connections: An experimental investigation. Proceedings of the International Conference on Steel Structures, Budva, Yugoslavia, 28 Sept.- 1 Okt. 1986.

SEMI-RIGID CONNECTIONS IN ISOLATION AND IN FRAMES

J B Davison, P A Kirby and D A Nethercot
Department of Civil and Structural Engineering
University of Sheffield
Sheffield S1 3JD
UK

ABSTRACT

The objective of the present Sheffield programme is the development of a full understanding of the behaviour of steel frames leading to design approaches that recognise the semi-rigid nature of steelwork connections. It has therefore embraced in-plane testing of connections, restrained columns in subassemblages and complete frames, as well as associated theoretical studies. As a result of this co-ordinated approach, a number of points have emerged concerning the conduct of M-Φ tests, which could not necessarily have been foreseen, nor would they have shown up in a programme of joint tests only.

INTRODUCTION

A project has been completed at Sheffield University to provide experimental data on the behaviour of columns and frames restrained by simple beam to column connections. The work has involved testing a series of joints to determine their in-plane rotational response, a number of restrained columns and two large three-storey two-bay frames. The action of the joints in the column and frame tests was monitored and compared with the isolated joint tests. Some interesting points concerning the action of joints in frames and the implications for conducting joint tests have emerged.

JOINT TESTS

All individual joint tests [1] were conducted with the specific objective of providing M-Φ data for use later in the programme. A cruciform arrangement was used with direct loading on the internal column. Rotations were measured by a set of 3 T-bars attached to the specimen on the column centre line and the beam centre lines as near to the connection as possible - normally within 100 mm. By monitoring the change in length of the wires attached to these bars, it was possible to determine the relative rotation within each joint. Such an arrangement could not, of course, measure the contributions to this rotation due to the different sources of flexibility within the joint, merely the overall effect.

The same light column and beam sections featured in all joint tests were subsequently used in the subassemblage and frame tests. A total of 25 tests, 8 of which included various types of deliberate lack of fit [2], were conducted. Each test did, of course, provide two M-Φ curves, one for each side of the specimen and an illustrative result is given as Fig 1.

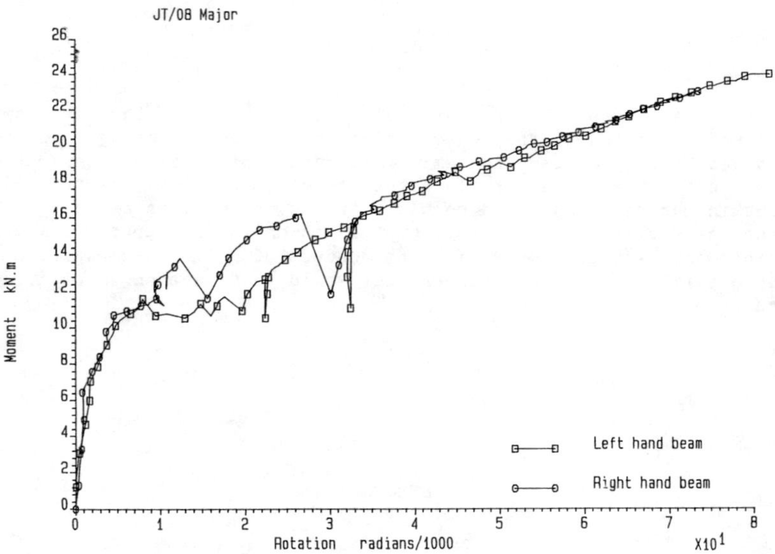

Figure 1. Moment-rotation curve from flange cleat joint test.

In some cases a limited number of unloading cycles were included in order to provide data on the unloading stiffness characteristics of the joints. The main variable in the programme was the type of connection - web cleat, flange cleat, end plate etc - with several of the joints fixed to the column web. Fig 2 summarises the complete set of M-Φ data in the form of average M-Φ curves for each connection type.

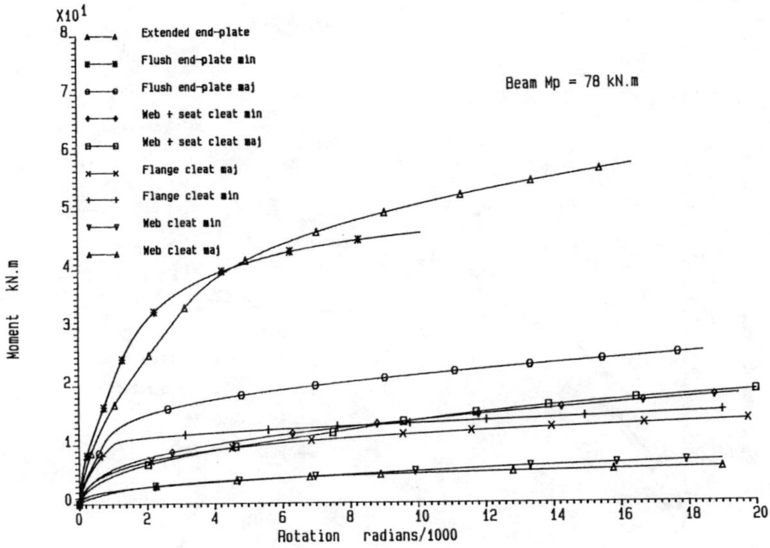

Figure 2. Collection of moment-rotation curves for various connections.

CONNECTION BEHAVIOUR IN FRAME AND SUBASSEMBLAGE TESTS

The performance of the connections in the frame and subassemblage tests was recorded. In the frame tests the bending produced in small spring steel strips, supporting heavy dumbells and secured to the beams and columns, as a result of rotation from the vertical was monitored [3]. A complete set of M-Φ curves for the joints in the frames was thus obtained. The subassemblage tests [4] were conducted in the horizontal plane and therefore methods of measuring rotation which rely on the effect of gravity could not be used. Limited access restricted the number of locations which could be monitored in the same way as the joint tests and a less satisfactory method using two LVDT's to record the relative location of the beam and column was adopted. A complete, reliable set of M-Φ curves for all tests in the subassemblage series is unfortunately not available. Therefore a complete comparison between joint subassemblage and frame behaviour is not possible.

Figure 3 compares M-Φ curves for notionally similar connections - top and bottom cleats with the same beam and column size, cleat size, bolt size and arrangement and bolt tightness - obtained from isolated joint testing, a subassemblage test and a frame test. The general form of each curve is very similar with the variation being approximately the same as that observed between repeat isolated joint tests. This suggests that the M-Φ characteristics measured in the basic joint tests provide an acceptable measure of the behaviour of similar connections when functioning as part of a framework.

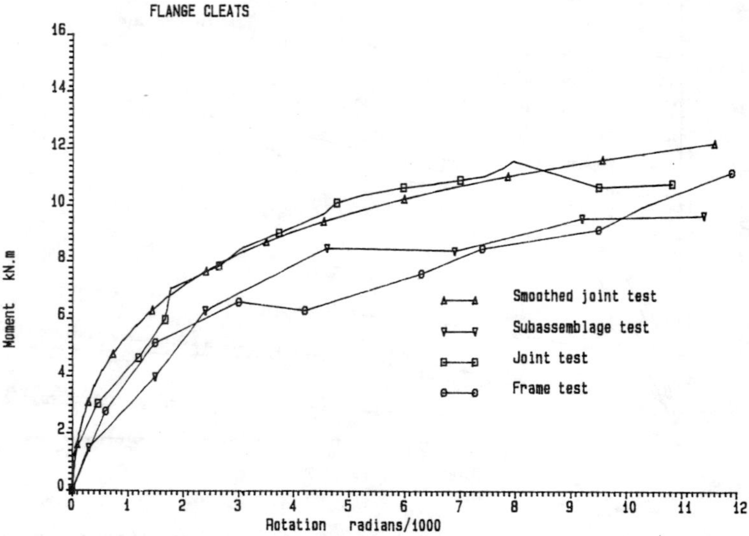

Figure 3. Flange cleat moment-rotation behaviour in isolated tests, subassemblages and frames.

Comparisons of the subassemblage and frame tests with finite element frame analysis programmes [5, 6] capable of including semi-rigid connections showed good correspondence between the predicted and actual response. Fig 4 illustrates the load deflection behaviour of a beam in frame 1 as measured in the test and predicted by the analysis program. Also shown in Fig 4 are the calculated deflections assuming the beam to be pinned at its supports. A reduction in deflection at service load of 25% of the simply supported estimate is evident. The analysis was conducted using an approximate M-Φ curve consisting of five straight lines closely fitted to the M-Φ curve obtained from the joint test - see Fig 5.

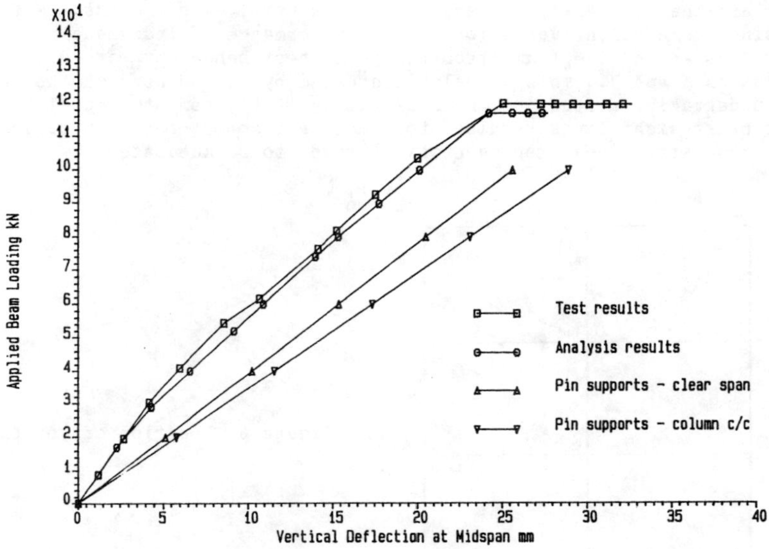

Figure 4. Beam deflection for test frame 1.

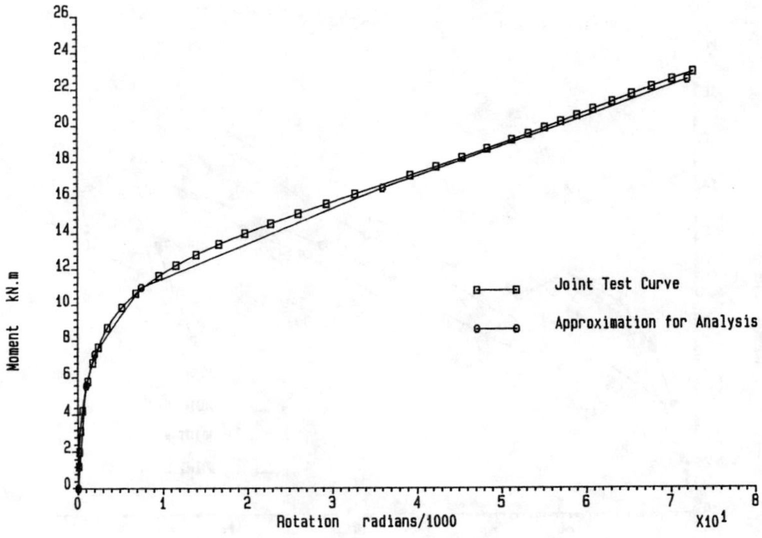

Figure 5. Approximation of moment-rotation curve for analysis.

It appears that the analysis is not overly sensitive to the exact M-Φ curve used since some slight variation of the performance of individual connections around the frame from the joint test behaviour was evident - see figures 6 and 7. This was also suggested by the authors of the program after undertaking a study of the sensitivity of the computer model to the number of straight lines required to simulate a moment-rotation curve. In many cases a tri-linear representation proved to be adequate [6].

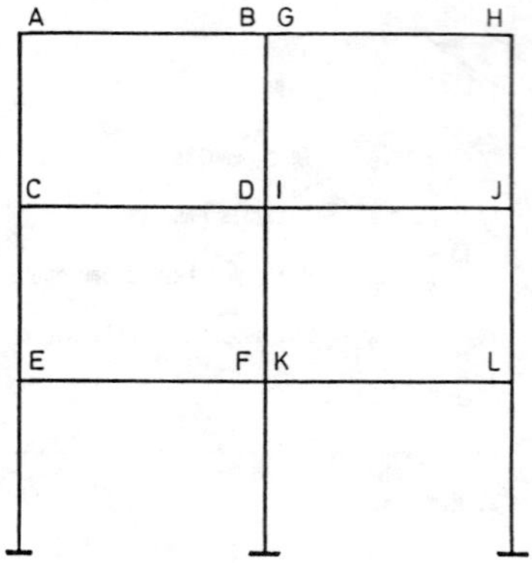

Figure 6. Location of joints.

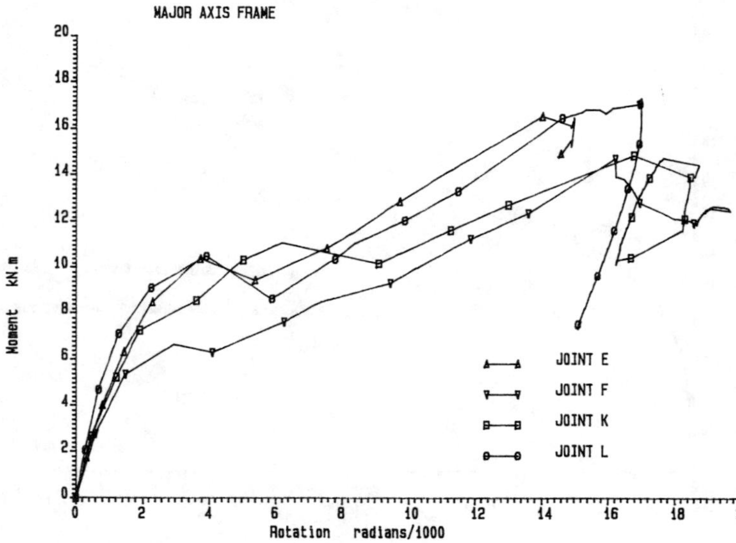

Figure 7. Behaviour of joints in frame test 1.

LESSONS FOR M-Φ TESTING DRAWN FROM FRAME STUDIES

The existence of a data base of the performance of various forms of structural framing employing semi-rigid connections permits information on connection behaviour to be extracted that can assist in the conduct of future joint M-Φ tests. Probably the most important item is the level of rotation likely to be experienced by the connections as the structure attains its ultimate load. The experience of the present work, which is limited to non-sway frames under static loading, is that these are generally far less than the range of rotations normally covered by published M-Φ data. In this context it is important to identify the type of component under consideration; axially loaded columns attain their ultimate load with very little end rotation - typically of the order of 0.005 radians, whereas beams provided with comparatively flexible connections may require end rotations of 0.02 radians; columns under primary bending e.g. due to unbalanced beam load, fall somewhere between these two extremes.

If Fig 2 is re-examined in the light of these requirements, it is clear that the most important part of the M-Φ curve is the early section, often the initial slope before the region corresponding to joint softening. It is therefore vital that the measuring systems for rotations are both sufficiently sensitive and are sampled sufficiently frequently during the initial stages of testing that this region be accurately determined.

A close examination of the behaviour of all the joints in the two frames as columns were brought up to failure showed that the unloading stiffness of the joints was of more benefit to the column than the loading stiffness of an adjacent joint. Figure 8 shows the behaviour of joints C and E as the external column was brought up to failure. The much larger unloading stiffness is readily apparent. It would therefore seem that the investigation of the unloading stiffness of a joint is equally important as the determination of the initial loading stiffness. The assumption that a connection unloads along a line parallel to its initial stiffness appears to be reasonable but needs to be conclusively established.

CONCLUSIONS

The work on connection behaviour conducted by Sheffield University has covered the response of joints in isolation, joints in subassemblages and in full size multi-storey frames. There appears to be no fundamental difference in the action of similar joints tested in the various environments. Studies of joint behaviour should pay particular attention to their action in the initial part of the M-Φ curve and also to their unloading behaviour. The response of connections at large rotations is likely to be of less importance in interpreting the action of semi-rigid frames.

ACKNOWLEDGEMENTS

The work reported forms part of a continuing investigation into the effects of semi-rigid joint action on the performance of steel frame structures supported by SERC, CIRIA and Constrado (now subsumed into the SCI) with assistance from the BRE. The authors are grateful for other help provided by staff of the Department particularly, Mr A M Rifai.

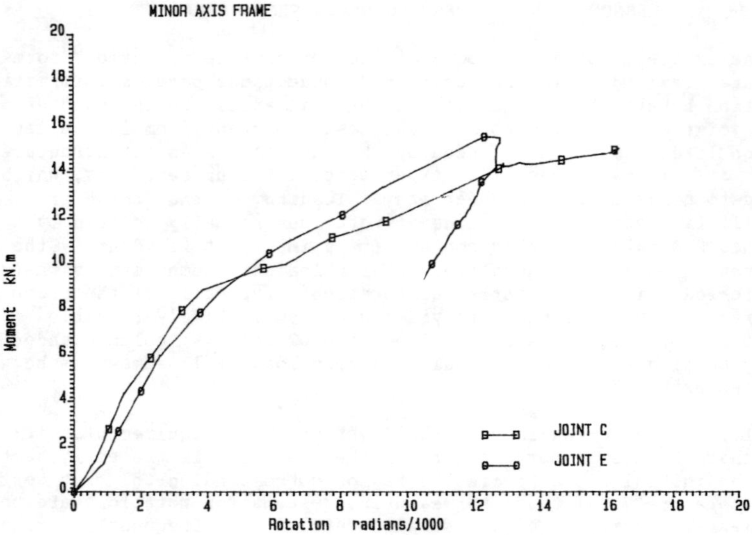

Figure 8. Comparison of joints C and E (frame test 2).

REFERENCES

1 Davison, J.B., Kirby, P.A. and Nethercot, D.A. "Rotational Stiffness Characteristics of Steel Beam to Column Connections", Journal of Constructional Steel Research, Vol. 7, 1987. (In Press).

2 Davison, J.B., Kirby, P.A. and Nethercot, D.A. "Effect of lack of Fit on Connection Restraint" Journal of Constructional Steel Research, Vol. 7, 1987, (In Press).

3 Jennings, D.A., Moore, D.B. and Sims, P.A.C "Instrumenting and Testing Bolted Steel Frame Structures" Proc., Determination of Dangerous Stress Levels and Safe Operation Conditions, Edinburgh, Aug. 1986.

4 Davison, J.B., Kirby, P.A. and Nethercot, D.A. "Column Behaviour in PR Construction - Experimental Behaviour", ASCE Convention, New Orleans, September 1986 (to be published in Journal of Structural Engineering, ASCE, 1987). SEP 87, p 2032-2050

5 Nethercot, D.A., Kirby, P.A. and Rifai, A.M. "Design of Columns in PR Construction - Analytical Studies", ASCE Convention, New Orleans, Sept. 1986 (to be published in Canadian Journal of Civil Engineering, 1987). V14, N4, AUG 87, p485-497

6 Poggi, C. and Zandonini, R., "Behaviour and Strength of Steel Frames with Semi-Rigid Connections", Connection Flexibility and Steel Frames, ed. W.F. Chen, ASCE, 1985, pp. 57-76.

STANDARDISED METHOD FOR MEASURING THREE-DIMENSIONAL RESPONSE OF SEMI-RIGID JOINTS

M Celikag and P A Kirby
Department of Civil and Structural Engineering
University of Sheffield
Sheffield S1 3JD
UK

ABSTRACT

Whilst procedures for in-plane M-Φ responses are becoming established little attention has been given to out-of-plane and torsional characteristics. The findings of exploratory tests are described in this paper which highlight a number of considerations which need to be resolved before any substantial experimental work is undertaken. There are several ways in which a beam column connection may be supported and various ways in which the rotation can be computed. It is suggested that the international research community might give attention to the production of a memorandum to develop a uniform approach to the problem.

INTRODUCTION

Since the 1930's many researchers have investigated the behaviour of beam-to-column connections but have mainly concerned themselves with the in-plane M-Φ behaviour. Very little experimental information is available on the out-of-plane and torsional response of semi-rigid connections. A standardised method for in-plane moment-rotation behaviour of beam-to-column connection has been drafted as a technical memorandum by Task Group 25 of the Structural Stability Research Council (S.S.R.C.) in 1986 [1].

It is obvious that, in a steel frame, there may be lateral loads acting out-of-plane on beams or unsymmetrical or one-sided floor loads which cause torsion in beams and therefore on to the beam and column connections. At the same time eccentric loads may well cause torsional effects. Such restraints are also important in the context of stability, e.g. lateral-torsional buckling.

This research is aimed at investigating the behaviour of connections due to out-of-plane and torsional loading. Because cruciform tests rely on symmetry, which can give rise to difficulties, a single connection cantilever arrangement was chosen (see inset of Fig. 1). The response of the test specimens was found to be more complicated than had been expected.

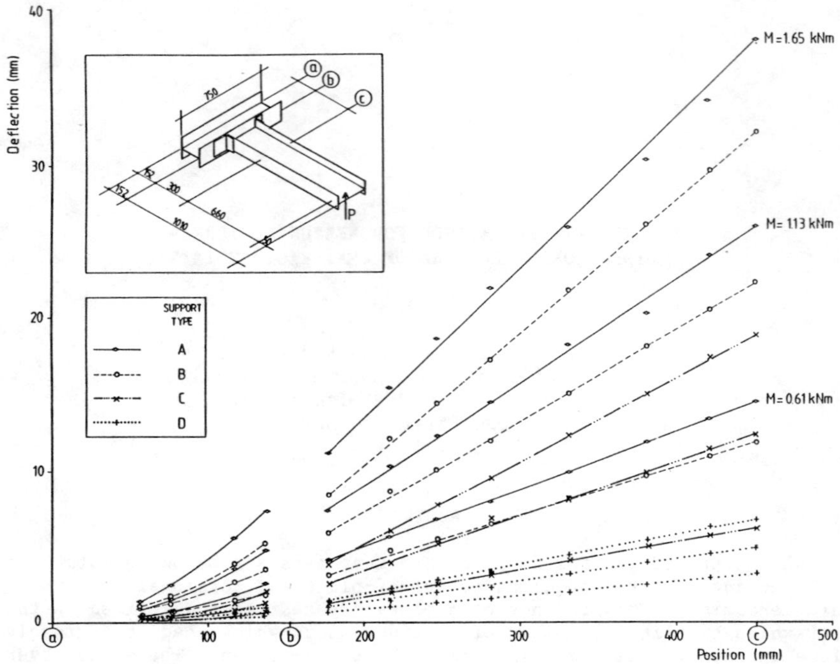

Figure 1. The deflection profile of the assemblage for various support conditions.

Some results of the exploratory tests are given and questions regarding the manner in which the M-Φ relationships should be determined are raised to initiate an agreed method similar to the S.S.R.C. Memorandum before substantial test work is carried out.

TEST SPECIMENS, APPARATUS AND INSTRUMENTATION

Flange cleat and web cleat connections were used for exploratory tests. These connections were designed according to an elastic design method of the Australian Institute of Steel Construction for in-plane loading. Previously similar connections had been tested for in-plane behaviour and this research was undertaken to find their out-of-plane and torsional M-Φ characteristics.

The beam and column elements were the same sections in all tests namely 254 x 102 x 22 UB, 1000 mm long and 152 x 152 x 23 UC, 750 mm long. The static yield stresses of the column material and the beam material were 266 N/mm² and 287 N/mm² respectively, and were found by testing coupons from their flanges. For the top and bottom flange cleat connections unequal angles were used; an 80 x 60 x 8 for the top cleat and a 125 x 75 x 8 for the bottom cleat. For the web cleat test again a pair of 80 x 60 x 8 angle sections was used.

M16 Grade 8.8 bolts with washers were used in drilled 18 mm diameter holes for all connections. Four bolts were used for the top cleat, six bolts for the bottom cleats and six bolts for the web cleat connection.

A self straining rig was constructed of steel channels and a 100 x 90 x 15 mm attachment plate was bolted on the vertical members to which the column was to be clamped. A hydraulic jack was used to apply the out-of-plane and torsional loading at the free end of the beam via a load cell. The bolts were tightened up to 160N.m. with a torque wrench. This torque was adopted as being consistent with the value expected of manual tightening with ordinary tools. A level was used to ensure the proper alignment of the specimen. For simplicity dial gauges were used for these exploratory tests which were intended primarily to examine overall response.

TEST PROCEDURES

3.1 Out-of-Plane Response

The single cantilever arrangement of the beam-column connection, as mentioned in section 1, was used. The beam was connected to one flange of the column with top and bottom flange cleats. For ease of application of load on the minor axis of the beam, the cantilever arrangement was turned through 90° and then the remote flange of the column was clamped horizontally against the attachment plate so that the hydraulic jack could be placed vertically to apply a point load at the free end of the beam web. Four different tests were carried out as listed below. The first, type B, represents the reference condition; the other types are variations of this situation.

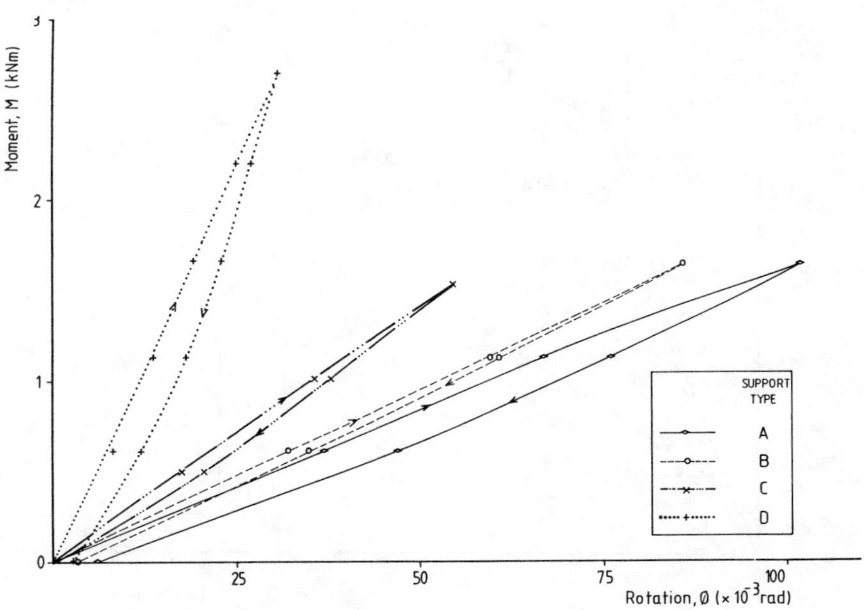

Figure 2. Beam rotations for various support conditions.

Type B: The rear flange of the column was clamped to the attachment plate over its entire length and the beam was bolted to the front flange of the column by top and bottom flange cleats (see inset of Fig. 3).

Type A: The same connection was used as in type B but an extra plate, equivalent to the depth of beam, was placed between the rear flange of the column and the attachment plate. The column was clamped in this region only. The aim of this arrangement was to simulate restraint over the depth of an equivalent beam rather than the clamping action being present over the full column length (see inset of Fig. 4).

Type C: The test arrangement used was as in type B but with additional stiffeners welded at four positions from the rear flange and up to the column centreline to substantially prevent the rotation of the rear half of the column web. The stiffeners were located on the column 125 and 250 mm away from the mid-length of the column in both directions (see inset of Fig. 5). This was designed to simulate symmetry about the column centreline.

Type D: The test arrangement in type C was used but, in addition, full stiffeners were welded to the column flanges and to the column web at positions to correspond to the beam flanges. This was intended to substantially eliminate both the web and flange distortions (see inset of Fig. 6).

Several series of deflection readings were taken along the mid-line of the beam and the corresponding mid-length of the column to produce a deflection profile of the assemblage for all types as is shown in Figure 1.

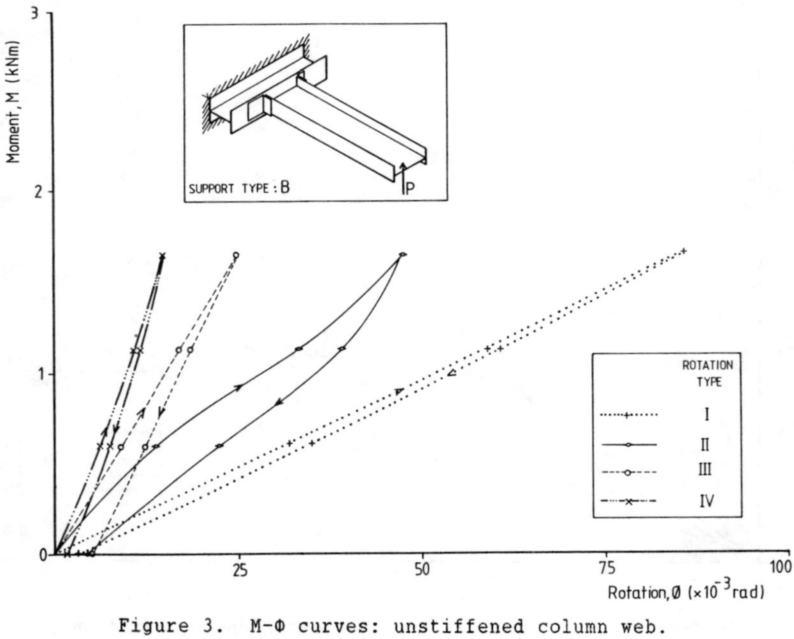

Figure 3. M-Φ curves: unstiffened column web.

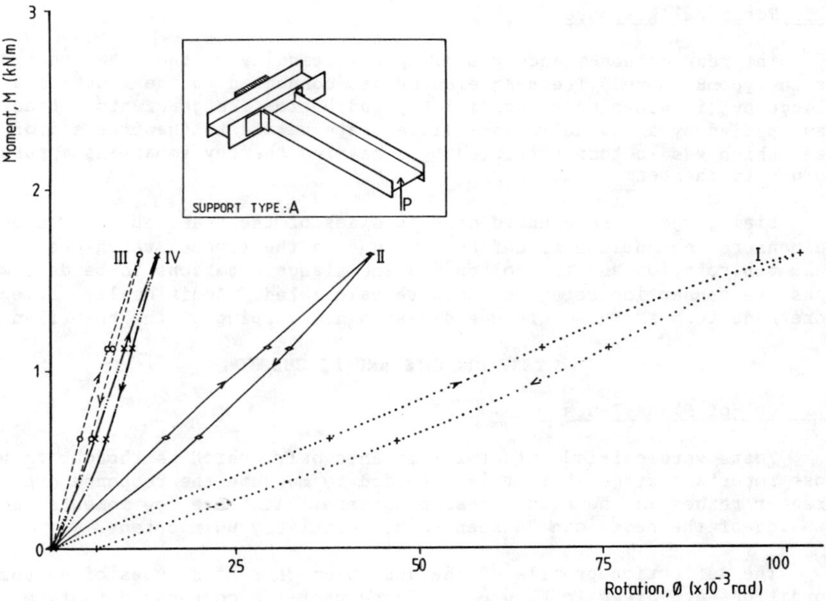

Figure 4. M-Φ curves: rear column flange partially fixed

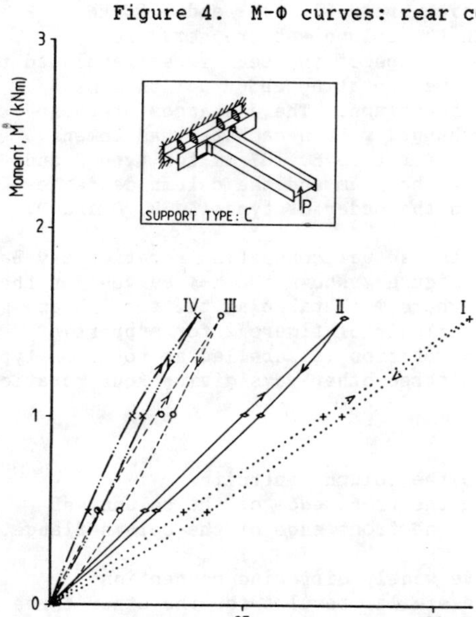

Figure 5. M-Φ curves: column web stiffened to centreline.

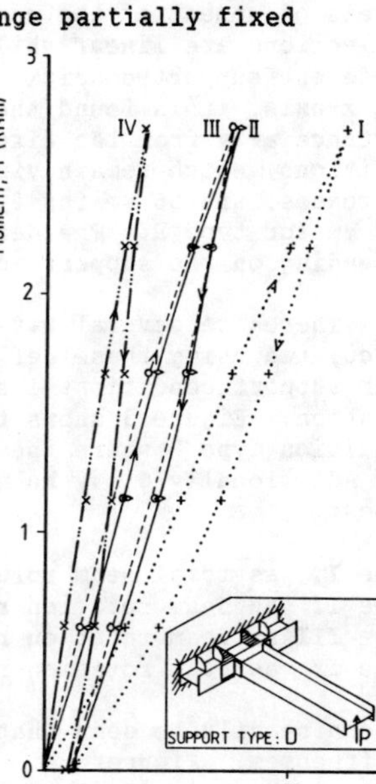

Figure 6. M-Φ curves: column web fully stiffened.

3.2 Torsional Response

The rear column flange was clamped vertically on the attachment plate as in type B above. The beam element was connected to the front column flange by (i) web cleats and (ii) top and bottom flange cleats. Loading was applied by a hydraulic jack via a plate welded to the free end of the beam which was in turn supported by a bearing thereby imparting a pure torque to the beam.

Dial gauges were mounted on both sides of the beam web and the column to measure the horizontal deflections due to the torque, which enabled the beam web rotation and the column web and flange rotations to be determined. Thus the connection rotation could be calculated. Additionally, attempts were made to measure the free and restrained warping of the beam flanges.

TEST RESULTS AND DISCUSSION

4.1 Out-of-Plane Tests

Tests were carried out at low moments only, which is thought to be the most important stage, but it is intended to measure the response over greater ranges in the actual test programme. At these low moments the bending of the beams can be seen to be relatively unimportant.

The deflection profile of the assemblage for four types of support conditions are given in Figure 1. These profiles correspond to three levels of moment at the connection which are 0.61, 1.13 and 1.65 kNm. Beam deflections are linear while those in the column web are not. For different support conditions, when the slope of the beam is extrapolated to the x-axis, it is found that the beam was rotating about points some distance away from the fixed end of the column. The distances of these positions, which remain virtually unchanged within each test as moment increases, are 66 mm for type A, 78 mm for type B, 107 mm for type C and 102 mm for type D. The deflections of the beam and the column decrease depending on the support conditions in the order of types A, B, C and D.

There are several ways in which the actual connection rotation may be calculated using these deflections. Figure 2 shows the M-Φ curves for the four support conditions listed above where Φ is taken as the total beam rotation. Figure 3 shows the characteristic of figure 2 for support condition type B where the total beam rotation is labelled as rotation type I. Additionally Φ may be measured in three other ways giving four rotation types:

Type I as total beam rotation
Type II as beam rotation relative to the column centreline
Type III as beam rotation relative to the front edge of the column web
Type IV as beam rotation relative to the front edge of the column flange.

It can readily be seen that these give widely differing connection stiffnesses. Figures 4, 5 and 6 give similar results for the other three support conditions. There are now 16 possible combinations of support and measurement types with 16 possible M-Φ characteristics for one connection.

Consider real structures. An exterior column usually does not receive symmetric loads from beams and floors. Type A and B support conditions could be representative in this case. An interior column often carries symmetrical loads coming from beams and floors. Type C and D support conditions could be appropriate in this case.

From the authors' point of view, considering the practical, analytical, and design requirements more than one type of support condition and type of connection rotation measurement might be needed. A change in the length of the column specimen used may be another factor which could affect the results. This would suggest that consideration needs to be given to the way in which such tests are carried out to ensure that enough information is provided to enable the appropriate M-Φ curve for any situation to be determined. The authors feel that the most conservative approach namely the combination of types A and I is perhaps the most appropriate for design. Alternatively it may be necessary to investigate two different conditions, one for interior columns and one for exterior columns for nominally identical joint conditions.

4.2 Torsion Tests

The results of torsion tests on two web cleat and one flange cleat connections are shown in Figures 7 and 8, from which it can be observed that the actual connection rotation is small compared with the twist occurring in the beam. Figure 7, shows that, at a moment of 2 kNm, the rotation of the web cleat connection (2) is 8 x 10^{-3} rad, whereas the twist occurring in a 1 m length of beam for the same condition is 235 x 10^{-3} rad,

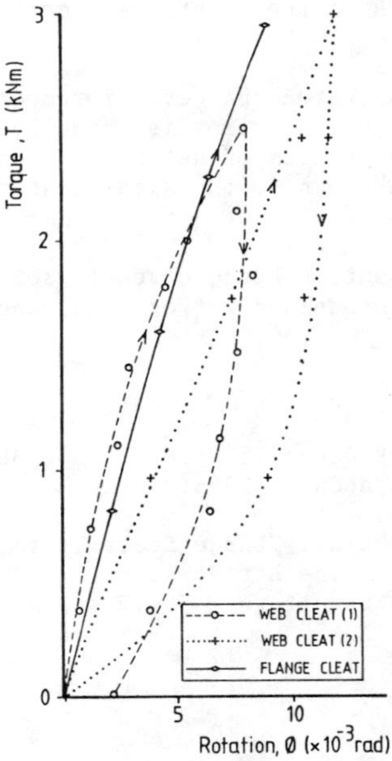

Figure 7. Connection torque-rotation curves.

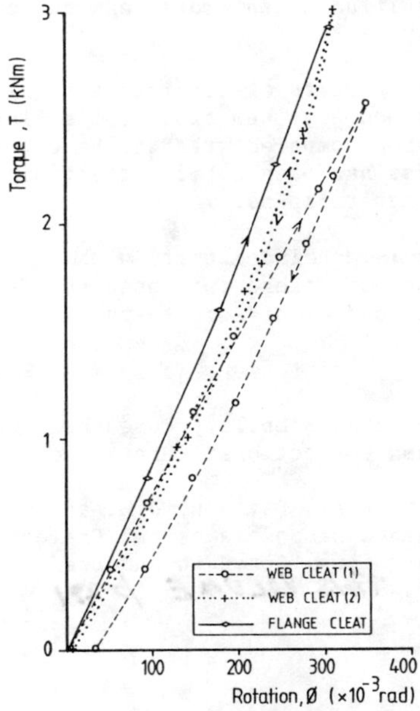

Figure 8. Beam torque-rotation curves

as can be observed from Figure 8. Thus even this modest connection is torsionally very stiff as compared with the beam and could be considered as infinitely stiff. However there could be other types of connections where this is not the case.

Attempts were made to measure the free and restrained warping distortions of the top and bottom flanges of the beam for web cleat and flange cleat connections. For theoretical analyses there is a need to measure the warping bi-moment and consideration needs to be given as to how this should be done.

4.3 General Observation

The work described above has concentrated on the first cycle of loading applied to a joint. However studies on the in-plane behaviour of subassemblages and frames [2] have shown, that as column buckling occurs, the direction of the moment applied at a joint often reverses. It is thus the unloading stiffness of a joint which is frequently most important in providing restraint to a column as the failure condition is approached. This occurs even with monotonic frame loading. The question of measuring unloading stiffnesses at various moment levels is another feature which requires attention.

CONCLUSION

Preliminary out-of-plane tests on beam-to-column connections have shown that there are a variety of ways in which the column may be supported and a number of ways in which the connection rotation might be measured. The resulting moment rotation characteristics of the joints vary quite markedly.

The results of torsion tests have demonstrated that even for modest connections e.g. web cleats, the flexibility of the joint is normally very small when compared with the bare beam. The problem of determining warping stiffness has been noted. Attention has been drawn to the significance of unloading stiffness.

It is concluded that, with the wide attention being given to semi-rigid action, there is a need to develop a recognised set of requirements for the conduct of such tests.

REFERENCES

1 Structural Stability Research Council Task Group 25, "Testing of Beam to Column Connections", Draft Technical Memorandum, 1986.

2 Kirby, P.A., Davison, J.B. and Nethercot, D.A., "Large Scale Tests on Column Subassemblages and Frames" State of the Art Workshop, Connections, Strength and Design of Steel Structures, E.N.S. Cachan, May 1987. THIS VOLUME p 291

DISCUSSION OF CLASSIFICATION PAPERS

There was general consensus among speakers and discussers alike that with the advent of limit states design, it is critical to arrive at a system of classification of connection response with regard to strength, stiffness and deformation capacity. In particular, this will facilitate the work of designers, to the effect that it may not be necessary to work with individual connection M-Φ-curves, for example.

Suggestions for simplified moment-rotation curves were made, to the extent that it would be preferable from an analytical standpoint to be able to deal with linear relationships rather than highly non-linear ones. Several discussers expressed concern over this as being a too liberal simplification, although analyses have shown that for most practical purposes, it is the magnitude of the initial connection stiffness that is the most important to the overall frame response. This applies to frame drift calculations as well as to stability analyses, although for the latter it is clear that the initial stiffness is of overriding importance.

Several questions dealt with the initial and subsequent connection stiffness characteristics. Thus, there appeared to be general agreement that loading and unloading stiffness magnitudes are crucial to the stability of the frame. Test results for structural subassemblages demonstrated that this was the case as well, and good correlation was found when the unloading stiffness was set equal to the initial stiffness. It was pointed out that for such problems, the overall rotation capacity of a connection is of lesser importance ; rather, the initial portion of the curve and the shape of the unloading part are of primary concern.

Several speakers emphasized that although it appears that the unloading stiffness of a connection is the same as the loading one, this should be examined for a range of connection types and sizes. Similarly, it was pointed out that an engineering definition of the connection stiffness is required, not the least because many test curves exhibit little or no initial linear range, or even a much softer connection response than what is eventually determined. This is certainly in part due to the nature of the physical testing in itself, but further emphasizes the need for a clear definition of the stiffness.

It was noted that although significant variations can be found in the stiffness values for different types of connections, their impact on the response of the frames is not as severe. In other words, with a given type of connection one may use quite different EI-values, and still arrive at substantially comparable strength and deflection results. This also holds true for stability analyses of individual columns and columns as part of subassemblages. Again, the importance of the connection unloading stiffness was stressed by discussers and speakers alike.

SESSION N°3

FRAME ANALYSIS

FRAME ANALYSIS INCLUDING SEMI-RIGID CONNECTIONS AND P-DELTA EFFECTS

Socrates A. Ioannides, Ph.d., P.E.
Structural Affiliates International, Inc.
2416 Hillsboro Road
Nashville, Tennessee 37212
USA

ABSTRACT

An analysis technique for Steel Frames that include Semi-Rigid connections is presented. The technique is easy to implement since it utilizes the stiffness method of analysis of structures. A typical beam element with six degrees of freedom(d.o.f.) is utilized together with a six d.o.f column element that utilizes stability functions to account for the P-Delta effect. The connection flexibility is incorporated into the model through a two d.o.f. spring element whose stiffness is the secant stiffness of the particular connection at the particular load level. Due to the non-linearity of the connection and the P-Delta effect the solution is iterative and the solution algorithm is presented.

INTRODUCTION

Analysis, design and drift calculations for steel frames are inter-linked processes that jointly dictate the final sizes of the individual members and connection details of steel frameworks. As a result each one of these has to be compatible in order that a consistent approach be achieved (i.e. the design approach used depends greatly on the analysis assumptions and techniques used). For the purposes of this paper only the analysis portion will be examined; design and drift considerations will be presented in a subsequent paper.

The modeling requirements that will make the analysis, design, and drift calculations consistent will be discussed in this paper. The use of this consistent approach will result in better understanding of structural behavior by demonstrating the actual force distribution and by avoiding such "design approximations" as the "Moment Magnification Factor". Specifically the inclusion of factors such as P-Delta and Connection Flexibility in the analytical model will be discussed and demonstrated.

The A.I.S.C. specification as presented in the Eighth edition of the Manual of Steel Construction[1] describes three types of construction:

Type I, commonly designated as "Rigid Framing",

Type II, commonly designated as "Simple Framing", and

Type III, commonly designated as "Semi-Rigid Framing".

It further stipulates that "the design of all connections shall be consistent with the assumptions as to type of construction called for on the design drawings". Further "type II construction is allowed as long as the wind moments are distributed among selected joints of the frame and certain provisions are met". Since 1964 Disque[2] has promoted the use of type II construction and has provided certain simplifying design assumptions such as the directional moment connections to facilitate the design engineer in utilizing this simple type of construction. Work done at the University of Colorado and R.P.I. by Ackroyd, Gerstle et.al.[3,4] has substantiated the use of type II construction in accordance with the Disque assumptions with certain restrictions. The purpose of this paper is to deal with type III (Semi-Rigid) not type II.

It can be argued that no connection is either fully rigid or totally pinned or flexible; every connection possesses some restraint no matter how flexible it is considered and even fully welded connections exhibit some flexibility. The Load and Resistance Factor Design (L.R.F.D.)[5] by A.I.S.C. recognizes this fact and distinguishes connections in two categories : Fully Restrained (FR) and Partially Restrained (PR). The technique described herein is suitable for analyzing frames that include PR connections, provided the Moment-Rotation characteristics of these connections are known.

MODELING

For the purpose of demonstrating the technique and in accordance with common construction practice the columns will be assumed continuous at the floor lines. However the technique could be extended to utilizing P.R. connections even between column stacks. The inclusion of Semi-Rigid connections connecting the beams through the column stacks necessitates a modelling technique that would allow relative rotation between the beam and column at a joint. As shown in Figure 1 this can be achieved by introducing additional rotational degrees of freedom at each joint; thus exterior joints will have a total of four degrees of freedom i.e. horizontal and vertical translation, and rotation of the column stack, and beam-end rotation, whereas interior joints will have five degrees of freedom, namely: the joint translation in the horizontal and vertical direction, the joint or column stack rotation and the two beam-end rotations. In addition to the beam and column elements spring elements are introduced connecting the joint rotation to each beam end rotation.

In developing a computer code for analyzing frames such as shown in Figure 1 the following stiffness matrices should be utilized:

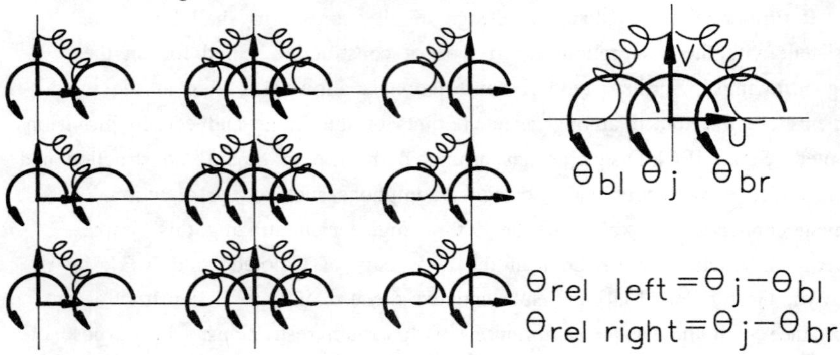

FIGURE 1: TYPICAL FRAME MODEL

A. Beam elements : The stiffness matrix utilized for the beam element should be the standard stiffness matrix found in basic structures texts[6].

B. Column elements : To account for P-Delta effects in the analysis the column stiffness matrix is formulated utilizing stability functions. The expressions for the stability functions can be found in basic structural analysis texts[6]. It should be noted that the stability functions only influence the first 4x4 portion of the stiffness matrix relating the rotational and shear degrees of freedom. These functions replace the standard stiffness and carry-over functions with ones whose values depend on the axial load present in a member.

C. Spring elements : The stiffness matrix for the spring element is a simple 2x2 matrix with the stiffness of the spring (Ks) in the diagonal terms and "-Ks" in the off-diagonal terms.

Frames with semirigid connections are generally more flexible than those designed with fully rigid connections; The P-Delta effects are, therefore, generally more pronounced and it is the author's opinion that any analysis that includes connection flexibility should also account for P-Delta. The column element described above, formulated with stability functions to account for the loss of stiffness with increasing axial load, has been found to work consistently well in the presence of semirigid connections with non-linear semirigid connections.

The proposed method is dependent on the availability of Moment-Rotation curves. In the past this seemed to be the bottleneck of analysis techniques that considered connection flexibility. Recently, however, efforts from various sources are beginning to make available such information both in the form of available experimental data and predictor equations[7,8,9].

A typical moment rotation curve is shown in Figure 2. The non-linear nature of the Moment-Rotation curve shown is typical of most connections tested in the past[7,8,9]. It should therefore be accounted for in the method of analysis. Three different stiffness values can be identified with any Moment-Rotation curve:

a) the initial stiffness,
b) the tangent stiffness at any point , and
c) the secant stiffness.

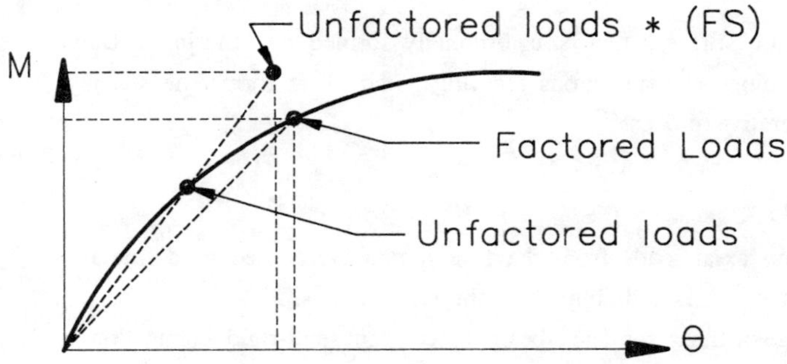

FIGURE 2: MOMENT ROTATION CURVE

Any of the three may be utilized in the analysis so long as it is consistent with the overall technique used; namely the tangent stiffness could be used with an incremental load application technique or, as recommended here, a secant stiffness should be used in an analysis where the full factored load is applied in one step. In the method proposed herein the secant stiffness is recommended for the following reasons:

a) the factored loads will be applied in one step,
b) the secant stiffness provides an integrated average of how the

connection arrived at the present level of loading,

c) regardless of the mode of loading and unloading at one time in the life of the structure the actual moment-rotation was followed to arrive at the present state of load.

d) if an initial stiffness is used the deflections derived will be erroneous.

It is further proposed that factored loads be utilized in the analysis, otherwise due to the non-linear nature of Moment-Rotation curves an incorrect factor of safety may be assumed to exist as shown in Figure 2. Using unfactored loads and allowable stress design leads to false factor of safety, underestimates total deflections, and thus underestimates the P-Delta effects.

NUMERICAL ALGORITHM

Having discussed the components of the model and the loads to be applied the numerical algorithm is hereby presented.

First Iteration

The structure stiffness matrix is originally formed by ignoring P-Delta effects and assuming all connections are fully rigid. This forms the starting point for the iterative process.

Second Iteration

The column axial loads from the first iteration are used to define new column stiffness matrices utilizing the stability functions.

The spring stiffness for each beam to column semi-rigid connection is obtained from the "available" Moment-Rotation curves for each type of connection by making an "educated guess" on the relative connection rotation. A method found by the author to give consistently good results for this initial guess is to superimpose the "beam line" on the moment-rotation curve(Figure 3). The beam line is a straight line connecting the point of fixed end moment at zero rotation to the point of zero moment and maximum rotation for the loading condition. This second point is the amount of rotation the beam would undergo if it was pinned. Although this value is dependent on the type of loading it varies within a narrow range of 0.016 radians for a point load at mid-span to 0.021 radians for uniformly distributed loads. The algorithm is not very sensitive to this initial value, so a value of 0.02 radians was found to work well for commonly encountered loadings. The spring stiffness for each connection is thus determined as the secant stiffness passing though the intersection of the Moment-Rotation curve and the Beam-Line.

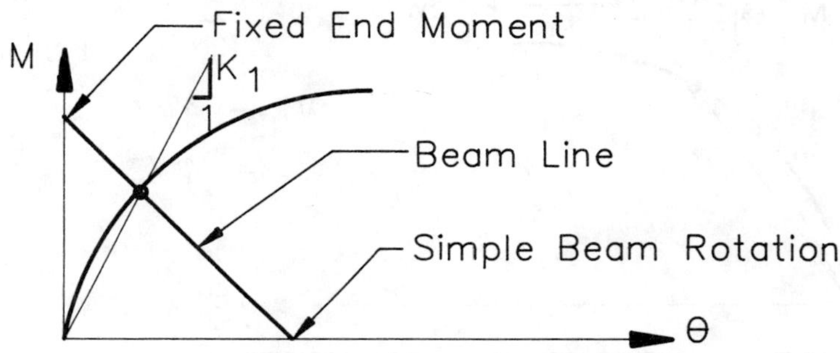

FIGURE 3: BEAM LINE

The structure stiffness matrix is then recalculated utilizing the new column and spring stiffnesses and the problem resolved.

Typical Subsequent Iteration

The column stiffness matrices are reevaluated based on the axial loads from the previous iteration.

The new spring stiffness for each connection is found by comparing the resulting moment and relative rotation at each spring location with the Moment-Rotation curve as shown in Figure 4. If the moment-rotation point from the previous iteration does not fall on the Moment-Rotation curve a vertical line is drawn until the M-Theta curve is intersected. The new spring stiffness is then defined as the slope of the secant passing through this new point. The convergence has been found to accelerate if instead of the new stiffness defined above a weighted average of this and the previous iteration is used as shown in Figure 4.

The new spring stiffness matrices are thus reevaluated and the new structure stiffness matrix formed.

The problem is solved again.

At this point convergence is checked by comparing the results of the present to the previous iteration. Specifically the relative rotation between the ends of each spring and the column axial loads are compared to the previous iteration. If convergence is detected the algorithm is stopped; if not the steps described in this section are repeated.

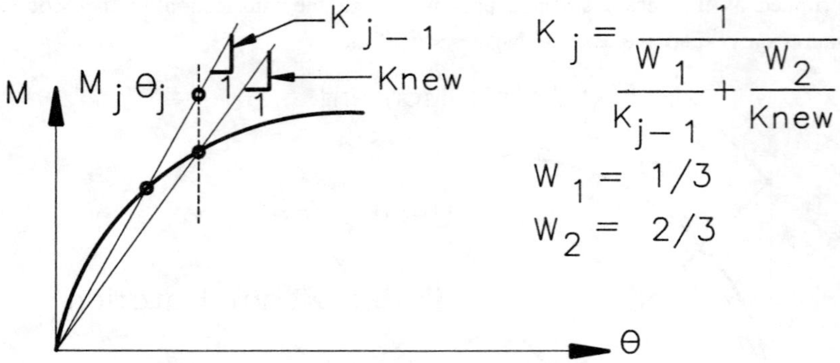

FIGURE 4: NEW SPRING STIFFENERS

CONCLUSIONS

A method has been presented for the analysis of steel frames utilizing semi-rigid connections. The basic components of the model are generally available so implementation of the algorithm should be straight forward for individuals with basic knowledge of the stiffness method of analysis. P-Delta effects are critical due to the increased flexibility of the frames and should be included. The resulting iterative problem is solved by iterating on both the P-Delta and connection flexibility at the same time. Convergence has been achieved in four to six iterations for a variety of frames.

REFERENCES

1. ______________, *Manual of Steel Construction,* American Institute of Steel Construction, Eighth Edition, Chicago, 1980.

2. Disque, R. O., "Wind Connections with Simple Framing," *Engineering Journal,* AISC, Vol. 1, No. 3, July 1964.

3. Ackroyd, M. H., and Gerstle, K., "Behaviour of Type 2 Steel Frames," *Journal of the Structural Division,* ASCE, Vol.108, No. ST7, July 1982, pp 1541-58.

4. Dewkett, K. A., Ackroyd, M. H., "Influence of Connection Behaviour on Steel Frame Designs at Factored Loads," paper presented at the ASCE Annual Conference 1984, San Francisco.

5. ______________, *Manual of Steel Construction,* Load & Resistance Factor Design, American Institute of Steel Construction, First Edition, Chicago, 1986.

6. Beaufait, F. W., Rowan, W. H., Hoadley, P. G., and Hackett, R. M., *Computer Methods of Structural Analysis,* Department of Civil Engineering, Vanderbilt University, Nashville, Tennessee, 37235, 1975.

7. Ioannides, S. A., "Flange Behaviour in Bolted End Plate Moment Connections," Thesis Presented to Vanderbilt University, Nashville, Tennessee, 1978 in Partial Fulfillment of the Requirements for the Degree of Doctor of Philosophy.

8. Goverdhan, A. V., "A Collection of Experimental Moment-Rotation Curves and Evaluation of Prediction Equations for Semi-Rigid Connections," Thesis Presented to Vanderbilt University, Nashville, Tennessee, 1983 in Partial Fulfillment of the Requirements for the Degree of Master od Science.

9. Nethercot, D. A., "Steel Beam to Column Connections - A Review of Test Data and Their Applicability to the Evaluation of the Joint Behaviour of the Performance of Steel Frames," CIRIA, London, 1985.

NON-LINEAR ANALYSIS OF PLANE FRAME STRUCTURES WITH SEMI-RIGID CONNECTIONS

Yvan GALEA
C.T.I.C.M. - B.P. 1
78470 Saint Rémy lès Chevreuse
FRANCE

1. INTRODUCTION - ABSTRACT

The essential objective of the present paper consists in exposing the procedure of taking into account semi-rigid connections with nonlinear behaviour, a posteriori introduced into the CTICM-analysis programme of plane frame structures (PEP).
After a succinct description of the initial state of the PEP-programme, the double spring-ended special member introduced is presented. The description of the taking into account of nonlinear behaviour of the semi-rigid connections follows and the paper is completed by results obtained on a two level-three span portal frame.

2. SUCCINCT DESCRIPTION OF THE INITIAL STATE OF THE PEP-PROGRAMME

A detailed description of the initial state of the PEP-programme is given in [1] and [2].
PEP is an analysis programme of plane frame structures the CTICM has developped for research purposes. It allows to proceed to four types of structural analysis :

- first order elastic analysis,
- second order elastic analysis,
- first order plastic analysis,
- second order plastic analysis.

The main characteristics of the PEP-programme are the following :

a) The plastification in the structural members is based on the concept of plastic hinges with bending moment-axial force interaction.

b) The determination of the states of equilibrium is based on the " displacement method ".

c) The loads on the structure, which are static loads, maintain their proportionality during the loading evolution. This leads to the definition of a parameter α, representing a multiplier of the loading provided by the input data of the programme.

d) Any non-linearity in the structural response to the applied loads, resulting from the occurence of either plastic hinges or second order effects, is dealt with by a step-by-step method without internal iterations (for example, similar to those of the Newton-Raphson method). The incremental behaviour at each loading step assumes the formulation :

$$|\Delta P| = [K_t] \cdot |\Delta D|$$

where,

$|\Delta P|$; $|\Delta D|$ - increments of the vectors "nodal loads" $|P|$ and "nodal displacements" $|D|$ during the step.

$[K_t]$ - tangent stiffness matrix of the structure as a function of the plastification state, of the displacements of the nodes and of the axial forces determined in the structure at the end of the previous step.

The nonlinear behaviour is therefore described by a succession of elementary linear behaviours with K_t being reactualized at each step. The accuracy is obtained by specifying limited amplitudes for the loading steps. For very divergent behaviours (close to instability), displacement control is possible.

At each step, the condition of the structure is determined by summing up the variations of condition calculated at the previous steps.

Should an instability occur (collapse mechanism, member buckling, global instability), the detection criterion is the sign of the determinant of the matrix K_t that changes from positive to negative.

e) Prior to the introduction of the present developments, the connections between the structural members were either perfectly rigid or perfectly hinged.

3. MODELLING OF SEMI-RIGID CONNECTIONS IN THE PEP-PROGRAMME

A detailed description of this modelling is given in [5]. In the framework of this study, neither the influence of the shear flexibility of the connections upon the global structural behaviour nor the possible influence of the shear force upon the intrinsic behaviour of the connections are taken into account.
As a consequence, each connection of semi-rigid type is modelled by two equivalent springs :

- an axial stiffness spring K_ε simulating the axial behaviour of the connection when the connected member is strongly compressed or in tension ;

- a spiral stiffness spring K_θ simulating the bending behaviour of the connection.

The interaction of these two springs reproduces the global behaviour of the connection. In case of nonlinear behaviour of the connection, the stiffnesses K_ξ and K_θ of the equivalent springs are modified at each step of the "step-by-step" procedure, according to the evolution of the parameters that have an influence over these stiffnesses (for instance, the forces upon the connection).

The thus modelled connection is then "attached" to a member whose stiffness matrix integrates K_ξ and K_θ. The modelled connection is therefore considered to have no length.

In this simple way, a special member element is made up of a standard member with the double spring just described at each extremity (figure 1).

This type of element allows one to treat similarily the members with semi-rigid connections at one or at both ends. For the end without semi-rigid connection, an infinite (immeasurable great) stiffness K_ξ and an infinite or null stiffness K_θ ,according as the connection is perfectly rigid or perfectly hinged respectively, are introduced.
In order to let appear the second order effects and the instability phenomena in strongly compressed members with semi-rigid connections, the tangent stiffness matrix of the member element represented en figure 1 has been determined.

Figure 1.

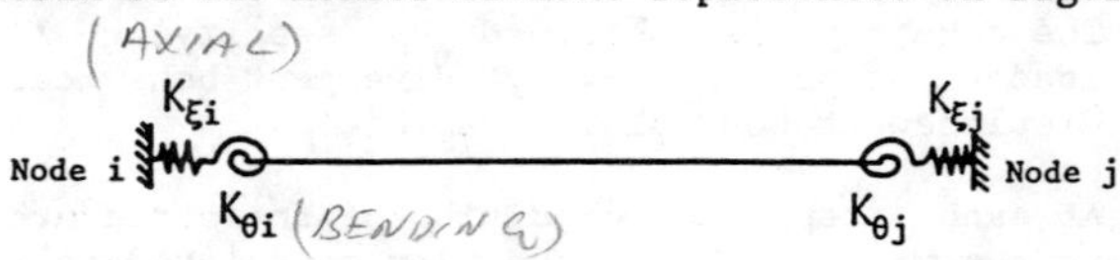

This tangent stiffness matrix R_t which relates, during a step, increments of forces at the ends i and j of the member to increments of displacements of the same ends according to the following relationship :

$$\begin{vmatrix} \Delta N_i \\ \Delta T_i \\ \Delta M_i \\ \Delta N_j \\ \Delta T_j \\ \Delta M_j \end{vmatrix} = [R_t] \cdot \begin{vmatrix} \Delta u_i \\ \Delta v_i \\ \Delta \omega_i \\ \Delta u_j \\ \Delta v_j \\ \Delta \omega_j \end{vmatrix}$$

N : axial force
T : force shear
M : bending moment
u : longitudinal displacement
v : transverse displacement
ω : rotation

is expressed as follows in the local system of the member :

$$[R_t] = \begin{bmatrix} R_1 & R_1\Omega & O & -R_1 & R_1\Omega & O \\ & R_2 + R_1\Omega^2 + \frac{N}{L} & R_4 & -R_1\Omega & -\left(R_2 + R_1\Omega^2 + \frac{N}{L}\right) & LR_2 - R_4 \\ & & R_3 & O & -R_4 & LR_4 - R_3 \\ & & & R_1 & R_1\Omega & O \\ & \text{symétrique} & & & R_2 + R_1\Omega^2 + \frac{N}{L} & R_4 - LR_2 \\ & & & & & R_3 - 2LR_4 + R_2L^2 \end{bmatrix}$$

Where : N - axial force as obtained at the end of the previous step (negative in compression).

Ω - $(v_j - v_i)/L$ - global rotation of the member as obtained at the end of the previous step.

L - length of the member between ends i and j.

E - Young's modulus.

I - inertia of the member cross section.

A - area of the member cross section.

And :

$$R_1 = \frac{EA}{\gamma L} \qquad \alpha = 1 + \frac{k}{K_{\theta_j}}$$

$$R_2 = \frac{k + \lambda k}{\alpha\beta L^2}\left[2 + (k - \lambda k)\left(\frac{1}{K_{\theta_i}} + \frac{1}{K_{\theta_j}}\right)\right] \qquad \beta = 1 + \frac{k}{K_{\theta_i}} - \frac{(\lambda k)^2}{\alpha K_{\theta_i} K_{\theta_j}}$$

$$R_3 = \frac{1}{\beta}\left[k - \frac{(\lambda k)^2}{\alpha K_{\theta_j}}\right] \qquad \gamma = 1 + \frac{EA}{LK_{\xi_i}} + \frac{EA}{LK_{\xi_j}}$$

$$R_4 = \frac{k + \lambda k}{\beta L}\left(1 - \frac{\lambda k}{\alpha K_{\theta_j}}\right)$$

. N<0 (compression) :

$$k = \frac{EI}{L} \cdot \frac{\eta(\sin\eta - \eta\cos\eta)}{2(1 - \cos\eta) - \eta\sin\eta} \qquad \eta = \sqrt{\frac{|N|\,L^2}{EI}}$$

$$\lambda k = \frac{EI}{L} \cdot \frac{\eta(\eta - \sin\eta)}{2(1 - \cos\eta) - \eta\sin\eta}$$

. N>0 (tension) :

$$k = \frac{EI}{L} \cdot \frac{\eta(sh\,\eta - \eta\,ch\,\eta)}{2(ch\,\eta - 1) - \eta\,sh\,\eta} \qquad \eta = \sqrt{\frac{NL^2}{EI}}$$

$$\lambda k = \frac{EI}{L} \cdot \frac{\eta(\eta - \eta\,sh\,\eta)}{2(ch\,\eta - 1) - \eta\,sh\,\eta}$$

. $N \simeq 0$: $k = 4\dfrac{EI}{L}$ $\qquad \lambda k = 2\dfrac{EI}{L}$

The assemblage of the stiffness matrices R_t of all the members, with or without semi-rigid connections, allows one to obtain the global tangent stiffness matrix K_t of the structure.

4. MODEL OF NON LINEAR BEHAVIOUR OF THE CONNECTION ADOPTED IN THE PEP-PROGRAMME

The model of nonlinear behaviour of the connection adopted at the present time in the PEP-programme is that worked out by COLSON and PILVIN [3,4]. Hereafter, the basic characteristics of their simplified model are recalled.
The ultimate limit curve of the connection is a bending moment-axial force interaction curve. In a first step, the curve has been linearized for the sake of simplification (figure 3).

M_u is the ultimate bending resistance of the connection ; it is assumed to behave in a similar manner whatever the sign of the bending may be (symmetrical connection).

N_u^- , N_u^+ are the ultimate axial resistances of the connection in compression and in tension respectively.

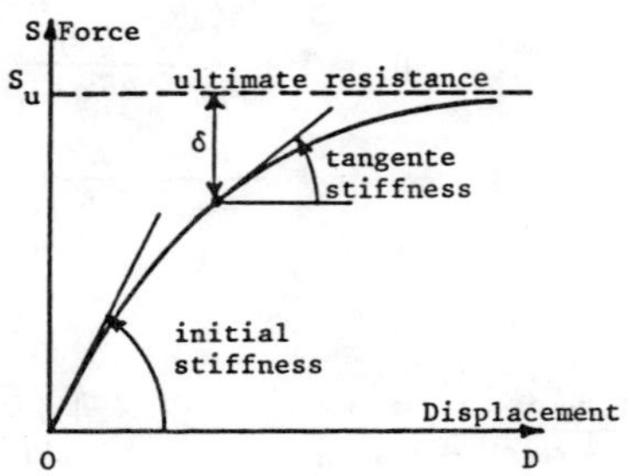

Figure 2.

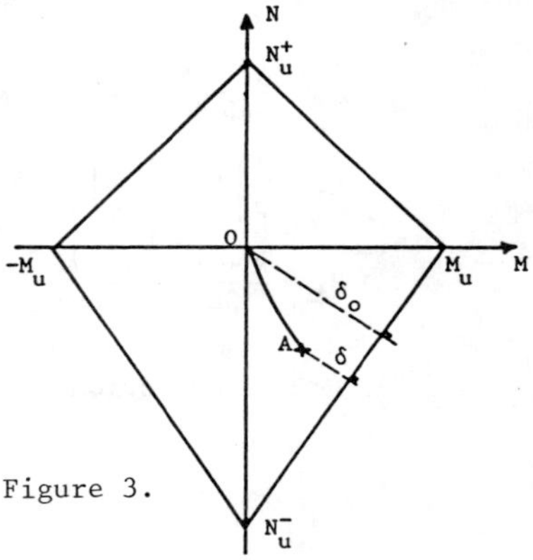

Figure 3.

Figure 2 shows a typical force-displacement curve characteristic of a semi-rigid connection under static loading.
Within the adopted model, the description of the nonlinear behaviour, during the loading phase, is based on knowledge of experimentally determined values of :

- the initial bending and axial stiffnesses of the connection,
- the ultimate bending and axial resistances,
- other parameters characteristic of the non linear behaviour of the connection,

and of a variable δ depending on the loading level that has been reached and characterizing the distance with regard to the ultimate limit curve (Figures 2 and 3).

The law of differential behaviour of the connection has been described in form of the following relationships :

$$\left.\begin{aligned} d\xi &= \frac{N_u}{K_n} \cdot \left[dn + g.\ (A_n.dn + B_n.\ |dm|) \right] \\ d\theta &= \frac{M_u}{K_m} \cdot \left[dm + g.\ (B_m.dn + A_m.\ |dm|) \right] \end{aligned}\right\} \qquad (1)$$

where :

. $d\xi$, $d\theta$: variations of elongation and rotation within the connection,

. dn, dm : variations of reduced forces

$m = M/M_u$ $\quad n = N/N_u$

$N_u = N_u^+$ if tension, $N_u = N_u^-$ if compression

. K_m, K_n : initial stiffnesses of the connection in bending and elongation

$K_n = K_n^+$ if tension, $K_n = K_n^-$ if compression

. $g = x(x + 2)$ with $x = (\delta_o - \delta)/\delta$

δ_o = initial value of δ (see figure 3).

. A_n, B_n, A_m, B_m : (experimental) coefficients for characterization of the non linear behaviour and M-N interaction in the connection.

The tangent stiffnesses of the connection : $K_\xi = \frac{dN}{d\xi}$ and $K_\theta = \frac{dM}{d\theta}$ are derived from the relationship (1).

Consequently, during a step *p* of the "step-by-step" procedure of the nonlinear analysis of the structure, the tangent stiffnesses input for a semi-rigid connection in the tangent stiffness matrix of a member are calculated by :

$$. \; K_\xi^{(p)} = K_n / \left[1 + g^{(p-1)} . \left(A_n + B_n . \frac{|\Delta m^{(p-1)}|}{\Delta n^{(p-1)}}\right)\right]$$

$$. \; K_\theta^{(p)} = K_m / \left[1 + g^{(p-1)} . \left(A_m + B_m . \frac{\Delta n^{(p-1)}}{|\Delta m^{(p-1)}|}\right)\right]$$

with : $g^{(p-1)} = x^{(p-1)} . (x^{(p-1)} + 2)$, $\quad x^{(p-1)} = (\delta_o - \delta^{(p-1)})/\delta^{(p-1)}$

$\Delta m^{(p-1)}$, $\Delta n^{(p-1)}$: variations of reduced stresses during the previous step.

$\delta^{(p-1)}$: distance with regard to the ultimate limit curve as determined at the end of the previous step.

If, during a load step, the ultimate limit curve is transcended, the amplitude of the load step is reduced in order to bring the point A (figure 3) back on the curve and the connection is replaced by a perfect hinge with regard to the later load increments. The use of this artifice is simply related to the numerical aspect of the "step-by-step" description of the non linear behaviour. Strictly speaking, the limit curve should never be reached but only approached asymptotically.

Particular cases : If $N \simeq 0$ $\quad K_\xi = \frac{1}{2}(K_n^+ + K_n^-)$; If $M \simeq 0$ $\quad K_\theta = K_m$

5. EXAMPLE OF APPLICATION

The chosen example of application concerns one of the calibrating frames defined by R. ZANDONINI (Polytechnic School in Milan), responsible within a subgroup of the ECCS - TC8 "Stability" for a study dealing with the stability of structures with semi-rigid connections.
Both the structure and the loading are defined in figure 4.

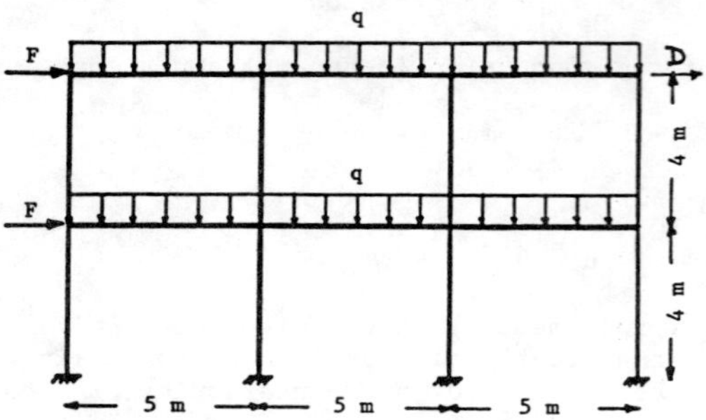

f_y = 235 MPa
q = 40 kN/m
F = 5 kN

beams : IPE 300 (strong axis)
columns : HEA 160 (strong axis)

Figure 4.

The structure is considered with an initial verticality imperfection ψ = 1/300. The type of beam-column connection is depicted in figure 5. The values of the connection parameters have been determined on a supplied bending moment-rotation curve ; they are :

M_u = 50,1 KNm K_m = 10 959 KNm rd^{-1} A_m = 0,113

Since the connections are essentially in bending, the following values have been adopted :

$A_n = B_m = B_n = 0$ $K_n^+ = K_n^- = \infty$ (very large)

$N_u^+ = N_u^-$ = plastic axial force of the beams

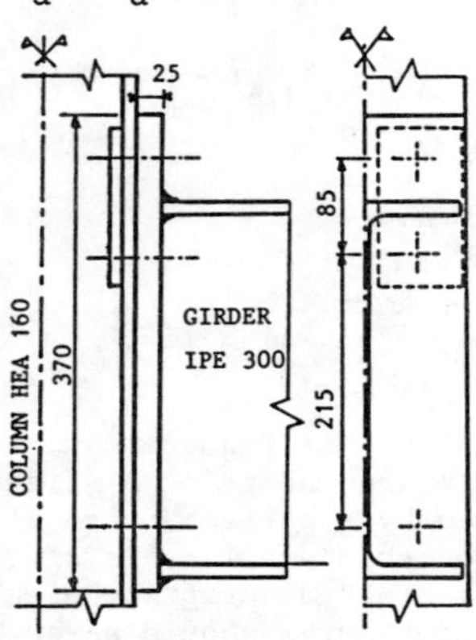

6 M16 bolts grade 10.9

Figure 5.

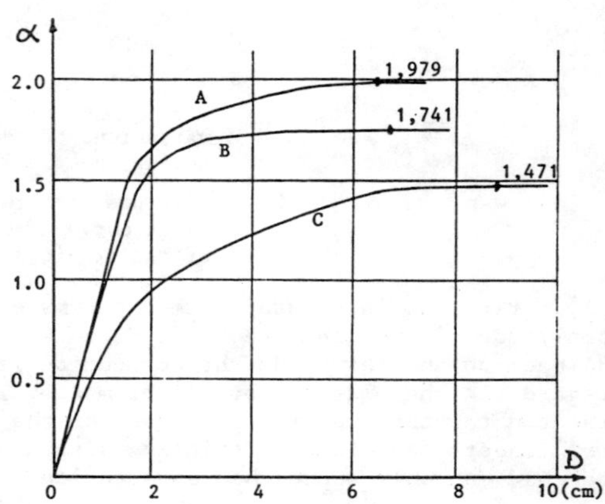

Figure 6.

In figure 6 , the evolution of the lateral displacement D at the top of the structure has been drawn as a function of the loads that increase proportionally with a multiplying factor α , and this for the three following analysis :

- first order plastic analysis + rigid connections (curve A)
- second order plastic analysis + rigid connections (curve B)
- second order plastic analysis + semi-rigid connections (curve C)

The amplitude of the loading step was limited by the most severe condition, i.e. $\Delta \alpha = 0{,}1$ or $\Delta D = 2$ mm. As concerns the curve C, that resulted in 48 steps before reaching the collapse and 4 mn 20 seconds calculation time on a IBM 4361-3 computer.

In order to take correctly the second order effects into account, the columns have been discretized in 3 sections. Because the double spring-ended element, described in 2, can not receive loads between its ends, the beams have also been discretized in 3 sections as shown in figure 7.

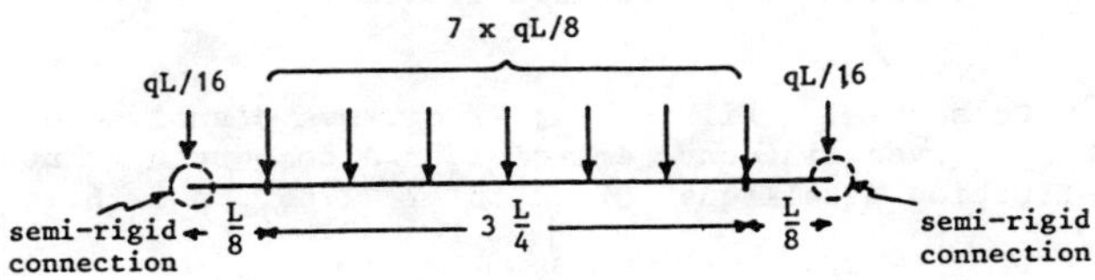

Figure 7.

The important divergence of behaviour depending on whether the semi-rigid connections have or have not been taken into account should be noted.
The difference between the collapse load by instability in the elasto-plastic range is about 16%. The lack of web stiffeners in the columns at the lower flange of the beams may explain the flexibility of this type of connection.

6. CONCLUSIONS

The fact of taking into account the semi-rigid nature of the connections in a pre-established programme resulted in adopting a simplified geometrical model. As a consequence, a certain number of assumptions have been admitted :

- connections without their own dimensions,
- shear is not taken into account,
- the ultimate limit curve is linearized.

Some other assumptions belong implicitly to the analysis method adopted for the PEP-programme (plastic hinges, "step-by-step" method without iterations).
Although a comparison with results obtained by other programmes used abroad has been satisfactory, a further stage is necessary that will consist in comparing a numerical simulation with large scale tests aimed at confirming the validity of the adopted assumptions.

7. REFERENCES

1. Lescouarc'h Y., Programme de calcul en élastoplasticité des structures planes. Revue Construction Métallique, CTICM, n° 1, 1976, pp. 40-50.

2. Galéa Y., Programme d'analyse élastoplastique non linéaire de structures planes à barres. Revue Construction Métallique, CTICM, n° 4, 1978, pp. 3-16.

3. Pilvin P., Modèle de comportement statique des liaisons de structures à barres. Revue Construction Métallique, CTICM, n° 2, 1984, pp. 43-50.

4. Colson A., Modélisation des conditions aux limites de liaisons et d'assemblages en mécanique des structures métalliques. Thèse de Doctorat d'Etat. Paris 6. 18 décembre 1984.

5. Galéa Y., Colson A., Pilvin P., Programme d'analyse de structures planes à barres avec liaisons semi-rigides à comportement non linéaire. Revue Construction Métallique, CTICM, n° 2, 1986, pp 3-16.

NONLINEAR INELASTIC ANALYSIS OF FLEXIBLY-CONNECTED STEEL FRAMES

M.H. Ackroyd
Techtrol, Ltd.
Montreal, Quebec

ABSTRACT

This paper describes a nonlinear computer analysis program that can simulate the behavior of unbraced steel frames that have flexible beam-to-column connections and/or partial composite girders. It summarizes the modeling capabilities and cites instances of how the tool has contributed to a better understanding of complex frame behavior and frame design.

INTRODUCTION

This paper describes an analysis program that has been developed for evaluation of structural steel building frames at their ultimate load level. The program was developed over a period of nine years in a research effort to evaluate the actual behavior of AISC "Type 2" frames (1), funded primarily by the American Iron and Steel Institute.

SCOPE

The program described herein is a research tool that has been specialized for very precise modeling of a large number of nonlinear phenomena that arise in the class of planar unbraced sway frames consisting of horizontal beams and vertical columns. No accounting is made for the possibility of out-of-plane effects such as lateral-torsional buckling of beams, out-of-plane frame buckling, or torsional buckling of members.

MODELING APPROACHES

In order to accurately predict the behavior of flexibly-connected steel frames, both material and geometric nonlinearities need to be accounted for in the mathematical models of the various structural components of the frame. The program described here includes models for inelastic beam-columns, beam-to-column connections having nonlinear moment-rotation curves (including unloading capability), and inelastic composite concrete-steel girders with slip.

Beam-columns

The model for steel I-shaped members treats the cross-section as three rectangular elements. All cross-section effects are computed as analytic expressions relative to these rectangular elements and then in terms of overall section response. Second order geometric effects are accounted for directly at the member level. Details of the development can be found in (2).

A. Cross-section Model

Residual Stresses. The residual stresses in the cross-section are represented as initial stress patterns that vary parabolically and symmetrically along the length of each rectangular component. The patterns in the flanges are taken to be identical, so that one need only specify the values of the residual stress at the tip of a flange and at the center of the web: the program automatcally computes the residual stress at the web-flange junction based on equilibrium of the unloaded member.

Force Resultants. When loads are applied to the section, a linear strain diagram is assumed, elastic-plastic fiber strains are evaluated as appropriate, and stress resultants are found that equilibrate the applied loads. The portions of the section that remain elastic are then used to compute the available bending rigidity for use in associated member stiffness calculations.

B. Spread of Plastification

When loads are applied to a member, the moment diagram is checked to see if any sections along the member length will experience plastic strains in the presence of axial load and moment. In the regions where plastification is predicted, the length of the plastic region is computed as the distance from the point where elastic behavior ceases to the point of maximum moment. At the point of maximum moment, the precise fiber model above is used to compute the bending rigidity of the cross-section. Then a linear variation of the bending rigidity is assumed from the linear elastic value at the start of the plastic region to the minimum value just computed. (The accuracy of this approach was corroborated by ancillary studies comparing finite difference solutions of the continuous problem to the displacements solutions predicted by the discretized equivalent taper model). A closed form stiffness matrix for the tapered portion of the member is then combined with the stiffness matrix for the elastic portion of the member, internal degrees of freedom are condensed out, and the total member stiffness matrix is obtained for use in a frame analysis.

C. Second-Order Geometric Effects

In formulating the stiffness matrix of columns in the frame, the classical stability functions are used for reducing the flexural stiffness coefficients due to the presence of axial thrust. Overall member P-Delta effects are accounted for by adding the usual member geometric stiffness matrix to the existing member stiffness matrix.

Nonlinear Beam-to-Column Connections

Flexible beam-to-column connections are modeled as two-degree-of-freedom spring elements having moment-rotation characteristics appropriate for the given stiffness. The frame analysis program uses a secant stiffness approach, so that the stiffness is simply the ratio of the current connection moment to the current angular distortion, plus any permanent angular offsets due to lack of fit or elastic unloading. Details of the development can be found in references 2 and 3.

A. Initial Loading Curves

The initial loading curves for the connection models can be represented as either a Frye-and-Morris polynomial or a Richard-Abbott exponential. The polynomial forms must be used with care so as not to choose a connection type and size that is beyond the extrapolation limits of the original Frye-and-Morris data. The Richard-Abbott expression, though, provides a stable curve-fitting of a wide range of responses, from elastic-plastic to stiffness-degrading.

B. Unloading Behavior

When the moment on a connection decreases, elastic unloading is assumed. In the secant stiffness formulation used here, this behavior can be handled in the same way that initial lack-of-fit problems are accounted for. The result is that one need only add an "initial force" vector to the applied load vector based on the permanent angular distortion of the connection and specify the initial elastic stiffness of the connection in evaluating the two-degree-of-freedom spring stiffnesses.

C. Finite Joint Size

When the depths of the connected members need to be considered in the frame analysis, the member stiffness matrix can be automatically adjusted to model rigid panel zones at the ends.

Partial Composite Girders

The model for partial composite girders predicts the nonlinear responses of concrete tensile cracking and compressive crushing, steel yielding in both compression and tension, and nonlinear slip/deformation of the shear studs bonding the concrete to the steel W-shape. Details of the development can be found in reference (4).

A. Material Behavior

For each of the three components of the composite girder (steel I-section, concrete, and Nelson studs), realistic material behavior is assigned according to accepted elemental force-deformation relations. The model for steel is the widely used elastic-plastic response for individual fibers within a cross-section. The model for concrete is taken as the well-known Hognestad continuously nonlinear response (5). The load-slip response of the shear stud connectors is that reported in Olgaard, Slutter, and Fisher (6).

B. Cross-section Model

If one assumes that no transverse separation of the concrete and steel occurs, but that longitudinal slip of the concrete can occur relative to the steel, then the curvatures of the concrete and steel will be identical. The strains induced by flexure are taken as linear, but with a possible jump in the value of strain at the concrete-steel interface due to slip, ϵ_d.

The stress fields associated with the above strains will produce a net compressive resultant in the concrete, C, and a tension resultant, T, in the steel, with a moment arm of z between them. This couple is then the internal reaction to the applied moment. Compatibility, stress-strain, and equilibrium then lead to a pair of first-order, coupled ordinary differential equations for the composite girder, as developed in detail by Yam and Chapman (7):

$$\frac{dY}{dx} = \epsilon_d = f(\ M(x),\ C(x)) \qquad \text{Slip Rate}$$

$$\frac{dC}{dx} = \frac{Q}{s} = g(\ Y\ ) \qquad \text{Shear Flow}$$

where Y is the amount of slip (e.g., inches) and Q is the force in the shear stud.

Solution of these coupled equations must be done by numerical integration techniques for any given girder and loading.

C. Composite Girder Model

The solution technique used for analyzing girders that are parts of frames consists of a preprocessing step of finite difference solution for isolated girders under gravity loads and a frame analysis that utilizes stiffness data generated in the preprocessing step.

In the preprocessing step, the desired gravity loads are specified on a simply-supported composite girder with the given cross-section. This girder is then analyzed for 4 to 6 levels of gravity load factor up to 1.3 or the collapse of the girder. For each load level, the coupled differential equations are solved numerically using a shooting method of forward numerical integration using the Euler-Cauchy finite difference equations. Details of this solution method are presented in Zaremba (4).

Once the differential equations are solved for any given problem, one has obtained the values of curvature at all station points along the span. These in turn can by integrated to obtain beam slopes and deflections along the span. The resulting data is then used to establish the load-deflection data to be used in the ensuing frame analysis.

In the frame analysis, an equivalent prismatic model is used for portions of girders in positive bending, where the value of EI is interpolated from reference responses of the prototypical partial composite girder analyzed in the preprocessing step.
In general, the response of a composite girder can encompass a range of stiffness regimes. For example, it could start with a large positive moment at the face of an attached column to result in a tapered end due to "necking down" of the effective slab width from the usual effective width along the span to the width of the column flange (under positive bending), an equivalent prismatic region in the region of positive bending, a region of steel-only in regions of negative bending, and possibly a region of tapered steel-only when steel yielding occurs under large negative bending moments. The model used in the frame analysis includes provisions for all of these possibilities, using a substructuring approach similar to that used for columns under partial plastification.

NONLINEAR FRAME ANALYSIS

All of the individual models described above are organized into a plane frame analysis program that allows combined gravity and wind loads. The loads are defined independently in terms of generic surface pressures over decks and cladding areas. Then they are applied to the frame according to tributary areas and user-defined gravity and wind load multiplier histories. The overall analysis proceeds in user-prescribed load-step increments, using a secant-stiffness formulation. When convergence is difficult (e.g., when approaching the collapse load), the program can automatically backstep and re-continue with reduced load step sizes. Results are played back through high resolution graphics sequences of displaced shapes to facilitate the identification of modes of collapse and progressive collapse scenarios.

RESULTS

Comparison of Frame Capacity to Test

The above program was used to analyze a frame that was experimentally loaded to collapse at Lehigh University (8) in order to corroborate its accuracy. It was necessary to include the nonlinear beam-column model and the rigid panel zone models to simulate the behavior of the rigid frame that was tested. Comparison of results showed that the analysis predicted a collapse load within 4% of that observed in the test results. Moreover, the sequence and pattern of cross-section plastifications were well predicted by the analysis.

Comparison of Frame Analysis to ANSYS

At the time of program completion, no experimental results were available for flexibly-connected frames. However a complete independent analysis check was performed by analyzing a flexibly-connected frame using the ANSYS nonlinear finite element package (9). Good agreement was observed.

APPLICATIONS

The major use of the above described program has been as a research tool for evaluating the adequacy of frames designed as Type 2 Construction. The most significant new finding that can be directly attributed to the program is the unsuspected mode of failure of most Type 2 frames in the so-called "zipper mode" of collapse (10), wherein the leeward column stack undergoes extensive plastification just prior to the onset of overall frame instability. This mode of frame failure occurs at load factors below those associated with the classical story shearing modes presumed in "rational" stability analysis approaches such as used in generating the effective length nomographs. Inadequacies of such design approaches have been remedied in modified design procedures that have been proposed as a result of studying the analysis results of both the original designs and modified re-designs (11).

REFERENCES

1. AISC, Specification for the Design, Fabrication, and Erection of Structural Steel for Buildings, Effective November 1, 1978.

2. Ackroyd, M.H., "Nonlinear Inelastic Stability of Flexibly-Connected Plane Steel Frames", PhD dissertation, University of Colorado, Boulder, Colorado, May, 1979.

3. Connolly, J.T., "Shakedown of Flexibly-Connected Plane Steel Frames", Master of Engineering Thesis, Rensselaer Polytechnic Institute, Troy, N.Y., May, 1982.

4. Zaremba, C.J., "Limit State Analysis of Partial Composite Girders and Application to Building Frames", Master of Engineering Thesis, Rensselaer Polytechnic Institute, August, 1986.

5. Park, and Pauley, Reinforced Concrete Structures.

6. Olgaard, J., Slutter, R., and Fisher, J., "Shear Strength of Stud Connectors in Concrete", AISC Engineering Journal, Vol. 8, No. 2, April, 1971.

7. Yam, L.C.P. and Chapman, J.C., "The Inelastic Behavior of Simply-Supported Composite Beams of Steel and Concrete", Proceedings of the Institution of Civil Engineers, London, Vol. 41, December, 1968.

8. McNamee, B.M. and Lu, L.W., "Inelastic Multistory Frame Buckling", Journal of the Structural Division, ASCE, Vol. 98, No. ST7, July, 1972.

9. Edinger, J.A., "Nonlinear Frame Analysis using the ANSYS General Purpose Computer Program", AISC Research Report No. 83-1, July, 1983.

10. Ackroyd, M.H., "Limit States of Type 2 Steel Building Frames", Research Report to Amercian Iron and Steel Institute, under Project No. 311, "Ultimate Capacity of Flexibly-Connected Steel Building Frames", Rensselaer Polytechnic Institute, Troy, N.Y., August, 1981.

11. Cronembold, J., "Evaluation and Design of Type 2 Steel Building Frames", Master of Architecture Thesis, Rensselaer Polytechnic Institute, Troy, N.Y., May, 1986.

A FINITE ELEMENT FOR THE ANALYSIS OF SEMI-RIGID FRAMES

Carlo Poggi
Department of Structural Engineering
Politecnico di Milano, Milano, Italy

and

Riccardo Zandonini
Department of Engineering
University of Trento, Trento, Italy

ABSTRACT

A realistic assessment of the frame behaviour requires the influence of joint flexibility to be taken into account in the analysis. The numerical approach highlighted in this paper makes use of the finite element technique. A beam-joint element was formulated, in which both joint deformation and joint finite dimensions are allowed for. The method is an extension of a highly refined formulation, previously developed for rigidly jointed frames. The validation of the approach has been obtained against available experimental data.

INTRODUCTION

The significant influence that joint flexibility may have on the performance of framed systems has been recognized since the early developments of modern structural analysis. Numerical methods which include this parameter in frame analysis were available more than fifty years ago [1] and approaches more and more refined were then developed [2]. Recent studies have indicated that design methods recognizing the joint semi-rigid frame action may allow an improvement in the cost effectiveness of the structure, due to a possible better balance between material and labour costs. Furthermore, the degree of flexibility of joints, usually modelled as rigid, may in some cases non negligibly affect the frame safety level.The interest in the semi-rigid frames as a realistic design alternative has then significantly increased. However, studies are still necessary to thoroughly understand how joint behaviour affects frame response. This knowledge should enable us to develop sensible design criteria and methods.
Sophisticated numerical approaches need to be set up in order to take into account all the parameters governing the behaviour of frames with semi-rigid joints. Several methods have been developed very recently [2], which make use of models at different degrees of refinement for simulating the joint and/or the member response.
This paper intends to highlight the basic aspects of a numerical method, in the frame of the finite element technique, which adopts a particular formulation [3] allowing for a combination of a refined modelling of the material and geometrical nonlinearities with computational effectiveness. In particular, the spreading of plasticity through the cross section and along the member is taken into account. This formulation was proved to be highly accurate in the case of steel frames with rigid joints [4,5,6].

A joint-beam element was developed, able to take into account both the joint nonlinear behaviour and the joint finite dimensions.
Two different solving strategies were adopted in order to optimize the computer time with reference to the cases of proportional and nonproportional loading paths.
Several comparisons to available experimental results are also reported and some conclusions on the accuracy of the proposed method drawn.

JOINT-BEAM ELEMENT MODEL

A joint-beam finite element model was developed, suitable for the analysis of semi-rigid frames (Fig. 1). The main characteristics of the model are highlighted here; detailed presentation is reported in [7]
The model is conceived as an assemblage of three different parts :
- a central nonlinear beam-column element
- two connection elements
- two rigid bars

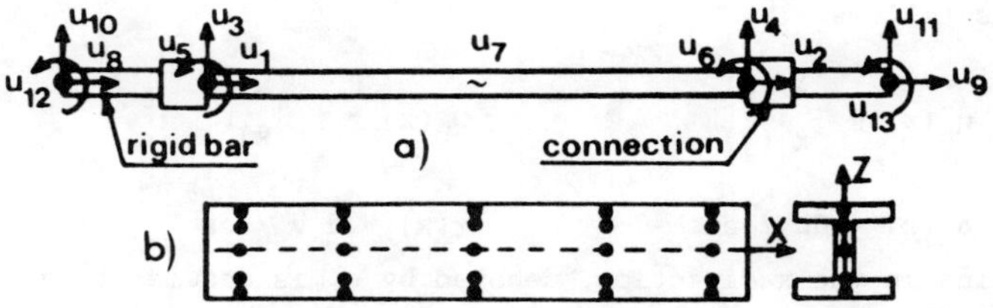

Fig. 1 a) Joint-beam element.
b) control points over the central beam element.

Central nonlinear beam element

The material costitutive law, characterized by a bilinear behaviour with cyclic hardening, is described by the following relations:

$$\sigma_\ell = E(\varepsilon_\ell - p_\ell - \varepsilon_{0\ell}) + \sigma_{r\ell} \quad ; \quad p_\ell = \underline{n}_\ell \underline{\lambda}_\ell \qquad (1a,b)$$

$$\underline{\Phi}_\ell = \underline{n}_\ell^t \sigma_\ell - \underline{h}_\ell \underline{\lambda}_\ell - \underline{r}_\ell \qquad (1c)$$

$$\dot{\underline{\lambda}} \geq \underline{0} \quad ; \quad \underline{\Phi} \leq \underline{0} \quad ; \quad \underline{\Phi}^t \dot{\underline{\lambda}} = \dot{\underline{\Phi}}^t \dot{\underline{\lambda}} = 0 \qquad (1d\text{-}g)$$

where:

$$\underline{n}_\ell^t = \begin{Bmatrix} 1 \\ -1 \end{Bmatrix} \quad ; \quad \underline{r}_\ell = \sigma_Y \begin{Bmatrix} 1 \\ 1 \end{Bmatrix}$$

$$\underline{\Phi}_\ell = \begin{Bmatrix} \Phi_{\ell+} \\ \Phi_{\ell-} \end{Bmatrix} \qquad \underline{\lambda}_\ell = \begin{Bmatrix} \lambda_{\ell+} \\ \lambda_{\ell-} \end{Bmatrix} \qquad \underline{h}_\ell = K \begin{bmatrix} 1 & -\beta \\ -\beta & 1 \end{bmatrix}$$

E is the Young's modulus of the material and σ_y its yield stress, p_ℓ is the plastic part of the total strain (expressed as the difference of two non-negative and non-decreasing plastic multipliers), ε_0 is the thermal part of the total strain and $\sigma_{r\ell}$ is the value of residual stress. $\underline{h}_\ell$ is an hardening matrix, K is a value related to the slope of the hardening branch and factor β is a constant to define the type of hardening rule and may vary from 1 (kinematic hardening) to -1 (isotropic hardening). Vectors $\underline{r}_\ell$ and $\underline{n}_\ell$ are constants and $\underline{\lambda}_\ell$ and $\underline{\Phi}_\ell$ are the vectors of plastic multipliers and yield functions. For simplicity, but without lack of generality, in the following the material is supposed to be elastic perfectly plastic, without thermal strains and residual stresses (this means $K = \varepsilon_{0\ell} = \sigma_{r\ell} = 0$)

The central beam element is supposed to be of constant cross section. Its centroidal axis is denoted by x and symmetry about z axis is assumed (Fig.1). If the shear effects are neglected, as in the classical beam theory, the elongation δ(x) and the curvature χ(x) of the beam may be assumed as generalized strain components governing the deformability of the cross section x. The axial force N(x) and the bending moment M(x) are the corresponding generalized stresses. Let vectors q(x) and Q(x) be defined as follows:

$$\underline{q}_b(x) = \begin{Bmatrix} \delta\ (x) \\ \chi\ (x) \end{Bmatrix} \qquad \underline{Q}_b(x) = \begin{Bmatrix} N(x) \\ M(x) \end{Bmatrix} \tag{2a,b}$$

where: $$\delta\ (x) = du\ /\ dx \qquad \chi(x) = d^2 v\ /\ dx^2 \tag{3a,b}$$

The strain in the x direction, denoted by ε, is assumed to vary linearly with z, so that:

$$\varepsilon\ (x,z) = \delta(x) - z\ \chi(x) \tag{4}$$

The model is based on the independent modelling of the displacements (u(x), v(x)) and of the plastic multipliers ($\lambda_{\ell+}(x,z)$, $\lambda_{\ell-}(x,z)$).
The displacements u(x) and v(x) (collected in vector $\underline{u}_b(x)$) are modelled as functions of the nodal parameters $\underline{u}$ by means of the following relation:

$$\underline{u}_b(x) = \underline{\Psi}\ (x)\ \underline{u} \tag{5}$$

where Ψ(x) is a matrix of shape functions.
Note that, if all the springs are active, only the first seven of the thirteen degrees of freedom of vector $\underline{u}$ (Fig. 1a) are necessary to completely define the displacement field. Otherwise, if several springs are fixed, a slave condition on the relevant displacements is imposed and the relation is modified by means of a condensation process. This was shown to be necessary to avoid any ill conditioning of the element stiffness matrix due to the introduction of fictitiously high spring stiffnesses.
Once eqs.(5) are defined, the strain distribution in the element is given by eqs (4) and (3). A constant distribution of δ(x) and a parabolic one for the curvature χ (x) is obtained because of the introduction of parameter u_7. It is an additional displacement parameter associated to a fourth order function that permits the use of a limited number of element even in the presence of geometric nonlinearities [4,5].

The plastic multipliers are modelled, indipendently of the displacements, on the base of their values at an established number of points (x_j, z_h) within the element and over the cross sections (Fig.1b).
To obtain an accurate modelling of the plastic zones it is often necessary that the distribution of plastic strains (p(x,z)) be controlled at a large number of points. This would produce a lack of compatibility between plastic and total strain distributions that causes self equilibrated stresses in the isolated element subject to plastic strains only. The procedure proposed in [3] is adopted to avoid the presence of these spurious terms. This is equivalent to an assumption for the distribution of N(x) and M(x) along the element that be consistent with the assumed displacement model. For this reason N(x) is constrained to be constant and M(x) to vary parabolically along x. The coordinates of the control points are those used to calculate an integral over the element volume by means of Gaussian Quadrature Rules. For doubly symmetric cross sections the values of z_h are reported in [4].
The above described relations together with eqs.1 are combined to obtain the beam elastic-plastic set of relations:

$$\underline{Q}_b(x) = \underline{d}_b(\ \underline{q}_b(x) - \underline{p}_b(x)\ ) \quad ; \quad \underline{p}_b(x) = \underline{n}_b \underline{\mu}_b(x) \qquad (6a,b)$$

$$\underline{f}_b(x) = \underline{n}_b^t\, \underline{Q}_b(x) - \underline{h}_b \underline{\mu}_b(x) - \underline{r}_b \qquad (6c)$$

$$\underline{f}_b \leq \underline{0}\ ;\ \dot{\underline{\mu}}_b \geq \underline{0}\ ;\ \underline{f}_b^t\, \dot{\underline{\mu}}_b = \dot{\underline{f}}_b^t\, \dot{\underline{\mu}}_b = 0 \qquad (6d,g)$$

The expressions of matrices $\underline{d}_b$, $\underline{n}_b$, $\underline{h}_b$ and $\underline{r}_b$ may be found in [7].

Connection elements

Three springs are located at each end of the beam to simulate the actual behaviour of the joints.
The piecewise-linearized cyclic costitutive laws governing the elastic-plastic spring behaviour are described by the following relations formally identical to the beam ones:

$$\underline{Q}_s = \underline{d}_s(\ \underline{q}_s - \underline{p}_s)\ ;\ \underline{p}_s = \underline{n}_s\ \underline{\lambda}_s \qquad (7a,b)$$

$$\underline{\Phi}_s = \underline{n}_s^t\ \underline{Q}_s - \underline{h}_s \underline{\lambda}_s - \underline{r}_s \qquad (7c)$$

$$\underline{\Phi}_s \leq \underline{0} \quad ; \quad \dot{\underline{\lambda}}_s \geq \underline{0} \quad ; \quad \underline{\Phi}_s^t\ \dot{\underline{\lambda}}_s = \dot{\underline{\Phi}}_s^t\ \dot{\underline{\lambda}}_s = 0 \qquad (7d,g)$$

Where $\underline{Q}_s$ and $\underline{q}_s$ are vectors collecting the axial, shear and bending stresses and strains of the springs. $\underline{d}_s$ is a symmetric, positive definite matrix collecting the elastic spring stiffnesses that may also take into account possible coupling of springs, and $\underline{h}_s$ is a symmetric hardening matrix. Without lack of generality, Koiter's hardening rule was adopted so that $\underline{h}_s$ becomes a diagonal matrix.

Rigid bars

Two rigid bars are located at the ends of the beam to account for the actual location of the joints.

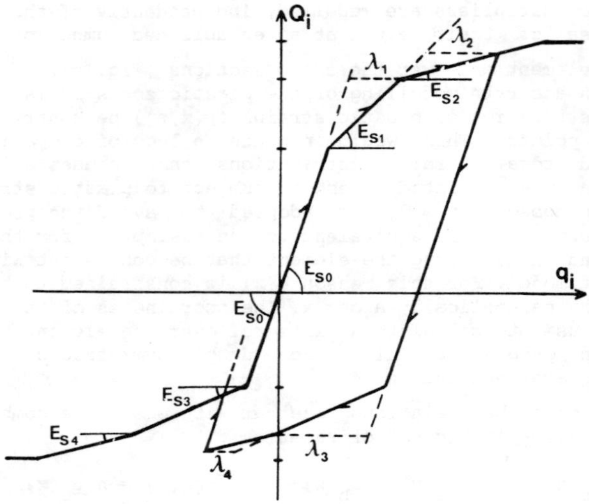

Fig. 2 Spring cyclic costitutive law.

Frame problem

After a usual assemblage procedure, the entire system relations may be obtained :

$$\underline{F} = \underline{K}_{uu}\underline{u} - \underline{K}_{u\lambda}\underline{\lambda} \quad (8a)$$

$$\underline{\Phi} = \underline{K}_{\lambda u}\underline{u} - \underline{K}_{\lambda\lambda}\underline{\lambda} - \underline{R} \quad (8b)$$

$$\dot{\underline{\lambda}} \geq \underline{0} \qquad \underline{\phi} \leq \underline{0} \qquad \underline{\Phi}^t\dot{\underline{\lambda}} = \dot{\underline{\Phi}}^t\dot{\underline{\lambda}} = 0 \quad (8c,f)$$

Where $\underline{u}$ and $\underline{F}$ are the assembled vectors of nodal displacements and forces, $\underline{\lambda}$ is a vector of plastic multipliers ; $\underline{\Phi}$ is a vector of plastic potentials of the same dimension of $\underline{\lambda}$. $\underline{R}$ is a vector of positive constants, $\underline{K}_{uu}$ is the usual elastic stiffness matrix while $\underline{K}_{\lambda u} = \underline{K}^t_{u\lambda}$ and $\underline{K}_{\lambda\lambda}$ are constant matrices. Such a set of equations is linear in the free parameters $\underline{u}$ and $\underline{\lambda}$ and must be respective to the sign restrictions and to the complementarity conditions (eqs. 8c,f). The problem may be solved by means of well known algorithms of mathematical programming or, as an alternative, following an incremental procedure. Two different incremental solving strategies were adopted. In the case of proportional loading the step amplitude is automatically calculated by the program while in the presence of non proportional or cyclic loading paths the step amplitude is imposed by the user.
In the case of non negligible geometrical nonlinearities the stiffness matrix is modified by the use of a geometric stiffness matrix [5]. The structure of the relations remains the same and the above mentioned solution procedures are still adequate but some iterations must be introduced in the solution of each step.

COMPARISON WITH EXPERIMENTAL RESULTS

The finite element approach to the nonlinear analysis of semi-rigid frames is an extension of a general formulation already validated [4,5,6] for the case of steel frames with rigid joints. Its verification against experimental results is nevertheless necessary in order to check the degree of accuracy of the model adopted for simulating the joint behaviour. Tests aimed at pointing out the semi-rigid frame action in the inelastic range were conducted recently and represent a very useful benchmark for the validation of a numerical method. In fact, data related to a series of stability tests carried out on column subassemblages under nonproportional loading [8], and to experiments conducted on a limited number of sway frames subject to cyclic lateral forces [9], allow a thorough check of the proposed formulation. The former set of experiments makes it possible to verify the capability of the method to predict the structural response, with reference to the complex interaction between column, beams and joints and to its influence on column stability. The latter experiments enable a verification of the assumed joint cyclic law.

The geometrical configuration and the general loading and restraint conditions of the column subassemblages tested by Davison and Nethercot are reported in figure 3. The same column and beam sections were used throughout the series. Tests differed for the type of beam to column connection, the loading and loading history and the axis of column bending. Instrumentation of the subassemblage allowed to measure, or to determine, the governing mechanical and geometrical parameters.

COLUMN AXIAL FORCE N AT COLLAPSE (KN)

TEST	BEND.* AXIS	CONN.+	N num	N exp	ERROR %
ST2	S	WC	665.10	675.87	-1.59 %
ST3	W	WC	556.90	519.91	7.11 %
ST4	S	FC	735.80	762.00	-3.44 %
ST7	W	FC	481.00	526.40	-8.62 %
ST8	W	WSC	491.40	518.25	-5.18 %
ST9	W	FEP	484.60	486.49	-0.39 %

* : S = STRONG AXIS W = WEAK AXIS

\+ : WC = WEB CLEATS FC = FLANGE CLEATS
WSC = WEB AND SEAT CLEATS
FEP = FLUSH END PLATE

fig. 3.a

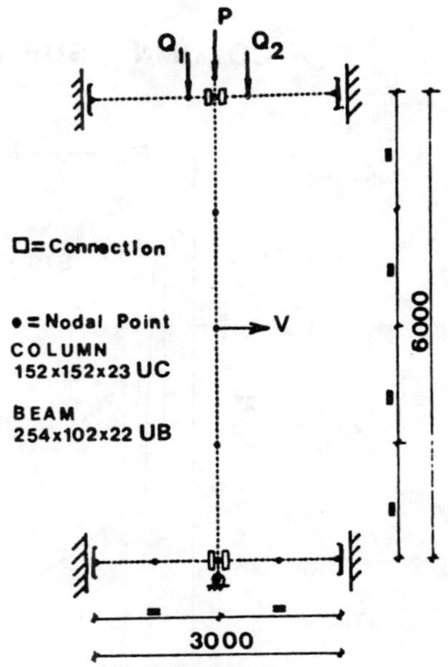

fig. 3.b

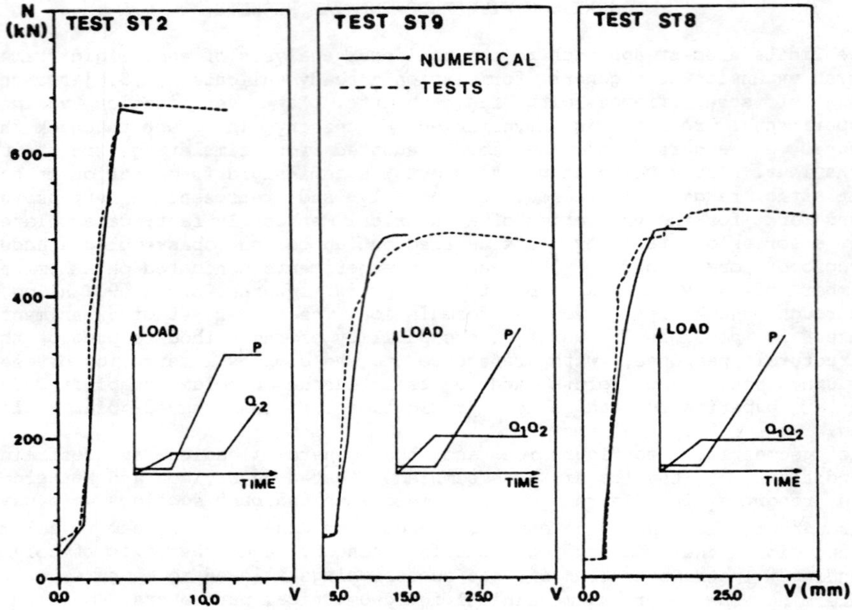

fig. 3.c

COLUMN SUBASSEMBLAGES AFTER [8]

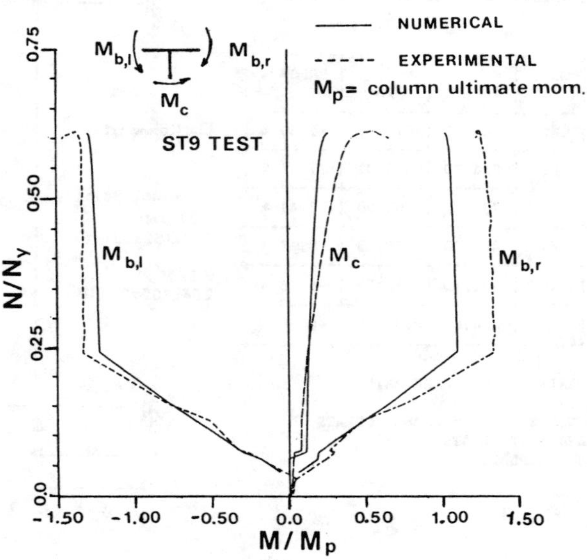

fig. 3.d

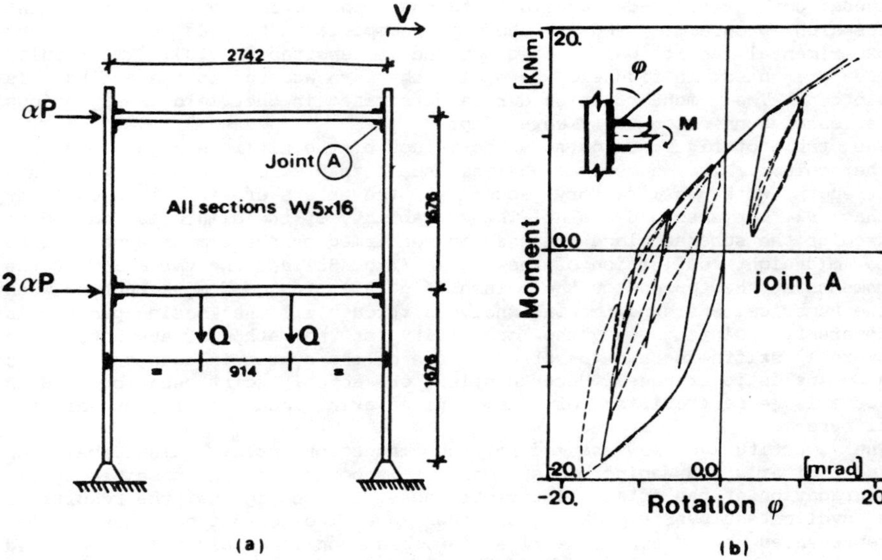

TEST N°10 AFTER [9]

Fig. 4

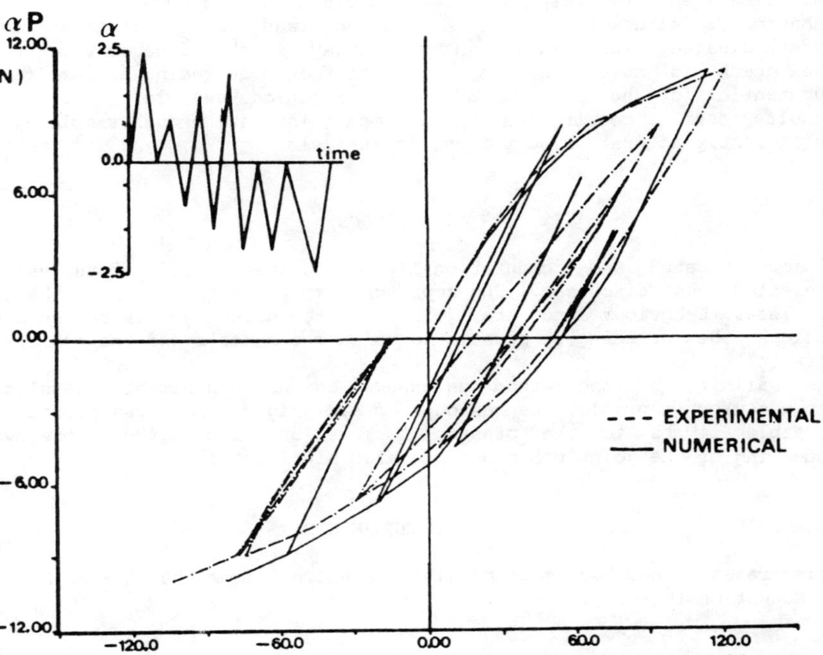

(c)

These data were made available to the authors, in the frame of a joint research programme, and a thorough comparison between numerical and experimental results was carried out and is reported in [10]. Some results are presented in figure 3 where also the mesh adopted in the analysis is plotted. The moment rotation curves determined in Sheffield in a previous series of connection tests were adopted.
The table of figure 3a gathers the values of the ultimate axial load N in the column. The numerical analysis seems to predict the ultimate column strength with satisfactory accuracy; the errors are in all cases lower than 9% in absolute value. The suitability of the simulation method to predict the subassemblage response is confirmed by the comparisons related to midheight deflection of the column (fig. 3c) and the variation of the moments at the top end of the column (fig. 3d). The good agreement between the numerical and experimental analysis through all the loading process is apparent, pointing out the capability of the method of evaluating the overall stiffness and the stress state of the system in both the elastic and inelastic ranges. Uncertainties on actual joint behaviour and on possible eccentricities of the applied axial load can well explain the differences.
The constitutive law assumed for the connection spring element, based on the Koiter's hardening rule, do not account for the possible cyclic degradation of the joint elastic stiffness. A check against the results of a cyclic test by Stelmack [9] was then made. The connection behaviour was represented by a non symmetrical law, based on the experimental measured response. Some results are presented in figure 4 related to the lateral frame drift and the performance of the top right connection. The effect of neglecting the stiffness degradation is apparent in both plots. However, the simulated frame response shows a satisfactory agreement with the test behaviour, although frame deflections and joint rotation may be underevaluated also significantly. Anyway, it should be considered that the check is severe because the stiffness degrading is particularly substantial for the top and seat angle connections used in the test. These results seem to confirm that the assumed joint law, though simple, may be sufficiently accurate also for cyclic analysis.

CONCLUSIONS

A sophisticated and computationally effective method of analysis was presented and discussed. The approach permits an inclusion of the joint nonlinear behaviour and the joint finite dimensions in the analysis, besides the spreading of plastic zones in the members and the geometrical effects.
The validity of the method was checked against experimental results and its adequacy to predict the response of semi-rigid frames was pointed out. Possible limits to the present formulation with regard to the cyclic modelling of the joint behaviour are also highlighted.

ACKNOWLEDGEMENTS

This research has been made possible by a grant from the Italian Ministry of Education (M.P.I.).

REFERENCES

1. Pippard A.J.S. and Baker J.F., The Analysis of Engineering Structures, 3rd Edition, Arnold, London, 1936.
2. Anderson D., Bijlaard F.S.K., Nethercot D.A. and Zandonini R., Analysis and Design of Steel Frames with Semi-Rigid Connections, IABSE Survey, to appear.
3. Corradi L., A Displacement Formulation for the Finite Element Elasto-Plastic Problem, Meccanica, vol. 18, 1983, pp. 77-91
4. Corradi L. and Poggi C., A Refined Finite Element Model for the Analysis of Elastic Plastic Frames, Int. J. of Num. Methods in Engng., vol. 20, 1984, pp. 2155-2174.
5. Corradi L. and Poggi C., An Analysis Procedure for Nonlinear Elastic-Plastic Frames Accounting for the Spreading of Local Plasticity, Costruzioni Metalliche, No. 1, 1985, pp. 1-14.
6. Poggi C. and Zandonini R., Behaviour and Strength of Steel Frames with Semi-Rigid Connections, Proc. Session on Connection Flexibility and Steel Frames, ASCE, Annual Convention, Detroit, October 1985.
7. Poggi C., A Finite Element Model for the Analysis of Flexibly Connected Steel Frames, submitted for publication to Meccanica.
8. Davison B., Kirby P.A. and Nethercot D.A., Column Behaviour in P.R. Construction - Experimental Behaviour, ASCE Structures Congress, New Orleans, September 1986.
9. Stelmack T.W., Analytical and Experimental Response of Flexibly Connected Steel Frames, Master Thesis, Report to AISI, CEAE Dept., University of Colorado, Boulder, 1983.
10. Davison B., Nethercot D.A., Poggi C. and Zandonini R., Analytical and Experimental Studies of Semirigidly Connected Steel Frames, to appear.

DISCUSSION OF FRAME ANALYSIS PAPERS

It is evident that the methods of frame analysis have advanced to the stage where all significant parameters can be taken into account, although the computer programs are still mostly research tools. This was addressed by several speakers and discussers, and it would appear that the development of additional design-oriented programs is critical in the acceptance and use by the profession of the principles of frames with semi-rigid connections.

Second-order geometric effects (P-Delta) were considered by all of the frame analysis methods that were presented. It was noted that the magnitude of the P-Delta contribution for many frames can run as high as 20 % when end-plate connections are used ; it may be substantially higher when more flexible connections are used. However, the designers that have made use of semi-rigid concepts in frame design reported that it was still possible to achieve economical solutions, but the frame drift is almost always the controlling factor.

A separate report was presented on a comparative study of various frame analysis approaches that are currently in use in Europe. Despite differences in modeling, such as in the way the global response of the connection was treated, the use of discrete plastic hinges vs.gradual spread of plastification, and so on, the results of the various programs for several types of frames correlate very well. This would seem to indicate that as long as first-and second -order material and geometric effects for the members and the frame are taken into account, along with a connection element that reflects a realistic M-Φ-relationship, the final results will not differ appreciably. For the moment-rotation relationship it may even be acceptable to use a very simple bi-linear form, as long as it is not unconservative relative to the real form (for example, the initial stiffness value should be close to that of the actual connection).

Numerous points were raised as regards the use of composite girders in the frame, and how to incorporate partial composite action, composite connection effects, reinforcing steel in the slab and around the column, and so on. For example, it was pointed out that research leading to the development of the composite girder criteria of Eurocode EC-4 had demonstrated that for long spans (up to 20 meters (67 feet)) with a low degree of shear connection, significant deflection problems could be encountered. However, it was also noted that the increased stiffness of a composite girder generally leads to the use of a smaller steel shape, with the result that a shallower, more flexible beam-to-column connection is utilized. Thus, the expected benefits of using the increased stiffness of composite girders may not be as substantial as originally thought, but it is also clear that further research work needs to address these questions.

Several discussers noted the importance of the increased connection stiffness that results when slab continuity is achieved (through the use of reinforcing steel bars). In the same way, it was reported that current studies of composite connection effects have shown that connections with very low stiffness in bare steel frames (double web angles, for example) are capable of providing significant local restraint when composite action is accounted for. This is clearly an area where additional research is needed.

SESSION N°4

FRAME STABILITY
AND
SIMPLIFIED METHODS

COLUMN BASE PLATE CONNECTIONS

A. Picard and D. Beaulieu
Département de Génie civil
Université Laval, Ste-Foy
Québec, Canada G1K 7P4

ABSTRACT

This paper summarizes the results of experimental investigations on the behaviour of steel column base connections. The connection is made of a steel base plate welded to the lower end of the column and attached to the footing by means of two or four anchor bolts. It is generally admitted in analysis that this type of connection behaves like a hinge. However, two totally different series of tests indicate that the base connection behaves more like a fixed connection.

INTRODUCTION

It is common practice in North America to consider column base connections as either pinned or fixed to the foundation in calculations. For a perfectly fixed base connection the flexibility ratio at the column base (G_L) is equal to 0.0, that is the concrete footing is infinitely stiff as compared to the column [$G = \Sigma$ (column/beam) rigidities]. The flexibility ratio is equal to infinity for a perfect hinged connection. In the Canadian Standard S16.1-M84 [1], the recommended values are $G_L = 1.0$ instead of 0.0 for a fixed connection to account for the difficulty of obtaining perfect fixity, and $G_L = 10.0$ instead of infinity for a hinged connection to account for the fact that there always exists a certain amount of rotational restraint in a hinged connection.

The column effective length factor (K) used in the evaluation of the column capacity is obtained from appropriate alignment charts. The theoretical relationship between the effective length factor and the nondimensionalized rigidity of the base connection is shown in Fig. 1 for a column that is pinned at the top. The nondimensionalized rigidity (R) is the inverse of G_L which is a function of the rotational flexibility of the base connection (λ : flexibility factor). It is important to point out that the curve drops steeply for low values of R, which means that near fixity conditions are obtained at relatively low values of column base connection rigidity.

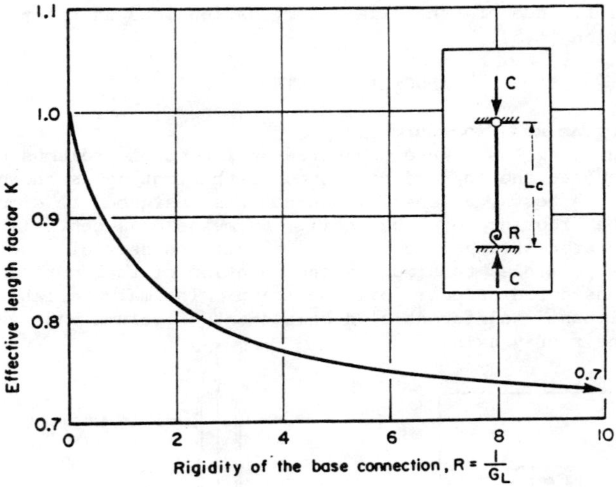

Perfectly hinged connection :	$G_L = \infty$	$R = 0.0$	$K = 1.0$
Standard S16.1-M84 :	$G_L = 10$	$R = 0.1$	$K = 0.975$
Perfectly fixed connection :	$G_L = 0$	$R = \infty$	$K = 0.7$
Standard S16.1-M84 :	$G_L = 1.0$	$R = 1.0$	$K = 0.875$

Fig. 1 -- Effective length factor as a function of the rigidity of the column base connection .

To evaluate G_L for a partial base restraint the following theoretical equation, which applies to non-sway conditions was suggested [2]:

$$G_L = \frac{2EI_c}{L_c}\left(\frac{\theta}{M}\right) = \frac{2EI_c}{L_c}(\lambda) \qquad (1)$$

In this equation, θ/M, which is the inverse of the slope of the moment-rotation curve, represents the rotational flexibility of the base connection (λ) and can be obtained experimentally. The rotation (θ) is the sum of the rotation between the footing and the surrounding soil and the rotation between the column base and the footing. When the column is free to sway, the factor 2 is replaced by 6 in Eq. (1). When $I_c = I_y$ and $\lambda = \lambda_y$ the G_L value obtained from Eq. (1) relates to the column weak axis (G_{LY}). The G_{LX} value is computed with $I_c = I_x$ and $\lambda = \lambda_x$.

The objective of this paper is to present the results of an experimental research project dealing with the behaviour of a simple standard column base connection frequently used in practice. The project was divided in two parts. Buckling tests were carried out on I section columns to check the adequacy of Eq. (1) and to study the stabilizing action of the base connection [3]. Flexion tests and compression - flexion tests were performed on column stubs to study the behaviour of the connection itself and to determine the influence of the axial compression load on the rigidity of the base connection [4,5].

BUCKLING TESTS

Experimental Program and Procedure

Nineteen buckling tests were conducted on I section columns of 5.5 m in length, hinged at the top and connected to the footing as shown in Fig. 2. In two of these tests, a special set-up was designed to simulate the rotation of the footing on the soil. By means of a centrally located roller and calibrated springs placed at the four corners of the concrete footing, it was possible to simulate the rotation of the footing on loose sand which is considered as poor soil condition. The main variable of the tests was the imposed differential displacement of column ends which were applied about the x or y axis.

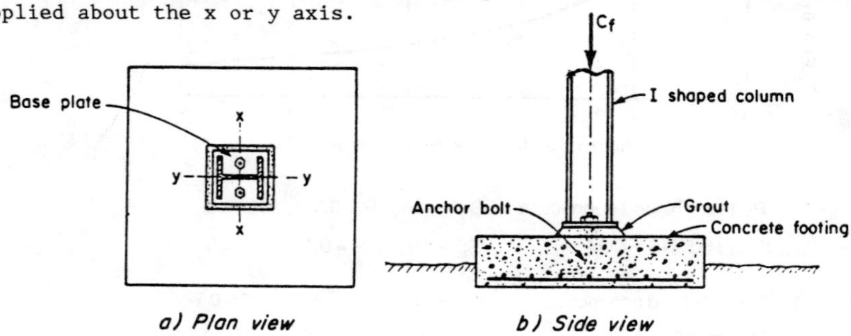

Fig. 2 -- Typical column base connection

The rotation of the steel column near the base plate was measured with LVDTs and the bending moment with electrical strain gauges. Lever arms welded to the lower end of the column were used to amplify the displacements measured by the LVDTs [3]. The rotation of the column was easily calculated from these measured displacements. In some tests, the rotation and the bending moment were also measured at the upper end of the column to evaluate the small restraint offered by the upper pinned connection. This small restraint was taken into account in the analysis of the test results.

Column initial curvatures were carefully measured before each test. Every significant parameter such as applied axial loads, moments and rotations at column ends, lateral deflections at mid-height, and rotations of the footing were recorded during the tests. The electronic data acquisition system was capable of recording up to twenty different measurements in less than forty milliseconds.

Buckling Tests Results

In these tests, all columns buckled about the weak axis. From a practical point of view, the main conclusions that can be drawn from these tests may be summarized as follows.

(a) For weak axis bending, the measured rotational flexibility at the column base ($\lambda = \theta/M$) ranges from 45×10^{-6} $(KN.m)^{-1}$ to 484×10^{-6} $(KN.m)^{-1}$.

With these minimum and maximum values of the flexibility factor, Eq. (1) gives $0.018 \leq G_{LY} \leq 0.196$. The effective length factor is then equal to: $0.71 \leq K_y \leq 0.76$. As shown in Fig. 1, these values are lower than the recommanded values in Standard S16.1-M84 for a fixed connection ($G_L = 1.0$, $K = 0.875$).

(b) Since the base connection behaved more like a fixed connection that a hinged connection, the measured buckling loads largely exceeded the theoretical buckling loads obtained for a pinned base column. The deformed shape of the buckled columns revealed the presence of a significant rotational restraint at the column base.

(c) For the tests where λ values were measured at both ends of the column, the measured buckling loads were 5.8% to 17.2% larger than the predicted buckling loads based on the measured λ values. It seems therefore that Eq. (1) used to evaluate G_{UY} and G_{LY} gives acceptable results. It should be noted that in practical cases where the connection at the top of the column is considered as pinned, the rotational restraint of the top connection is larger than the rotational restraint of the mechanical device used for the tests.

(d) The moment-rotation curves of the column base plate connections, for weak axis bending, measured during the buckling tests, show that the connection behaves elastically and the rigidity is constant as long as the base plate remains in full contact with the footing. A most interesting observation is that separation is initiated only after buckling when the bending moment becomes large with respect to the axial load. This was the fact in all cases even when large relative displacements of the column ends were imposed.

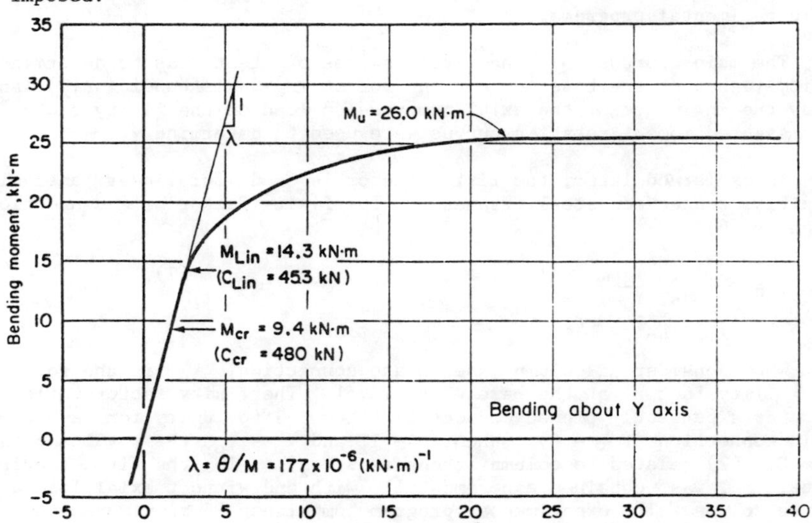

Fig. 3 -- Typical moment-rotation curve measured at the column base .

This conclusion is illustrated in Fig. 3 which shows one of the moment-rotation curves measured during the buckling tests. The moment developped in the base connection at buckling (M_{cr}) was smaller than the linear limit (M_{Lin}) in all tests. Consequently the connection was behaving elastically at buckling. Moreover, since the base plate remained in contact with the footing, the presence of more anchor bolts should have a limited influence on the behaviour of the connection at buckling.

Another interesting observation can be made from Fig. 3. At ultimate, the base connection developed 78% of the plastic moment capacity of the column about the weak axis (M_{py} = 33.3 kN.m). This is quite a large moment for a generally considered pinned connection. It seems therefore appropriate to provide a welded connection capable of developing M_{py} between the base plate and the column.

(e) Buckling tests with a special set-up designed to simulate the rotation of the footing on the soil, have shown that no significant rotation of the footing occurred even though very poor soil conditions were simulated. The moment developed in the column base connection at buckling and even after was too small to force the footing to rotate.

LOCAL BEHAVIOUR OF THE CONNECTION

Experimental Programs

Two series of tests were carried out on column stubs to study the local behaviour of the base plate connection [4,5]. In this short paper, it is not possible to present the full extent of the experimental programs and procedures. The reader is therefore referred to the published papers. The presentation will be limited to a brief description of some aspects of the experimental programs.

The main purpose of the first series of tests was to determine the fixity factor of the base connection for strong axis bending (γ_x) and to study the influence of the axial compression load on the fixity factor [4]. The measured moment-rotation curves were used to determine γ_x.

As explained later, the fixity factor is used for linear analysis of flexibly connected steel frames. The fixity factor of a connection is given by [6]:

$$\gamma = \frac{1}{1 + \dfrac{3EI_c\lambda_o}{L_c}} \qquad (2)$$

When considering column base plate connection, λ_o is the rotational flexibility factor at the base plate level. The fixity factor varies from zero for a perfectly pinned connection ($\lambda_o = \infty$) to unity for a perfectly rigid connection ($\lambda_o = 0$). When $I_c = I_x$ and $\lambda_o = \lambda_{ox}$ the γ value obtained from Eq. (2) relates to column strong axis (γ_x). In the first series of tests, λ_{ox} was obtained experimentally with and without axial load acting on the column. The experimental program comprised 15 specimens tested to failure: 7 flexion tests and 8 compression-flexion tests performed on 1220 mm long column stubs.

The main purpose of the second series of tests was to determine the influence of the axial compression load on the value of G_L obtained from Eq. (1). Since the buckling tests have shown that the base connection behaves elastically when buckling occurs, the moment-rotation curves were measured in the elastic range. Each column was tested at six, seven or eight axial load levels and was bent successively about both principal axes. More than 150 moment-rotation curves were obtained from tests performed on 2000 mm long column stubs [5].

Main Conclusions of the Tests

The results of the first series of tests have clearly shown that the flexural stiffness of the base connection is a function of the applied axial compression load in the column. A significant increase of the flexural stiffness was observed when a compression force was applied to the column. The test results also indicated that the fixity factor of the base connection tends towards a limit value when the axial load in the column increases. It was found that $0.517 \leq \gamma_x \leq 0.873$ for $0.24\ C_y \leq C \leq 0.45\ C_y$, where C is the axial compression load in the column and C_y is the nominal axial plastic capacity of the column ($C_y = AF_y$ where A is the area of the cross-section and F_y the yield stress).

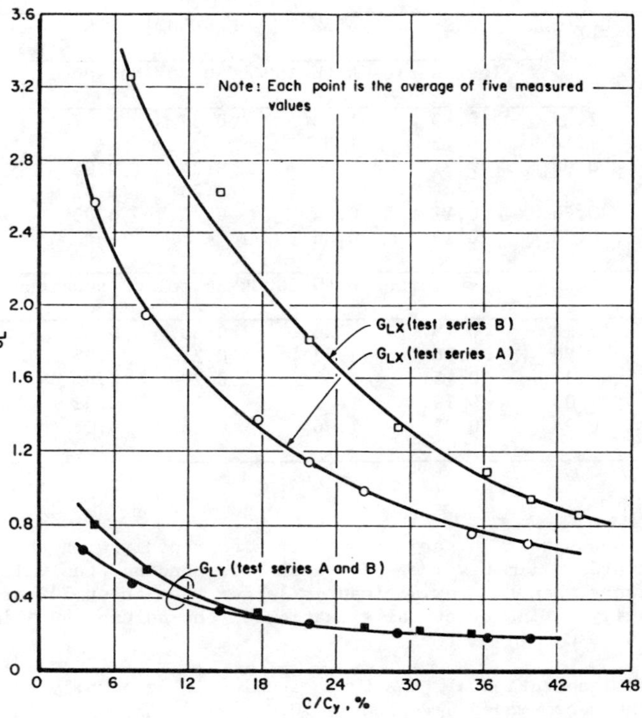

Fig. 4 -- Variation of G_L with axial load .

The axial compression load has also an effect on the slope of the moment-rotation curve. When the base connections were not subjected to an axial load, the measured moment-rotation curves were nonlinear almost from the start of loading. The slope of the M-θ curves decreased continuously with the applied moment. In that case the rotational flexibility factor ($\lambda = \theta/M$) and the fixity factor (γ) are not constant. With an axial compression load in the column the measured moment-rotation curves were linear at least up to 70% of the maximum applied moment. In that case, the flexibility factor and the fixity factor are constant for a large loading range.

Most of the results of the second series of tests are summarized in Table 1 and in Fig. 4. The mean values and standard deviations for G_L, evaluated at different axial load levels for weak axis or strong axis bending, are given in Table 1. Each mean value given in that table [G_L (50%)] is the average of five measurements and the standard deviations (S) were computed with the unbiased formula. The G_L values having only a 5% chance of being exceeded, assuming a normal distribution, are also given in the table.

TABLE 1
Statistical parameters of experimental G_L values

C/C_y (%)	G_{LX}(50%)	S	G_{LX}(5%)	G_{LY}(50%)	S	G_{LY}(5%)
	Test series A (W 150 x 30 column section)					
17.6	1.37	0.19	1.68	0.32	0.07	0.44
22.0	1.14	0.23	1.52	-	-	-
26.4	0.98	0.21	1.33	0.24	0.05	0.32
30.8	-	-	-	0.23	0.02	0.26
35.2	0.76	0.11	0.94	0.20	0.03	0.25
39.6	0.70	0.10	0.86	-	-	-
	Test series B (W 200 x 36 column section)					
14.6	2.62	0.39	3.26	0.33	0.09	0.48
21.8	1.80	0.26	2.23	0.25	0.05	0.33
29.1	1.33	0.15	1.58	0.20	0.04	0.27
36.4	1.09	0.19	1.40	0.19	0.02	0.22
40.0	0.94	0.07	1.06	0.18	0.03	0.23

Fig. 4 shows that G_L at first decreases rapidly with the axial load but tends towards a limit value. For weak axis bending, the value of G_L is almost constant when the axial load is larger than about 30% of the axial plastic capacity. In practical situations, the column buckling load is larger than 0.30 C_y.

The base connection of the type studied is generally considered as pinned and the recomemnded G_L value is 10.0 [1]. All G_L values given in Table 1 are much lower than 10. The test results indicate that for weak axis buckling a conservative value is $G_{LY} = 0.50$, and for strong axis buckling $G_{LX} = 1.50$, if there is no relative displacement of column ends (sway-prevented condition). In most practical situations, sway is prevented when considering weak axis buckling, therefore $K_y \leq 1.0$. According to Eq. (1), where the factor 2 is replaced by 6 when the upper end of the column is free to sway, $G_{LY} = 1.50$ and $G_{LX} = 4.50$ for the sway-permitted condition.

The effective length of the column is reduced when the flexural stiffness of the base connection is taken into account. Design examples show that the increase in column strength varies between 10% and 30% in current situations (slenderness ratios between 60 and 120).

FRAME ANALYSIS

It is well known that connection flexibility affects the overall behaviour of a frame. As the connection flexibility increases, the frame response changes from shear-type behaviour to flexural cantilever behaviour. Two approaches can be used to analyse flexibly connected steel frames: the linear and nonlinear approaches.

The linear approach is well suited for current office practice since it requires only minor modifications of the commonly used displacement or stiffness method. This approach is satisfactory if the rotational flexibility of the connections is almost constant from the start of loading up to the bending moment which corresponds to factored loads. In our opinion, the simplest linear method of analysis is obtained by inclusion of fixity factors at the member ends. The member stiffness matrices are then modified by a correction matrix which includes the fixity factors [6].

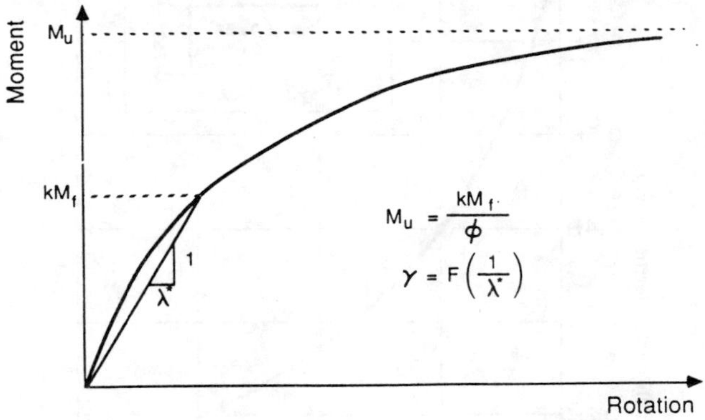

M_f = maximum moment at the connection due the factored loads, obtained from the analysis assuming perfectly rigid connections.

kM_f = design moment ($k < 1.0$)

λ^* = secant rotational flexibility

ϕ = resistance factor of the connection

Fig. 5 -- Linearisation of connection behaviour .

The nonlinear approach is used when the behaviour of the connections of a frame is inelastic and nonlinear almost from the start of loading. It should also be used if the ultimate strength of the frame and the conditions at failure are to be determined. In this latter case, three different nonlinear effects should be considered: connection nonlinearity, member plastification and geometric nonlinearity leading to member and frame instability.

Since nonlinear inelastic analyses are too arduous for most designers and unsuitable for design office practice, some representative linearisation of the nonlinear connection behaviour should be adopted for design calculations. To linearise the connection behaviour, the secant rotational flexibility can be used, as shown in Fig. 5. The secant is a straight line from the origin to the point on the curve corresponding to the design moment (kM_f).

The following design procedure is suggested. The structure is first analysed assuming perfectly rigid connections to obtain the maximum moments in the connections. Since the connections are partially rigid, these moments will never be reached. Then the designer selects a percentage of the maximum moment to design the connections, for instance 75% (k = 0.75). As shown in Fig. 5, the secant rotational flexibility corresponding to the design moment is obtained from the M-θ curve and is used to compute the fixity factor of the connection. Finally the fixity factors of the connections are introduced in a computer program, which incorporates the correction matrices [7], and the structure is reanalysed to obtain new bending moments.

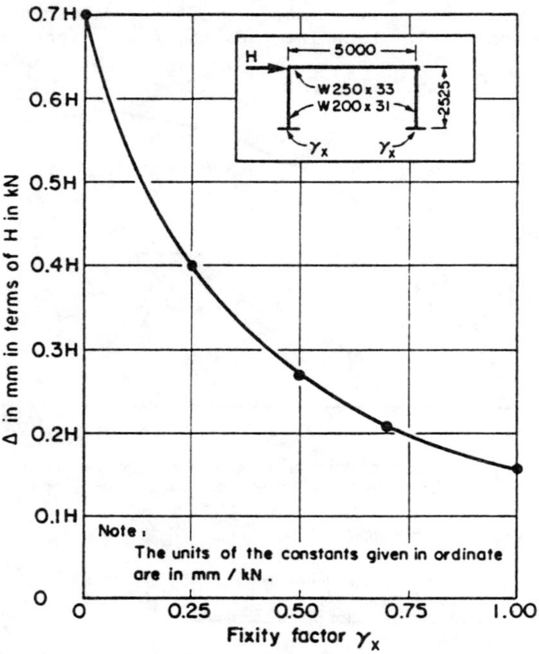

Fig. 6 -- Linear analysis of a simple frame .

If the bending moment in each individual connection is smaller than the design bending moment (kM_f), the design is satisfactory. If the bending moment is larger, the value of k should be increased. On the other hand, if the bending moment is much lower than the design moment, the value of k should be decreased. Of course, this design procedure assumes that sufficient experimental information is available for the type of connection used in the structure.

Fig. 6 shows an example of linear analyses considering fixity factors. The beam-to-column connections are assumed perfectly rigid (γ_x = 1.0) but the column base fixity factor varies between 0.0 and 1.0. For the selected frame, the first-order horizontal displacement (Δ) may be expressed in terms of the applied horizontal load (H), taking into account frame geometry and cross-section properties of beams and columns.

If it is assumed that the column base connection is a perfect hinge, the horizontal displacement is equal to 0.7 times the horizontal load. If the fixity factor is equal to 0.25, the horizontal displacement is reduced by 43%, when compared to the displacement obtained neglecting base fixity. Tests have shown that γ_x = 0.25 is quite a conservative value for a simple column base plate connection [4].

It is important to point out that the curve drops steeply for low values of the fixity factor. In other words, the fixity factor does not have to be high in order to reduce the lateral displacement and the corresponding sway effects in a significant manner. The advantages of considering the actual flexural stiffness of the column base connection is clearly shown by the results of the analysis of this simple frame.

CONCLUSION

The test results summarized in this paper indicate that the flexural stiffness about both principal axes of a simple column base plate connection has a very beneficial effect on column stability and frame behaviour. With these results, the effect of base restraint can be accounted for in design.

REFERENCES

1. Canadian Standards Association, Steel Structures for Building (Limit States Design). CAN3-S16.1-M84. Rexdale, Ontario, 1984.

2. Nixon, C.D., The Design of Light Industrial Building. Ph.D. thesis, Dept. of Civil Eng., University of Alberta, Edmonton, Alberta, 1979.

3. Beaulieu, D. and Picard, A., Contribution des assemblages avec plaque d'assise à la stabilité des poteaux. Construction Métallique, 1985, no. 2, pp. 3-19.

4. Picard, A. and Beaulieu D., Behaviour of a simple column base connection. Can. Journal of Civil Eng., 1985, vol. 12, no. 1, pp. 126-136.

5. Picard, A., Beaulieu, D. and Pérusse, B., Rotational Restraint of a Simple Column Base Connection. Can. Journal of Civil Eng., 1987, vol. 14, no. 1, pp. 49-57.

6. Monforton, G.R. and Wu, T.S., Matrix Analysis of Semi-Rigidly Connected frames. ASCE Journal of the Structural Division, 1963, vol. 89, no. 6, pp. 13-42.

7. Picard, A. and Beaulieu, D., SUPREM: logiciel d'analyse des structures planes. Rapport GCT-85-09, Génie civil, Université Laval, Québec, 1985 (rev. 1986).

SIMPLIFIED DESIGN APPROACH FOR END-RESTRAINED COLUMNS IN FRAMES

Reidar Bjorhovde
Professor and Chairman
Department of Civil Engineering
University of Pittsburgh
Pittsburgh, Pennsylvania 15261
U.S.A.

ABSTRACT

Recent studies of the strength, stability and design of columns have focused on the need for giving a more accurate representation of such members as parts of realistic frames. In particular, the research has addressed the influence of the connections between the columns and the adjacent parts of the structure, recognizing the stiffness characteristics of the semi-rigid beam-to-column connections. The paper details one of the approaches that may be used to determine the buckling load of columns in frames, by taking into account the initial stiffness of the beam-to-column assemblage. It is demonstrated that this property is crucial to the response and capacity of the member. To facilitate the incorporation of the method into current design procedures, use is made of the effective length concept, modified to reflect the actual restraints at the column ends. Finally, it is noted that the solution is equally applicable in allowable stress and limit states design formulations.

INTRODUCTION

Research and development studies dealing with the strength and behavior of columns have been conducted for many years. Although the basic problem may appear to be relatively simple, the many factors that influence column stability make the topic a highly complex one. In addition, solutions that apply to the individual column have limited applicability to members that are parts of structures. This is due to the interaction of all of the framing members and the overall response of the frame. For example, second order frame effects can be significant in a number of cases.

The original solutions for compression members dealt with pinned-end, perfectly straight columns with elastic behavior, which were subsequently modified to take into account material and cross-sectional non-linear effects. Thus, the works of Euler, Engesser, Considere. Shanley, and Dutheil, among many others [1], are well known, and need not be reiterated here. Similarly, the advent of the computer made possible the incremental, iterative numerical solutions for the strength of members with imperfections in the form of residual stress and initial crookedness. This took place much later than the above investigations, but the findings have now been incorporated into design standards around the world [1,2,3,4].

Over the past several years it was recognized that although the strength and behavior of the pinned-end column was an important and convenient reference case, somehow the influence of the actual support conditions of the member should be accounted for. At the outset it was felt that the restraining effects of the connections between the column and the other components of the surrounding structure would be likely to raise the actual capacity of the member. Although the connections would respond in a manner that depended on their rotational stiffness, it was deemed realistic to expect that buckling could be delayed to a certain degree, even in the case of the most flexible beam-to-column joints.

This paper will provide a brief review of some of the investigations that have examined the capacity of columns with semi-rigid end connections. The primary strength parameters are identified, along with the development of a design procedure that allows the engineer to incorporate semi-rigid concepts into current methods of analysis.

INDIVIDUAL COLUMNS WITH END RESTRAINT

Before examining the behavior and strength of end-restrained columns in frames, it is important to understand the response of the individual member when a measure of restraint is present at the ends. Specifically, this clarifies the parameters that influence the capacity of the column, as follows:

(1) Restraint characteristics of the column end supports

(2) Column length or slenderness ratio

(3) Material and geometric imperfections

In the following the influence of these parameters will be explained in some detail.

Restraint Characteristics of the Column End Supports

For most practical purposes this identifies the influence of the beam-to-column connection that joins the column to the adjacent structural elements. The most significant measure of its restraint effect is the moment-rotation or M-ϕ-curve, and Fig. 1 shows representative samples for some typical structural connections. In particular, the C-measure, as indicated in the figure, which reflects the initial rotational restraint of the connection, is important. This will be described in some detail later; at this point it is sufficient to note that the connection stiffness provides the actual column buckling restraint, along with that of the beam to which it is also connected.

A perfect pin has a C-value of 0 and a rigid connection provides infinite restraint. All real beam-to-column joints lie somewhere between these extremes, with the simple or shear-type connections offering a very low moment capacity and stiffness. However, it has been shown that even such connections can lend significant additional buckling strength to end-restrained columns [5,6].

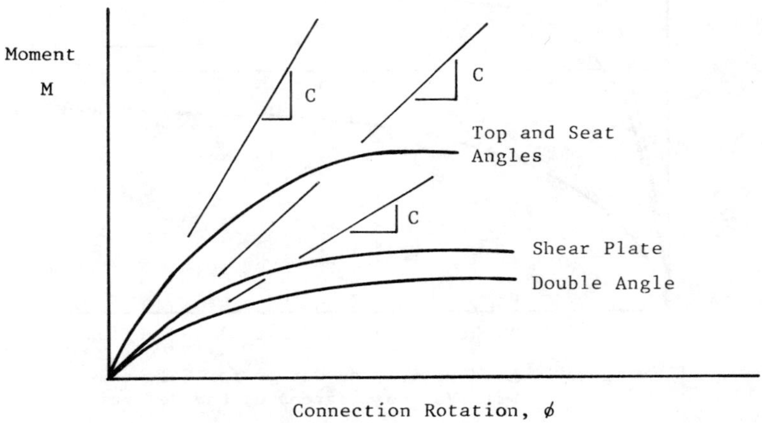

Figure 1 Some Typical Moment-Rotation Curves for Beam-to-Column Connections

Additional information regarding the influence of the connection stiffness on the column buckling strength is provided by Fig. 2. Using data from a study of Nethercot et al. [7], the figure shows the load-deflection curves that result when the same column shape is used with three different end connections. Specifically, the curves represent the response of a W-shape column with pinned ends, double angle connections, and top-and-seat angle connections. It is noted that the column in question had a slenderness ratio of 120, and an initial out-of-straightness of L/1000, where L is the length of the member. The large slenderness ratio is the primary cause of the magnitude of the strength increases, as will be discussed in the following section.

Influence of Column Length

The column strength studies have demonstrated clearly that the longer the member, the larger the relative strength increase that is created by having a semi-rigid end connection. In other words, for one and the same cross section and end restraint, the longer the member, the larger will be the percentage increase in the maximum capacity. Examining various types of beam-to-column connections, researchers have reported strength increases of more than 200 percent for certain very long columns [5,8]. However, it is noted that these lengths are well in excess of practical values, as has been shown by other findings [6,9]. As a further illustration of this observation, Fig. 3 gives the maximum strength column curves for members with three types of end connections, and the middle of the SSRC Column Curves [1,10] is also included for comparison (it is based on pinned ends).

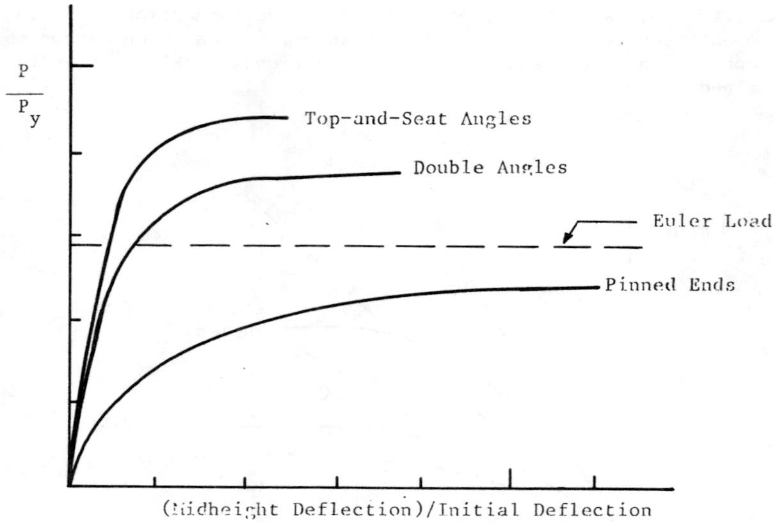

Figure 2 Load-Deflection Curves for Columns with Various Degrees of End Restraint

The vertical axis in Fig. 3 gives the relative maximum column strength, and the horizontal is based on the non-dimensional slenderness term. Also shown is the Euler curve, as well as the ranges of practical slenderness values for two common steel grades (yield stresses of 36 ksi = 248 MPa and 50 ksi = 345 MPa) that are in extensive use in North America. Although the strength of the end-restrained columns is still significantly higher than that of the pinned-end member, it is not on the order of the 200 % that has been indicated. Nevertheless, the benefits are clear, and advantage should be taken of this contribution to the performance of the structure.

Effects of Residual Stresses

Research has demonstrated that the end restraint effects are not as significant when residual stresses are present in the cross section [7]. Specifically, the strength increase that is prompted by the end restraint will be less for columns that buckle in the inelastic range of behavior.

Effects of Initial Out-of-Straightness

Studies have also shown that the end restraint effect is more important when the column is initially perfectly straight [7]. Therefore, for the more realistic maximum column strength models, which incorporate the influence of the crookedness (as well as the residual stresses), the benefit again will be less than what has been reported, but it still valuable to include the contributions in some form in design approaches.

END-RESTRAINED COLUMNS IN FRAMES

The preceding discussion has focused on the various parameters that influence the strength of individual end-restrained columns. The studies

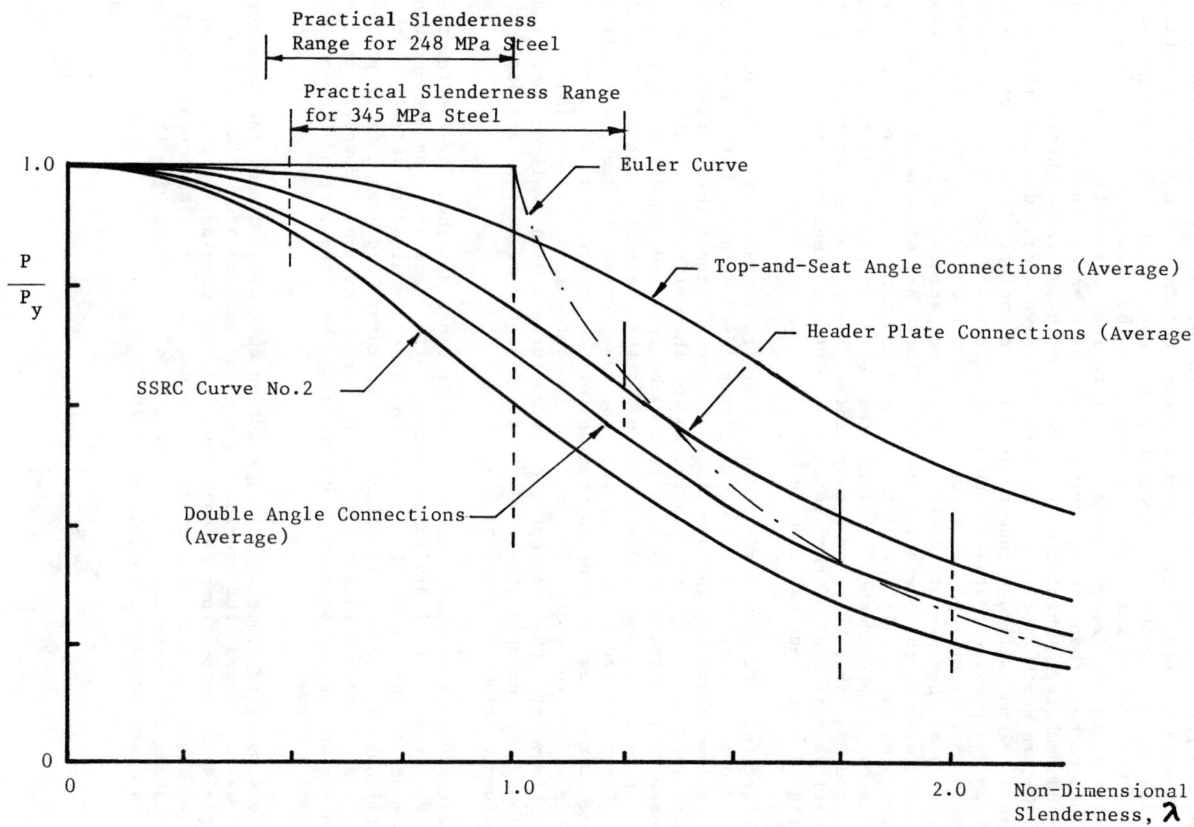

Figure 3 Column Curves for Members with Different End Connections

that led to these findings were crucial to the complete understanding of the problem of the stability of such members, but only limited information was gained regarding the implementation of the concepts into design code formats. Much debate was conducted regarding the potential for utilizing column curves in the design standards that reflected specific levels of end restraint, rather than having curves that were based on the traditional pinned-end members. The latter topic will not be addressed here, short of noting that in North America it appears to have been decided that the pinned-end column will continue to be the focal point. This has been done on the express premise that although it is recognized that true pinned-end members do not exist in real structures, such a model reflects accurately the capacity of the member itself, and is not encumbered by the factors that are related to the overall structure. There are obvious advantages to this approach, especially when it is borne in mind that frame stability solutions can be provided that reflect the restraint conditions between the columns and the other structural components. The following section of this paper will detail one of the potential solutions that has been found to offer significant practical promise [6].

Rather than using complete second order frame stability solutions, it is preferable in many cases to utilize simplified methods that can account for the column stability through other devices. Among the latter, North American practice for years has relied on the effective length concept, which essentially transforms a column with real end conditions into one that is pinned-end, but whose strength is the same as that of the real member. The procedure is well known, and has served designers adequately for some time. However, the assumptions of the analysis and the manner in which the effective length factor, K, is determined, have limited the applicability of the concept to structures with well-defined end conditions. For example, the traditional K-factor solutions are based on rigid beam-to-column connections [1].

Research work by Chen at al. [5,8] provided a method for finding the effective length of individual columns with realistic beam-to-column connections at the ends. This solution also examined the effect of the beam stiffness on the overall column end restraint, and gave data on the K-factors for members with a variety of support conditions. However, it did not address the larger problem of determining the stability of the column as part of a frame.

The simplified frame stability solution that makes use of the K-factor concept was developed from considering a subassemblage of an elastic frame, using rigid beam-to-column connections [1], and considering the sway- and non-sway cases. The characteristic equation for the subassemblage incorporated the K-factor and functions of it, along with the relative stiffness distribution factor, G, given as

$$G = \frac{\sum E_c I_c / L_c}{\sum E_b I_b / L_b} \tag{1}$$

where the numerator term expresses the combined column stiffness at a joint in the subassemblage, and the denominator gives the combined beam stiffness at the same joint. Briefly, Eq. (1) reflects the magnitude of the moment that will be transmitted to the column at the instant of buckling.

Conceptually, Eq. (1) actually shows that <u>the magnitude of the moment</u>

that is transferred to the column is a function of the stiffness of the restraining member or assembly. The higher the value of G, the more moment the column will be asked to carry. In the most extreme case the columns are infinitely stiffer than the beams (i.e. G approaches infinity), which means that the beams will transfer no moment to the column. This, in turn, is an expression of the fact that the beam in this case cannot offer any stabilizing effect for the column. In other words, a value of G equal to infinity indicates that the column has a pinned end connection.

In real framed structures the beams, which have a certain bending stiffness EI and span L, are attached to the columns of the frame through beam-to-column connections that have a certain rotational stiffness. The latter is given by the initial slope of the moment-rotation curve, C, as illustrated in Fig. 1. As shown by Bjorhovde [6], the combined rotational restraint of the connection and the beam that is attached to the column is given by the effective rotational restraint, C^*. This is exactly the same as the denominator term in Eq. (1), but now revised to take into account actual restraint characteristics of the assembly.

On the above basis it has been shown [6] that the stiffness distribution factor, G, for a realistic frame should be replaced by G_r, the end-restraint stiffness distribution factor. Equation (2) therefore takes the place in the effective length solution procedure, rather than Eq. (1). The former gives the expression for G_r as

$$G_r = \frac{\sum E_c I_c / L_c}{C^*} \qquad (2)$$

and it is noted that a single C^* is used for each column end. The reason for this can be explained as follows:

(a) Behavior of interior columns: In the general frame subassemblage, the beams are loaded and deflected a certain amount prior to column buckling. The connections are therefore rotated a certain amount. As the column buckles, two of the connections will unload elastically (stiffness = C) and the other two will continue loading, with a stiffness equal to the slope of the moment-rotation curve at a point equal to the beam moment before buckling took place. For the typical interior column, which has two beams framing into each end, this means that if the stiffness of the loading connection is ignored (which is usually a near-correct assumption, especially for low-moment joints), then only one connection at each end will provide rotational restraint. Hence, the use of a single C^*-value in the G_r-equation.

(b) Behavior of exterior columns: Following the above evaluation of the behavior of an interior column prior to and during buckling, it is clear that when an exterior column fails, one end connection will load and the other will unload. In the solution procedure that was developed [6] it was therefore determined that the values of G_r to use should be based on one C^* at one end, and at the other end a pin could be assumed as the support.

It is noted that in order for the above approach to be utilized, data for the moment-rotation characteristics of typical beam-to-column connections must be available. Such data collection efforts are currently underway in Britain and the United States.

At this point in time the effective restraint procedure that has been outlined above should only be applied to the design of columns in sway-prevented frames. The reason for this is that further research is needed to resolve the questions that arise with respect to the moment that is transferred into the column from the beams in a sway-permitted structure. The latter uses moment-resistant connections, and the moments that can develop are substantially larger than those of the types of joints that have been described in the preceding.

Design examples that have been worked out [6] show column size savings on the order of 5 to 15 percent, depending on the specifics of the member itself.

SUMMARY

The paper has presented an examination of the strength and behavior of columns with realistic end-restraint conditions, such as can be encountered in framed steel structures. The individual column strength parameters are given, along with their relative influences.

A solution procedure is given for finding the effective length factor for end-restrained columns in frames, taking into account the connection rotational restraint and the bending stiffness of the beams. It is shown that this leads to the effective rotational restraint, which in turn can be used in a traditional subassemblage K-factor procedure. An expression is given for the end-restraint stiffness distribution factor, which replaces the well-known G-term of the elastic stability equations.

It is noted that column economy on the order of 5 to 15 percent can be gained through this solution method. However, data are needed on the moment-rotation curves and their parameters for many actual beam-to-column connections. Further research should also be conducted to resolve the applicability of the method to sway-permitted frames.

REFERENCES

1. Johnston, B. G., Editor, "Guide to Stability Design Criteria for Metal Structures", 3rd Ed., Wiley-Interscience, New York. 1976.

2. American Institute of Steel Construction (AISC), "Specification for the Load and Resistance Factor Design, Fabrication and Erection of Structural Steel Buildings", AISC, Chicago, Illinois, September 1, 1986.

3. Canadian Standards Association (CSA), "Steel Structures for Buildings - Limit States Design", CSA Standard No. CAN3 - S16.1 - 1984, CSA, Rexdale, Ontario, Canada. 1984.

4. European Convention for Constructional Steelworks (ECCS),"Eurocode 3: Recommendations for the Design of Steel Structures for Buildings", ECCS, 1986.

5. Sugimoto, H., and Chen, W. F., "Small End Restraint Effects on Strength of H-Columns", Journal of the Structural Division, ASCE, Vol. 108, No. ST3, March. 1982 (pp. 661-681).

6. Bjorhovde, Reidar, "Effect of End Restraint on Column Strength-Practical Applications", AISC Engineering Journal, Vol. 20, No. 1, First Quarter, 1984 (pp. 1 - 13).

7. Jones, S. W., Kirby, P. A., and Nethercot, D. A., "Effect of Semi-Rigid Connections on Steel Column Strength", Journal of Constructional Steel Research, Vol. 1, No. 1, September, 1980 (pp. 35-46).

8. Lui, E. M., and Chen, W. F., "End Restraint and Column Design Using LRFD", AISC Engineering Journal, Vol. 21, No. 1, First Quarter, 1985 (pp. 29 - 39).

9. Bjorhovde, Reidar, "Discussion of "End Restraint and Column Stability" by W. F. Chen", Journal of the Structural Division, ASCE, Vol. 107, No. ST8, August 1981 (pp. 104 - 120).

10. Bjorhovde, Reidar, "Deterministic and Probabilistic Approaches to the Strength of Steel Columns", Ph. D. Dissertation, Lehigh University, Bethlehem, Pennsylvania, May, 1972.

INFLUENCE OF SEMI-RIGID CONNECTIONS
ON THE OVERALL STABILITY OF STEEL FRAMES

F.M. MAZZOLANI
Istituto di Tecnica delle Costruzioni
Engineering Faculty
University of Naples (Italy)

ABSTRACT

This paper summarizes the main results of recent parametrical analysis on the load carrying capacity of semi-rigid frames. In particular, the effect of partial restraint conditions of sway frames is analyzed with respect to the elastic and inelastic buckling. In addition to the effect of connections, the influence of geometrical imperfections on the ultimate behaviour of "industrial frames" is also examined.

INTRODUCTION

The behaviour of semi-rigid frames is influenced by several parameters. The main determinant aspects of the general problem have been analyzed from different points of view. Starting from the assumption of a mathematical model for joints under alternate loading conditions [1], an appropriate calculation method has been worked out [2]. The first results of its application have been mainly devoted to the elastic and inelastic buckling of multi-story frames [3], also including the effect of geometrical imperfections [4]. This led to a new definition of "industrial frames", in which the actual restraint conditions of joints together with the manufacturing tollerances are considered as generalized imperfections [5].
The problem of the overall ductility has been also analyzed in view of the assesment of the structural coefficients (the so-called q-factors) in the design of seismic resistant structures [6].

ELASTIC BUCKLING

By means of the following parameters:

α_k critical multiplier of the frame with semirigid connections;

α_∞ critical multiplier of the frame with perfectly rigid connections;

α_0 critical multiplier of the frame with pinned connections;

R ratio of girder-to-column flexural stiffness;

K ratio of joint-to-girder flexural stiffness;

fig. 1 shows some results referred to a portal frame and to a one bay-three story frame .
Curves of fig. 1,a) allow to define three regions which characterize the unstable behaviour of the semi-rigid frames:
- $0 < K < 3$, where even for small values of connection stiffness the critical multiplier considerably increases and the case of pinned joints can be safely assumed;
- $3 < K < 10$, where the actual value of the joint restraint must be introduced in the analysis without allowing any simplification;
- $K > 10$, where the case of perfectly rigid joints can be assumed provided 80 percent of elastic critical multiplier is safely adopted.

Fig. 1,b) shows that the obtained results are located with a good approximation on lines passing through the origin, which are given by the equation:

$$(\alpha_K - \alpha_0) = (\alpha_\infty - \alpha_K) \cdot A \cdot K \qquad (1)$$

Fig. 1,c) provides the value of the costant A for a given R ratio and a given frame typology. These results have been confirmed by a parametrical analysis on multy-bay (up to three) and multy-story (up to six) frames [3,4] and, therefore, the linear relation-ship (1) can be proposed for general applications.

INELASTIC BUCKLING

With reference to the $M-\varphi$ curves of connections and girders, the three cases of fig. 2 can arise: case a) corresponds to a full strength joint and the global relation-ship can be assumed as elastic-perfectly plastic with a good approximation.
The influence of partial strength joints (case c) produces important alterations in the plastic behaviour of the structure due to the early formation of plastic hinges (see dotted curves of fig. 2). It causes a strong reduction of the ultimate carrying capacities together with a change of slope of the decreasing branch of the curves.
Curves of fig. 2 show also the influence of the vertical loads for the given two-story frame with different joint restraint (K). Going from the case $0.1\, N_c^\infty$ to $0.25\, N_c^\infty$ (being N_c^∞ the critical load of the rigid joint frame) we find that:
- the collapse multiplier is reduced to about a half;
- the decreasing branch increases its slope, leading to a sensible reduction of the global ductility.

In both cases the effect of restraint is sensible in the increasing branch, whereas all curves are connected to only one decreasing branch, as it can be aspected because the rigid-plastic mechanism is not influenced by the joint flexibility.

IMPERFECTION SENSITIVITY

The so-called "industrial frames" [5] are today considered in the modern codes which introduce geometrical imperfections (as out-of-plumb in columns and eccentricities in joints), according to the usual erection tollerances.
The influence of such imperfections on the elastic-plastic behaviour has been analyzed for two portal frames of fig. 3, respectively fixed and hinged. The different cases are characterized by the column slenderness $\lambda = H/r$ = 40, 100, 160. Collapse multipiers are given in fig. 4. For λ = 40 we observe that all curves of the fixed frame practically reach 95 percent of the squash load N_y , excepted for K = 0.1. In the case of hinged frame, on the contrary, the effect of semi-rigid connections is more evident and only 60 percent of N_y is reached for K = 10.000 (which corresponds in practice to a rigid joint).As the column slenderness increases, it is evident that also in the case of fixed frame the effect of semi-rigid connection becomes more and more important. The percentage of collapse load is always decreasing up to the case of hinged portal with λ = 160, in which only 5 percent of the squash load of column is reached.

CONCLUSIONS

The above results show that it is possible to provide for several structural typologies the values of the elastic critical multiplier of semi-rigid frames by means a simple linear relation-ship.
The influence of the restraint conditions on the overall behaviour in elastic-plastic range plays a capital rôle on both strength and ductility of imperfect frames, which must be properly analized according to the definition of "industrial structures".

ACKNOWLEDGMENTS

This research is supported by the Italian Ministry of Education. The research team is composed by Edoardo Cosenza, Antonello De Luca, Attilio De Martino and Ciro Faella.

REFERENCES

1 Cosenza, E., De Luca, A., De Martino, A., Faella, C., and Mazzolani, F.M., Inelastic behaviour of framed structures under cyclic loads. Euromech Colloquium 174, Palermo, october 1983.

2 Cosenza, E., De Luca, A., Faella, C., Nonlinear behaviour of framed structures with semirigid joints. Costruzioni Metalliche, n. 4, 1984.

3 Cosenza, E., De Luca, A., Faella, C., Inelastic buckling and post-critical behaviour of steel structures with semirigid joints. Regional Colloquium on Stability of Steel Structures, Budapest, september 1986.

4 Cosenza, A., De Luca, A., Faella, C., Elastic and inelastic stability of semirigid frames. SSRC Annual Technical Session on Stability of Plate Structures, Washington, april 1986.

5 Cosenza, E., De Luca, A., Faella, C., and Mazzolani, F.M., Imperfection sensitivity of "industrial" steel frames. International Conference on Steel and Aluminium Structures, Cardiff,july 1987.

6 Cosenza, E., De Luca, A., Faella, C., and Mazzolani, F.M., On a simple evaluation of structural coefficients in steel structures. Eight European Conference on Earthquake Engineering, Lisboa, september 1986.

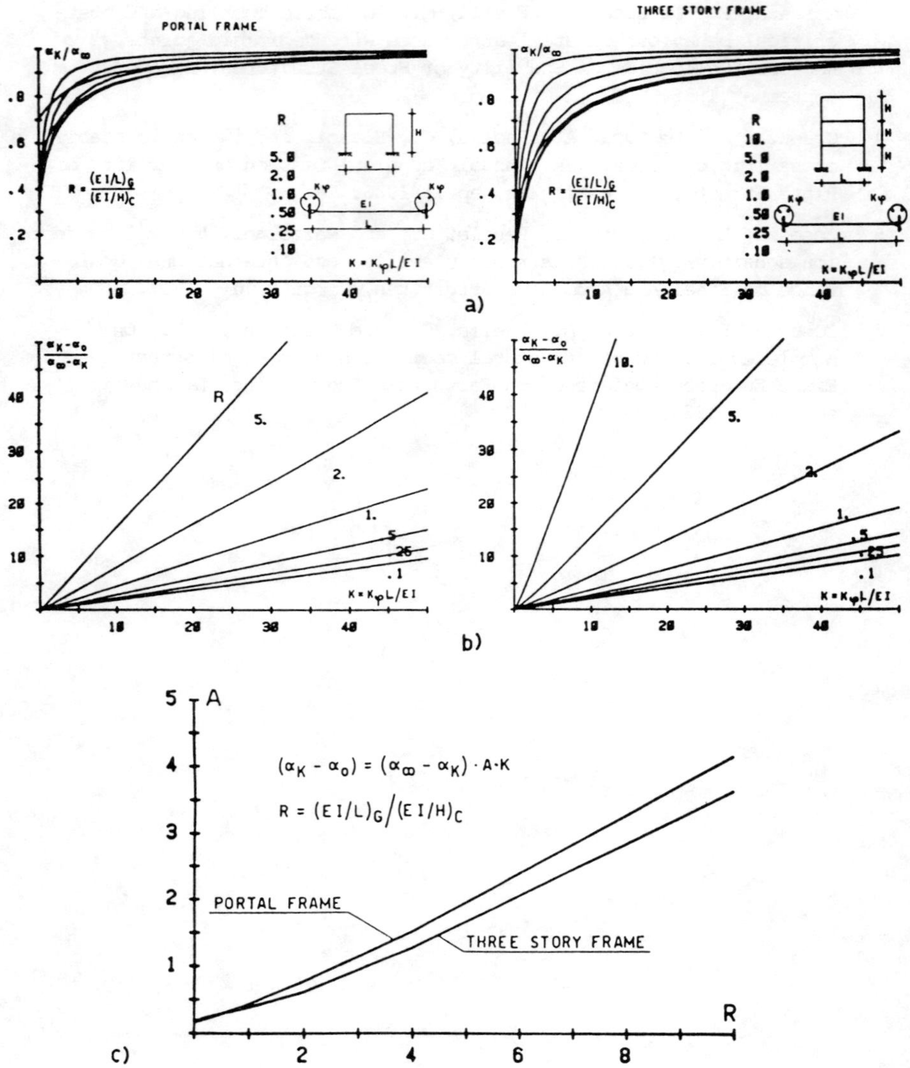

Figure 1 - Results of the parametrical analysis for the evaluation of the elastic critical multiplier of semi-rigid frames.

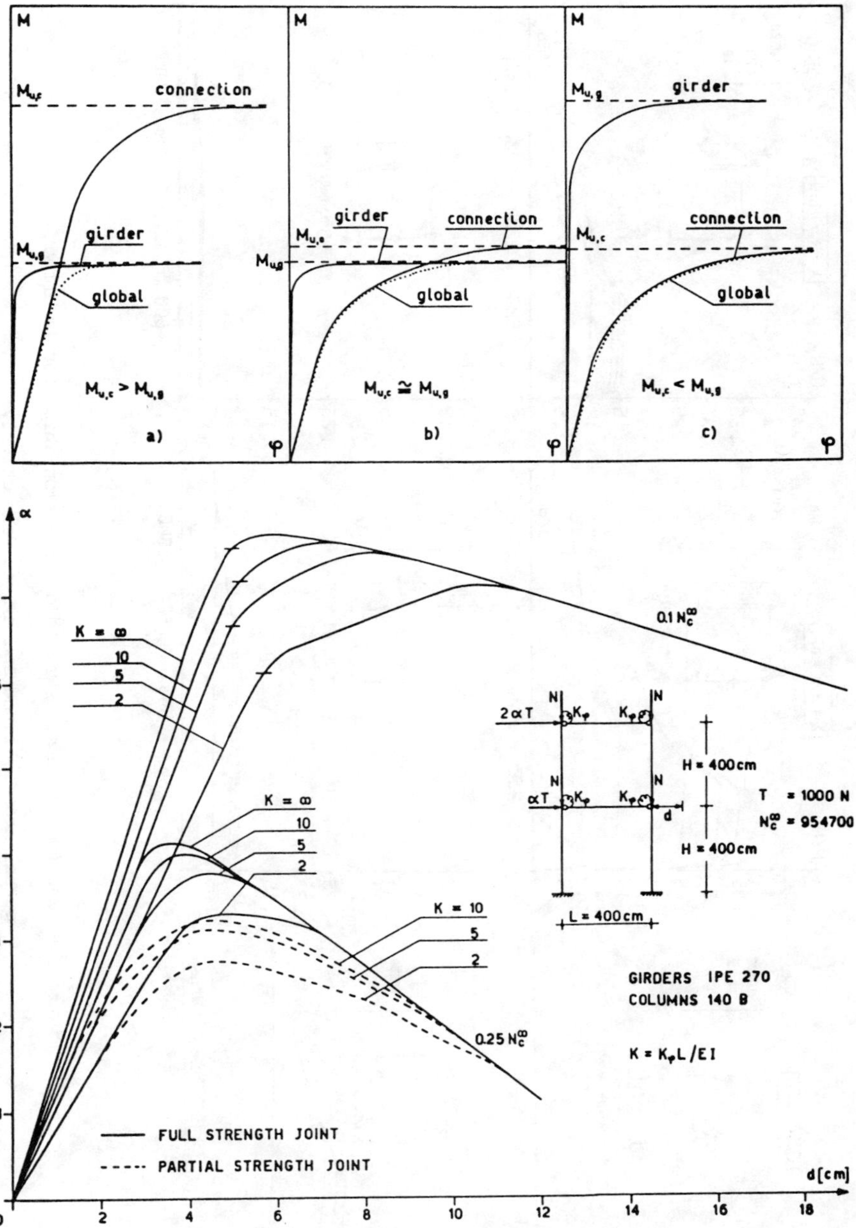

Figure 2 - Influence of the partial strength joints and of the intensity of vertical loads on the load carrying capacity of semi-rigid frames under horizontal forces.

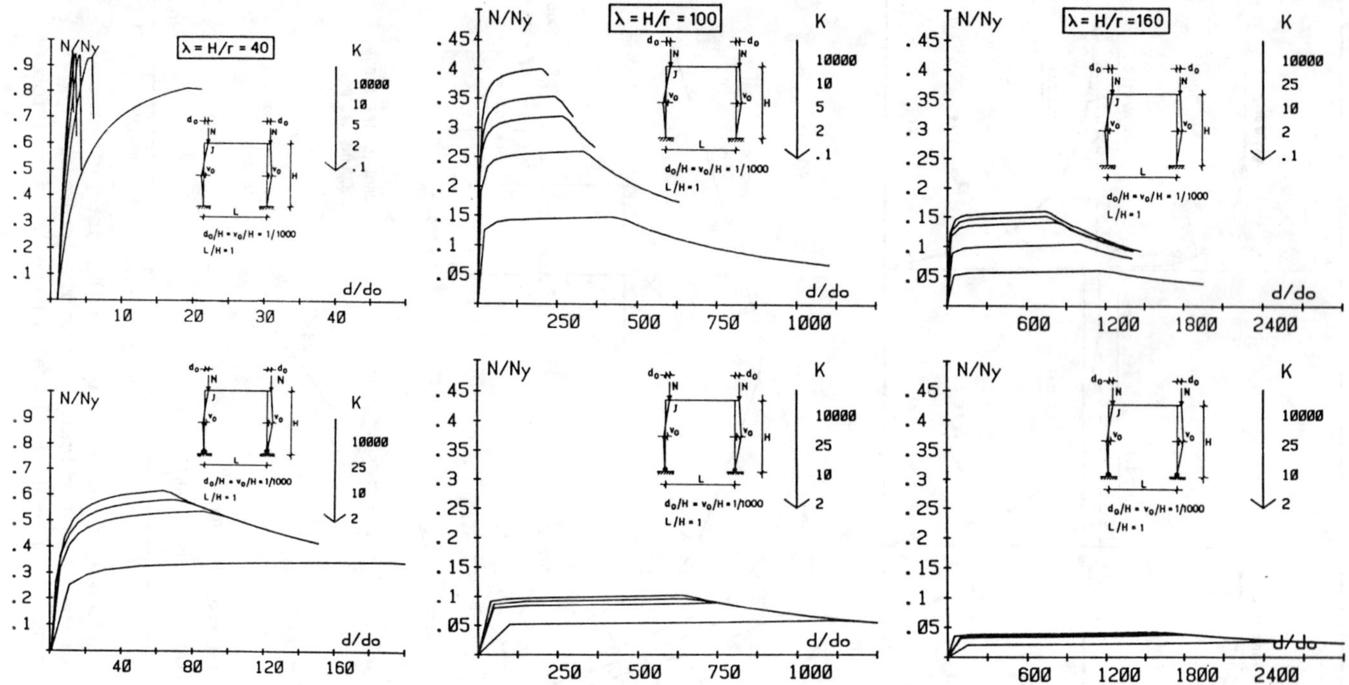

Figure 3 - Load (N/N_y) versus sway (d/d_o) relation-ships for imperfect portal frames, fixed and hinged respectively.

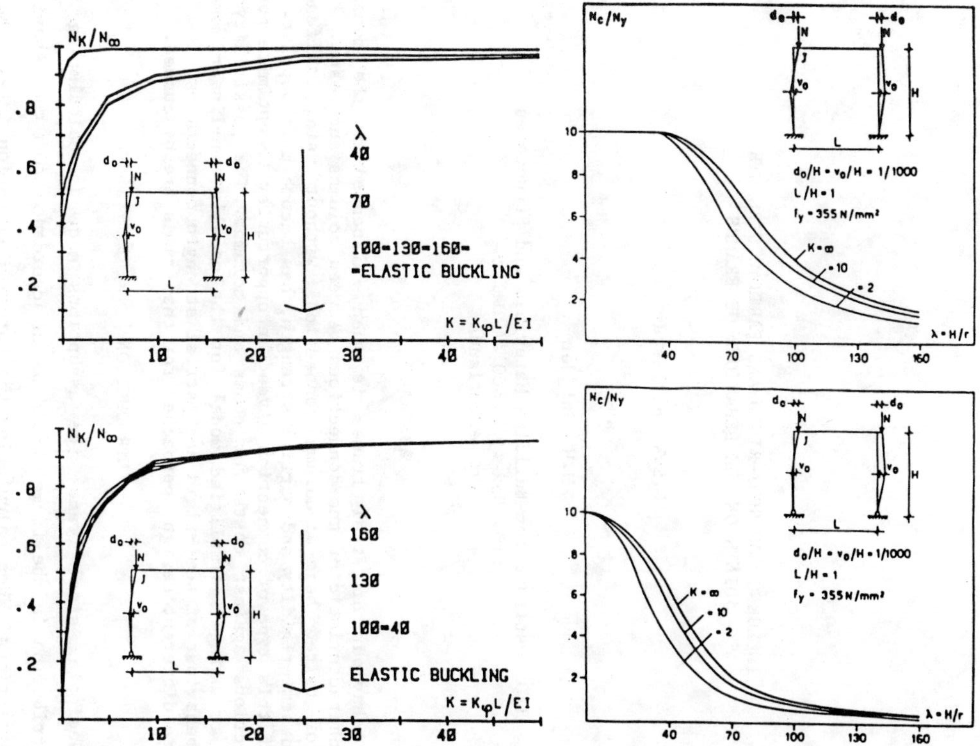

Figure 4 - Collapse multipliers of semi-rigid portal frames.

INFLUENCE OF SEMI-RIGID AND PARTIAL STRENGTH JOINTS ON THE BEHAVIOUR OF FRAMES

F.S.K. Bijlaard

and

H.H. Snijder

TNO Institute for Building Materials and Structures
P.O. Box 49, 2600 AA Delft
The Netherlands

ABSTRACT

The behaviour of steel frames is much influenced by the moment-rotation characteristic of the connections between columns and beams.
The use of bolted beam-to-column connections without using stiffeners leading to semi-rigidity and partial strength, has become more popular.
The use of this type of connection gives the opportunity to optimize costs for connections against costs for beams and columns. The possibility is discussed of using a simplified model for the usually non-linear moment rotation behaviour of semi-rigid and partial strength connections.
Furthermore different design approaches for the frames are discussed.

INTRODUCTION

In designing steel frames the assuptions made in calculating the force distribution in the frame should be in accordance with the behaviour of the connections. When calculating the force distribution in a frame in accordance with elastic theory, the specific joint behaviour with respect to stiffness, such as hinged, rigid or semi-rigid, has to be taken into account. In all cases, the connections should be able to resist the calculated forces. Hinged connections should be able to undergo the necessary rotation without producing substantial moments. Rigid connections have to transfer moments and the corresponding deformation in the connections has to be small, so that they do not influence the calculated force distribution. Semi-rigid connections must also transfer moments, but the corresponding deformations have an influence on the force distribution in the structure.

When the force distribution in a frame is calculated by means of plastic theory, the specific joint behaviour with respect to strength, such as full-strength or partial-strength, has to be taken into account. Due to the complex behaviour of connections, it is only possible to calculate their strength and stiffness with relatively modest accuracy. It is, however, necessary to know the influence of the strength and stiffness of connections on frame behaviour if calculations are to give a reliable estimate of the actual structural response of the frame.

STABILITY OF FRAMES WITH SEMI-RIGID AND PARTIAL STRENGTH CONNECTIONS

In order to study the influence of the joint characteristic on the structural response of frames, the basic behaviour of a joint expressed in the moment-rotation characteristic must be known. Three important aspects are the stiffness, the strength and the rotational capacity of the connection (see Fig. 1). In [1] these aspects are discussed.

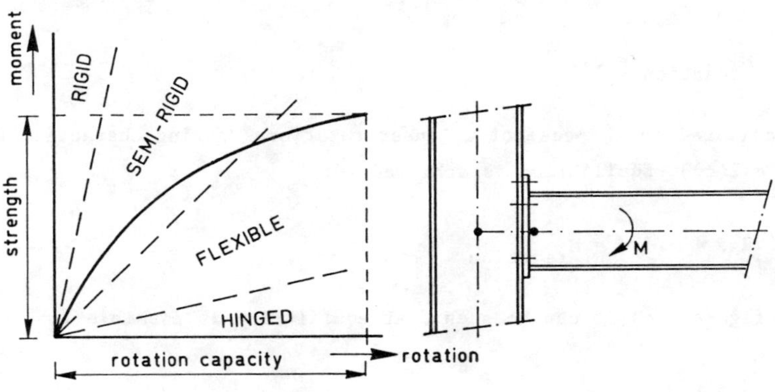

Figure 1: Moment-rotation characteristic of connections.

In calculating the stabilility of frames, an important question is whether or not a simplified model for the, in reality, non-linear connection behaviour can be used. Non-linear connection behaviour, as shown in figure 2(a), can be dealt with in a geometrical and material non-linear calculation where the load is applied incrementally. However, if a geometrical and material non-linear calculation is not used to obtain the force distribution in the frame and to verify the stability of the frame, the non-linear connection characteristic must be schematized. Use of the initial stiffness C_{in} leads to overestimating of the stability of the frame. When the bi-linear approximation is used, the secant stiffness of the connection is taken into account in calculating the frame stability. Suppose, however, that due to the design load on the frame, the connection is loaded up to a certain moment, say the calculated moment capacity $\hat{M}_v$ of the connection. In that case the actual tangent stiffness is far less than the secant stiffness used in the calculation. It can be shown, however, that using the secant stiffness is a safe approximation to the connection behaviour in calculating the stability of the frame. Assume a vertical bar with infinite bending stiffness, connected to the ground by a rotational spring, as shown in figure 2(b). At the top of the bar a horizontal and a vertical load act, producing an overturning moment:

$$M_{action} = H.\ell + F.\ell.\phi \qquad \text{..... (1)}$$

In the rotational spring a reaction moment:

$$M_{reaction} = c.\phi \qquad \text{..... (2)}$$

is activated in the case of a linear rotational spring characteristic (see figure 2(c)). Equilibrium is achieved if:

$$H.\ell + F.\ell.\phi = M_{reaction} \qquad \text{..... (3)}$$

From figure 2(c) it can be seen that equilibrium is possible if:

$$c > F.\ell \qquad \text{..... (4)}$$

The possibility of reaching equilibrium is also dependent on the value of the ultimate moment capacity $\hat{M}_v$. In figure 2(d) the lines a, b and c indicate three different values for $H.\ell$. Line b indicates the maximum value for $H.\ell$ at a certain value for $F.\ell$ for which equilibrium is possible. In the case of a non-linear rotational spring characteristic, the same procedure holds (see figure 2(e)). The maximum load line b does not reach the ultimate moment capacity of the connection characteristic, but touches that curve. Figure 2(f) shows a non-linear rotational spring characteristic and a bi-linear approximation to this curve. It can be seen that the use of the bi-linear approximation leads to a safe calculation of the stability of the frame, because it produces a lower value for $H.\ell$ than in the case of the non-linear curve at the same value for $F.\ell$. The same holds for the case of a constant value of $H.\ell$ (see figure 2(g)). In that case, using the bi-linear approximation leads to a lower value for $F.\ell$ than using the non-linear curve.

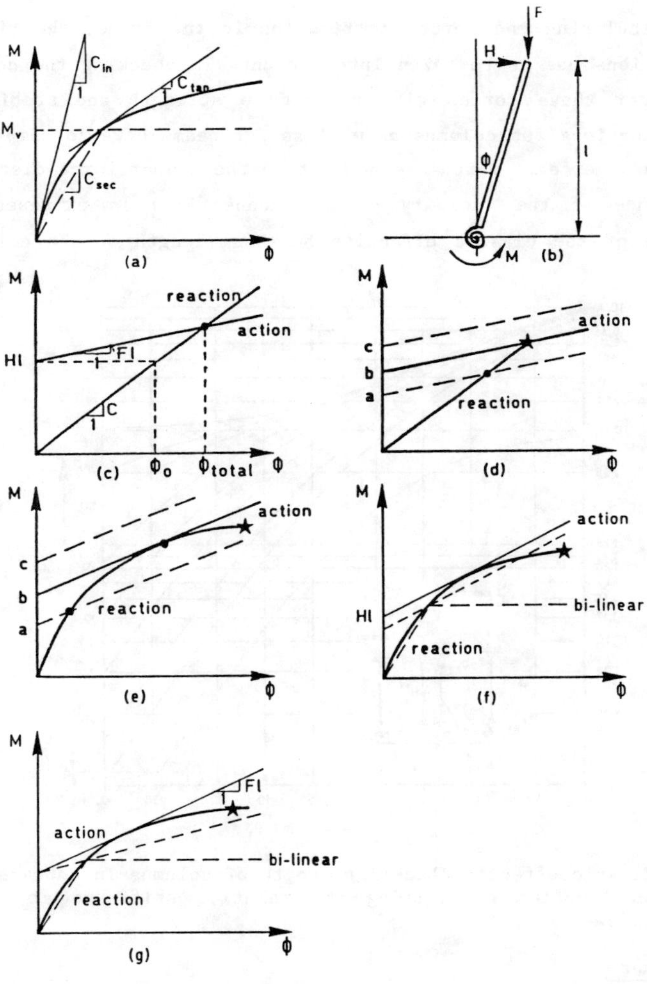

Figure 2: Influence upon frame stability of a bi-linear approximation to the moment-rotation characteristic of a connection.

It can be concluded that it is safe to use a bi-linear approximation to the non-linear moment-rotation characteristic for semi-rigid and partial-strength connections for the purpose of calculating the stability of frames.

ELASTIC DESIGN

Unbraced frames

In calculating the force distribution in the frame, the rigidity of the connections has to be taken into account. In checking the columns and members under these forces with respect to strength and stability, the strength functions for columns as well as for beams have to compensate for the non-linear effects which are not yet in the linear force distribution. The influence of the regidity of the connections is expressed in the calculation of the elastic effective buckling length of the columns (see Fig. 3).

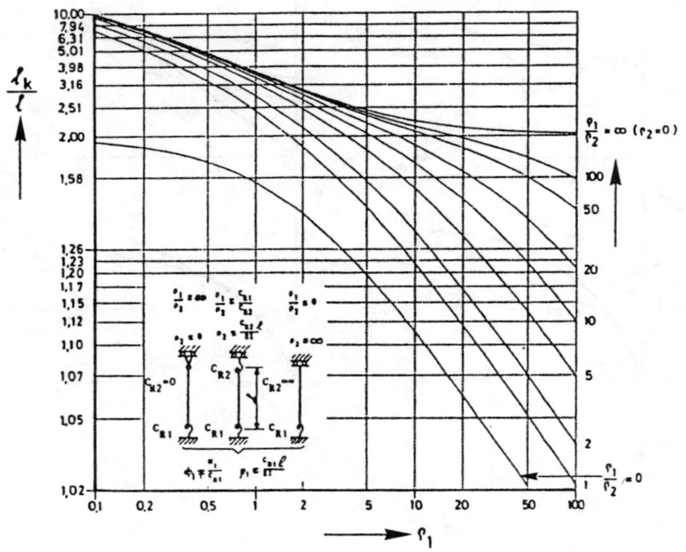

Figure 3: Elastic effective buckling length of columns in unbraced frames as function of ρ, being the relative stiffness at the column ends.

Braced frames

Elastic design of braced frames can also be referred to as "weak-column, strong-beam" design [2,3]. After making a first order elastic calculation to obtain the force distribution of the braced frame, the columns are designed first. Then, the beams are designed in such a way that they do not collapse prior to the columns. The same holds for the semi-rigid connections.

In [4] interaction formulae have been presented for pin-ended beam-colomns under combined axial compression and bending. To obtain the magnitude of the imperfection e^*, the amplification factor n/(n-1) and the buckling factor χ,the buckling length concept is used. The stability check of the total braced frame is replaced by stability checks of the individual columns. The columns are "cut" out of the frame and the resultant pin-ended columns, with the imposed bending moments and axial forces, are checked for stability. When the individual column approach is consistently applied, the buckling length should be taken equal to the system length [5]. In actual fact, the columns form part of the braced frame and it can be shown that the use of the system length as buckling length is conservative in many cases [6]. Therefore, checks on column stability in braced frames are acceptable on the basis of the elastic effective length. This, however, has consequences for checking of the beams. Use of the elastic effective length as buckling length in the interaction formulae, leads to safe estimations for the maximum strength of columns in braces frames, if the beams and the connections are strong enough and column collapse characterizes the frame behaviour [7,8,9,10]. The moment distribution at collapse is different from the moment distribution using the first-order elastic analysis. At collapse, the bending moments on the beam are greater than $\frac{1}{8}\,qb^2$ (Fig. 4). This is caused by the restraining moment M_r that has to stabilise the column.

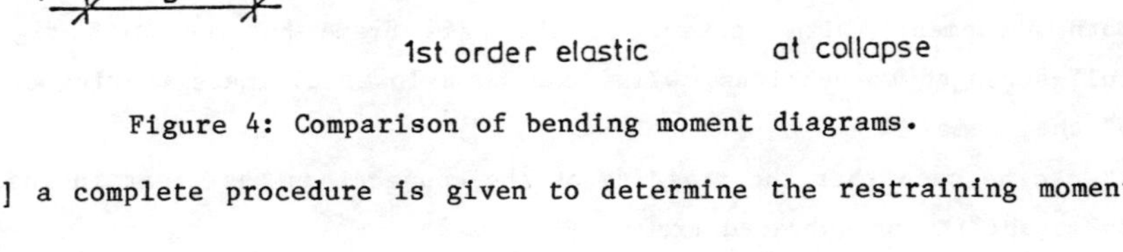

Figure 4: Comparison of bending moment diagrams.

In [6] a complete procedure is given to determine the restraining moment M_r.

Unbraced frames

Designing unbraced frames with plastic theory is rather simple when using the formation of a collapse mechanism for calculating the first order plastic collapse load. The ultimate limit bearing capacity of the frame, which is less than the plastic collapse load, due to second order effects, can easily be determined by using the Merchant-Rankine formula. Figure 5(a) shows an unbraced frame with rigid and full-strength connections while Fig. 5(b) depicts the same frame but with simi-rigid and partial-strength connections. In this figure, the load-deformation curves are plotted

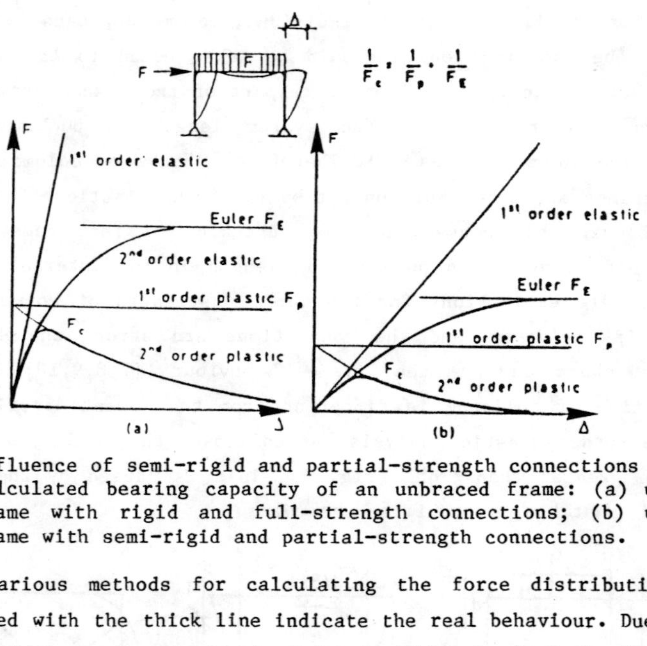

Figure 5: Influence of semi-rigid and partial-strength connections on the calculated bearing capacity of an unbraced frame: (a) unbraced frame with rigid and full-strength connections; (b) unbraced frame with semi-rigid and partial-strength connections.

using the various methods for calculating the force distribution. The curves plotted with the thick line indicate the real behaviour. Due to the flexibility of the connections, the Euler buckling load of the frame will decrease. As a result of the partial strength of the connection, the bearing capacity, based on the first-order plastic theory, will decrease. Both phenomena, with respect to the same frame but now with rigid and full-strength connections, will lead to a lower ultimate bearing capacity of the frame.

It can be seen that the rigidity of the connections has great influence on the stability of unbraced frames.

Braced frames

Plastic design can also be referred to as "strong-column, weak-beam" design [2,3]. The beams are designed first, making use of the first order plastic force distribution (beam mechanism). Then, the columns are designed in such a way that they do not collapse prior to formation of a beam mechanism. When the beam fails, a mechanism is formed with a plastic hinge in the mid-section of the beam and plastic hinges at the supports of the beam. When the moment capacity of the connections is smaller than that of one of the connecting parts, the plastic hinges will form in the connection itself. If that is not the case, the plastic hinge will form in the beam just aside the connections. In knee-connections it is also possible that a plastic hinge is formed in the column when, due to normal forces, the reduced moment capacity of the column section is smaller than the moment capacity of the beam section [11].
If plastic hinges occur at the top of the column, rotational capacity at the top of the column is necessary [12] if these plastic hinges are formed prior to the hinge at mid-span.
To reach a beam mechanism a redistribution of moments is necessary.
A redistribution of moments can only be attained when the parts which yield first have sufficient deformation capacity. In many cases this deformation capacity has to be brought up by the semi-rigid and partial-strength connections [1,6]. The columns are loaded with normal forces and moments equal to the plastic moment capacities of the connections being part of the beam mechanisms. The bearing capacity of the beam mechanism is only reached if the columns can withstand the force distribution belonging to the relevant beam mechanism. Therefore, columns must be checked using interaction formulae [4]. Now the question arrises which buckling length has to be used. Since the beam can not offer restraint to the column, because of the plastic hinge near the column end, the system length can be used as buckling length safely. However, this approach is on the conservative side if used for the frame with rigid supports. In general, the elastic effective length based on the remaining elastic structure may be used as buckling length in plastically designed braced frames.

DESIGN BASED ON SECOND ORDER THEORY

Using the second order theory, two calculation methods are available:

- geometrical non-linear calculation with linear elastic material behaviour;
- geometrical and material non-linear behaviour.

A geometrical non-linear calculation with a linear elastic material behaviour results in a force distribution which is in equilibrium in the deformed situation with the factored design load. Imperfections have to be taken into account. The magnitude of the imperfections is given in [4].
The moment-rotation characteristic can be taken as realistic as possible in the calculations, however, the semi-rigid connections can also be modelled with the secant stiffness of the bi-linear approximation.
This method of analysis results in the correct force distribution in the whole structure. Correction factors on the moment distribution are superfluous [13]. Plasticity effects need to be checked separately using interaction formulae for cross-sections loaded with normal and shear force and bending moment. Formally, because of the assumption of elastic material behaviour in the analysis, the cross-section should be checked in such a way that the yield stress is reached in the outer fibres of the cross-section only. However, the use of a plastic interaction formula for checking the cross-section is allowed because this procedure allows only one plastic hinge to be formed in the structure or more than one but at the same time.

The geometrical and material non-linear behaviour can also be obtained within one calculation [e.g. 14, 15]. In that case no seperate check on cross-sectional behaviour is necessary.

In most cases, when non-linear theory is used in calculating frames, only the planar behaviour of the members is considered. Therefore seperate checks on local instability and lateral torsional buckling are still needed to be carried out.

EVALUATION OF DESIGN METHODS

All three design methods discussed in the previous subsections, elastic design, plastic design and design based on second order theory, are acceptable design methods. Elastic design and plastic design are based on first order force distributions which are relatively easy to obtain. The post-processing of these results, to take the second order effects and plasticity into account, has been formalized in specific design rules. Second order theory results directly into a second order force distribution, but consequently asks more of the user's knowledge.

The methods of analysis for evaluating the ultimate limit state of frames is so much refined, that deflections and side sway become more important and determining. There is a need for criterea for the deflections and side sway dependent on the purpose of the building and the deformation capacity of the secondary structures attached to the supported frames.

REFERENCES

1. Stark, J.W.B., Bijlaard, F.S.K. "Structural properties of connections in steel frames", paper to be presented at "State of the art workshop on connections, strength and design of steel structures", Cachan, 25-27 May, 1987.

2. Vinnakato, S. "Design of columns in planar frames - A few comments", Proc. of the National Conference on Tall Buildings, held at New Delhi, 1973.

3. Vinnakota, S. "Design of columns as part of frames - Some remarks", Paper submitted to TG3: SSRC, 1983.

4. Eurocode 3, "Common unified code of practice for steel structures", Commission of the European Communities, Brussels 1984, Report EUR 8849 DE, EN, FR.

5. Second International Colloquium on Stability. Introductory Report, Liège, April 13-15, 1977, ECCS

6. Analysis and design of steel frames with semi-rigid joints. ECCS Technical Committee 8 "Structural Stability" (to be published).

7. Snijder, H.H. and Bijlaard, F.S.K. "The influence of end restraints on the maximum strength of centrically loaded columns in braced frames", J. Construct. Steel Research, Vol. 5 (1985) pp. 137-147.

8. Snijder, H.H., Bijlaard, F.S.K. and Stark, J.W.B. "Use of the elastic effective length for stability checks of columns and consequences for checks on beams in braced frames, Instability and plastic collapse of steel structures", Proc. of the Michael R. Horne conference, Edited by L.J. Morris, London, Granada, 1983, pp. 152-163.

9. Snijder, H.H., Bijlaard, F.S.K. and Stark, J.W.B. "Strength requirement for beams in braced frames in relation to column stability", Third international colloquium, Stability of metal structures, Final report, Paris, 1983, pp. 247-51.

10. Snijder, H.H., Bijlaard, F.S.K. and Stark, J.W.B. "Design of braced frames using the elastic effective length method", Festschrift Roik, Technisch-wissenschaftliche Mitteilungen, Institut für Konstruktiven Ingenieurbau, Ruhr-Universität Bochum, Mitteilung Nr. 84-3, September 1984, pp. 316-332.

11. Witteveen, J. "Frame analysis including column hinges", Stability of steel structures, Final report, pp. 309-310, Second International Colloquium, Liège, April 13-15, 1977, ECCS.

12. Manen, S.E. van "Plastic design of braced frames allowing plastic hinges in the columns", Heron 27 (1982), No. 2, pp. 1-20.

13. Vogel, U. "Gedanken zum Sinn und Zuverlässigkeit des Tragsicherheitsnachweises am Ersatzstab bei Stabsystemen", Festschrift Roik, Technisch-wissenschaftliche Mitteilungen, Institut für Konstruktiven Ingenieurbau, Ruhr-Universität Bochum, Mitteilung Nr. 84-3, September 1984, pp. 333-346.

14. Borst, R. de, Kusters, G.M.A., Nauta, P. and Witte, F.C. de "DIANA - a comprehensive but flexible finite element system", Finite element systems, third edn., ed. by C.A. Brebbia, Berlin, Springer-Verlag, 1983.

15. Foeken, R.J. van, Snijder, H.H. "Steel column and frame stability analysis using finite element techniques", Heron, Vol. 30 (1985) No. 4, pp. 1-29.

LARGE SCALE TESTS ON COLUMN SUBASSEMBLAGES AND FRAMES

P A Kirby, J B Davison and D A Nethercot
Department of Civil and Structural Engineering
University of Sheffield
Mappin Street
Sheffield S1 3JD
UK

ABSTRACT

The preliminary findings of eight beam and column subassemblage and two three storey two bay frame tests are reported. It is noted that the behaviour of joints as measured in the frame tests is not appreciably different from that determined from an individual joint test. It is seen that the semi-rigid joint continuity results in enhanced column capacity, reduced maximum beam moments and deflections. Existing analysis programs developed at Sheffield and Milan can accurately simulate this behaviour.

INTRODUCTION

When compared with the normal design assumption of pinned joints, actual semi-rigid joints can be conceived as having two potential benefits. The first involves restraint provided by the beam to the columns, thus reducing effective lengths and mid-height deflections and tempering P-Δ effects. The second occurs when considering continuity provided by the columns at the ends of the beam so as to reduce midspan moments and deflections. Much analytical work [1 - 3] has recently been accomplished and computer programs with varying degrees of sophistication exist which can predict the response of individual columns, beam and column subassemblages and complete structures. The complementary experimental programme of work needed to verify and validate these capabilities however has lagged far behind. This paper presents the initial findings of the only known reasonably extensive test programme designed to remedy this deficiency. As mentioned in a companion paper [4] eight column subassemblage tests and two complete three storey two bay frame tests with joints nominally identical to those of a specially conducted series of connection tests have been undertaken. The joint and subassemblage tests were carried out at the University of Sheffield, whilst the frame tests were conducted collaboratively at the Building Research Establishment (BRE). In addition, two similar frames fabricated with extended end-plates have been carried out by Hatfield Polytechnic and BRE and a fifth test is yet to be conducted.

TESTS ON SUBASSEMBLAGES

A series of eight column subassemblage tests have been carried out and a full description of the procedures and results are presented in ref 5. In all cases tests were conducted in the horizontal plane. The columns were 6.5m long with 1.5m long beams framing in at either end 6.0m apart as indicated in Fig 1. In some cases the beams framed into the column flanges and in other cases into the column web. A variety of connections were employed ranging from flexible web cleats to almost rigid extended end plates as listed in Table 1. The ends of the beams remote from the column were mounted in special bearings which were designed to prevent rotational movement but to freely permit movement parallel to the direction of the column. To ensure the desired in-plane response the column was restrained from out of plane movement by bracing in the form of guides, which were provided with sliding bearings so as not to inhibit in-plane movements.

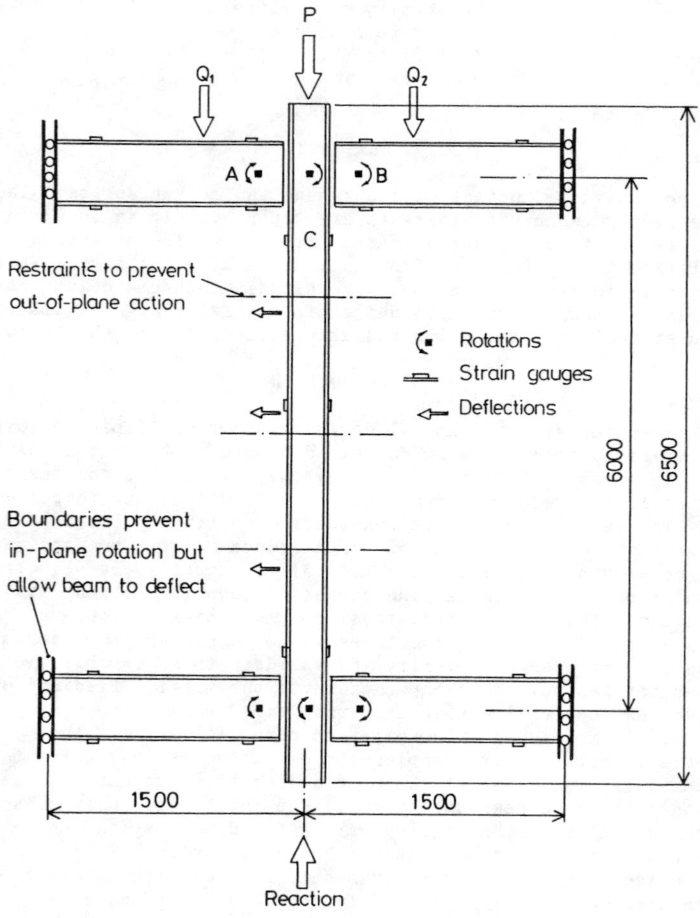

Figure 1. Subassemblage test arrangement.

TABLE 1
Summary of subassemblage tests

TEST	CONNECTION TYPE	AXIS BENDING	Q_1 kN	Q_2 kN	P kN	TOTAL COLUMN FAILURE LOAD ANALYSIS kN	EXPERIMENT kN	ANAL/EXP
ST2	Web cleats	Major	0	73.3	598	671	671	1.00
ST3	Web cleats	Minor	41.0	44.0	452	537	520	1.03
ST4	Flange cleats	Major	110.0	110.0	463	683*	760	0.90*
ST6	Flange cleats	Minor	41.4	43.0	510	594	520	1.14
ST7	Flange cleats	Minor	0	40.0	484	524	526	1.00
ST8	Web and seat	Minor	67.0	70.0	362	499	518	0.96
ST9	Flush end plate	Minor	83.0	85.0	343	511	486	1.05
ST10	Extended end plate	Major	93.0	100.6	403	597*	743	0.80*

* Numerical divergence

Loading could be applied independently to the beams and also axially at the head of the column via load cells. The other end of the column was prevented from translating and a load cell was employed at this location. Correspondence between the reactive force and the applied loads indicated that the sliding bearings worked well with only minimal friction losses. Column deflections were measured by LVDT's and strain gauges were positioned as indicated. On the beams the gauges were connected to permit bending moments to be monitored whilst on the column both bending moments and axial loads could be determined.

All columns showed evidence of having been roller straightened, an observation later confirmed by residual stress measurements, and, in order to induce a representative initial lateral deformation, a small force was applied at midheight. Loading then occurred in two stages. In the first beam loads Q_1 and Q_2 were applied up to the levels indicated in Table 1 and these were then held constant whilst axial column load P was applied to failure. After the subassemblage tests, material yield stresses were obtained from coupons cut from the column flanges. Also shown in Table 1 are failure loads predicted by the computer program. Apart from the two failure loads indicated by asterisks the correspondence is good and, in these two cases, it is thought that the analysis was terminated prematurely due to numerical divergence.

A typical load deflection plot is shown in Fig 2 together with the responses predicted by computer programs developed at the University of Sheffield and reported in ref 6. The close correspondence is apparent.

Table 2 shows the results of some simple 'design' calculations which assess the capacity of the column as an axially loaded member i.e. ignoring the presence of moments transmitted to the column. The capacity is first calculated assuming that no rotational restraint is provided at the connections giving a pin-ended column. Such a computation dramatically underestimates the actual capacity even for a column with simple web cleat connections. Also shown is the capacity determined in a similar manner but with an additional step which utilises the method detailed in ref. 7 to

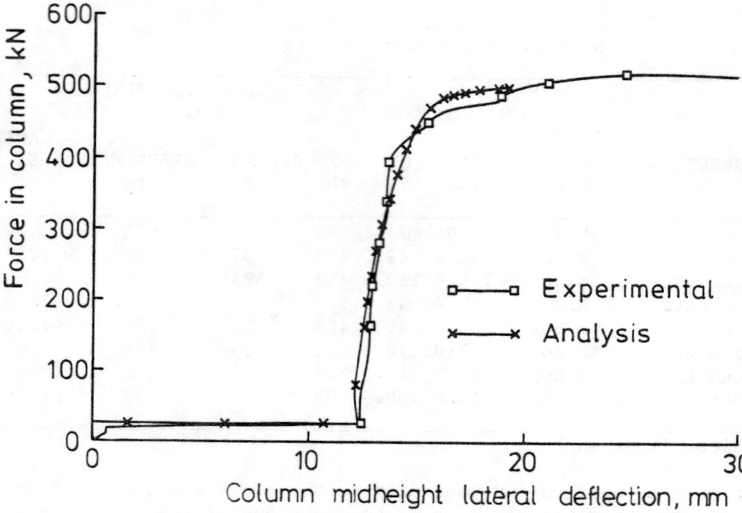

Figure 2. Load against deflection - test ST8.

TABLE 2
Test and 'design' loads for subassemblages

TEST	BASIC PROPERTTIES			AXIALLY LOADED COLUMN CAPACITY PIN-ENDS				MODIFIED CONTINUOUS				
	p_y (N/mm²)	A (mm²)	P_{test} (kN)	λ (k=1.0)	p_c (N/mm²)	P_{des} (kN)	$\frac{P_{test}}{P_{des}}$	k	λ	p_c (N/mm²)	P_{des} (kN)	$\frac{P_{test}}{P_{des}}$
ST2	290	2822	671	92.4	159.5	450.1	1.49	0.67	62.4	225.5	636.4	1.05
ST3	265	2914	520	164.8	57.7	168.1	3.09	0.54	89.0	141.0	410.9	1.26
ST4	273	2983	760	92.3	155.2	463.1	1.64	0.64	59.1	221.1	659.5	1.15
ST6	288	2874	520	164.4	59.1	170.0	3.06	0.53	87.1	151.9	436.6	1.19
ST7	278	2824	526	165.3	58.1	164.1	3.20	0.53	87.6	146.9	414.8	1.27
ST8	279	2837	518	166.2	57.6	163.4	3.17	0.54	89.7	142.9	405.4	1.28
ST9	271	2927	486	166.7	56.9	166.6	2.92	0.53	88.4	145.0	424.4	1.14
ST10	272	2845	743	92.7	154.1	438.3	1.69	0.59	54.7	226.9	645.5	1.15

p_y material yield stress from flange coupon tests
A cross-sectional area calculated from measured dimensions
P_{test} load at failure in test; $\Sigma(P = Q_1 = Q_2)$
slenderness (effective length/appropriate radius of gyration)
p_c reduced material compressive strength from Table 27 BS 5950
P_{des} $A.p_c$
k effective length ratio

obtain restraint coefficients which are then used in the chart of Fig. 23 of BS 5950 [8] to obtain reduced effective length ratios. It can be seen that these ratios, taken as unity for the previous calculation, now lie between 0.53 and 0.67. Perhaps somewhat surprisingly all of the test loads still exceed the enhanced design loads even though no allowance has been made for the presence of bending moments - indeed the values show a remarkable correspondence. Caution must however be exercised as there are a number of features which must be borne in mind namely the beams were very short, the columns were initially very straight and contained minimal residual stresses.

Fig 3 shows the variation in the bending moments observed in the beams and the column ends close to the load application joint (points A, B and C of Fig 1). In all cases the values have been nondimensionalised by dividing moments M by the fully plastic moment of the beam Mp and the total axial load in the column P by the column squash load P_Y. The values predicted by the computer simulation show reasonable correlation. It is interesting to observe that, as the beam load is applied, the moment transmitted through to the column attains a value of approximately $0.25M_P$. As subsequent axial load is added, this moment does not grow considerably; indeed over the next phase of loading it actually relaxes. Similar behaviour has consistently been observed with, in some cases, the moment relaxing to zero and in others actually reversing in sign as the beams are called on to restrain the column as it tries to buckle.

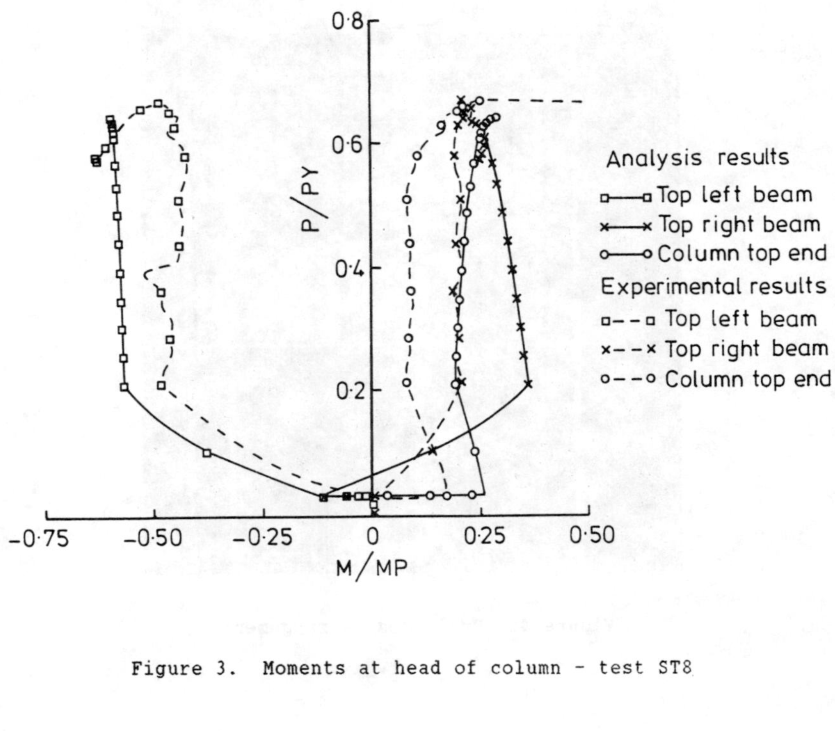

Figure 3. Moments at head of column - test ST8

FRAME TESTS

Two three storey two bay frames have been tested in the Building Research Establishment as part of the Sheffield test program. One frame involved the column members being bent about their major axes and the other about their minor axes. In both cases flange cleat connections, which are towards the more flexible end of the range, were employed as two similar frames with stiff flush and extended end plate connections formed the Hatfield/BRE part of the same sequence of tests employing a degree of common loading, bracing and monitoring equipment. Fig 4 shows a Sheffield test in which moments were measured close to column ends, at column midheights, beam ends and adjacent to loading positions. Deflections were measured at column midheight, at beam load points with rotations also being recorded on the columns at the beam column centre line intersection points and as close to the ends of the beams as was practicable (approximately 125mm away from the beam end). Beam loads were applied through saddles suspended from the beams by hydraulic jacks mounted under the laboratory floor via Macalloy bars. Column loads were similarly applied via substantial caps at the head of column extensions which protruded above the head of the frame.

Figure 4. Frame test arrangement.

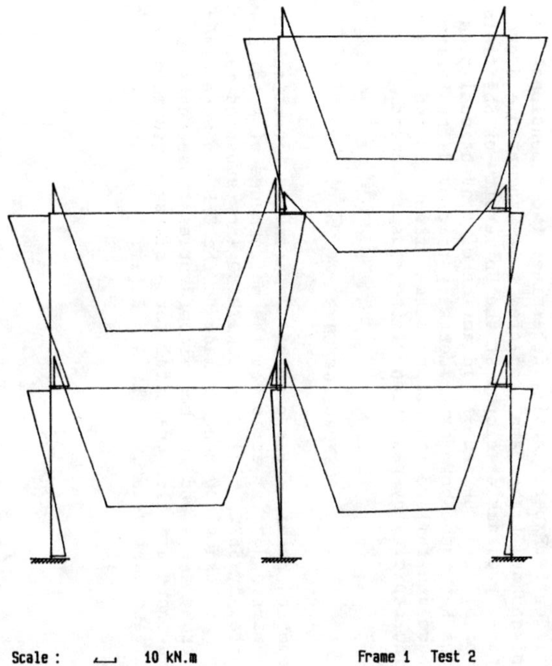

Figure 5. Measured moment distribution around frame 1

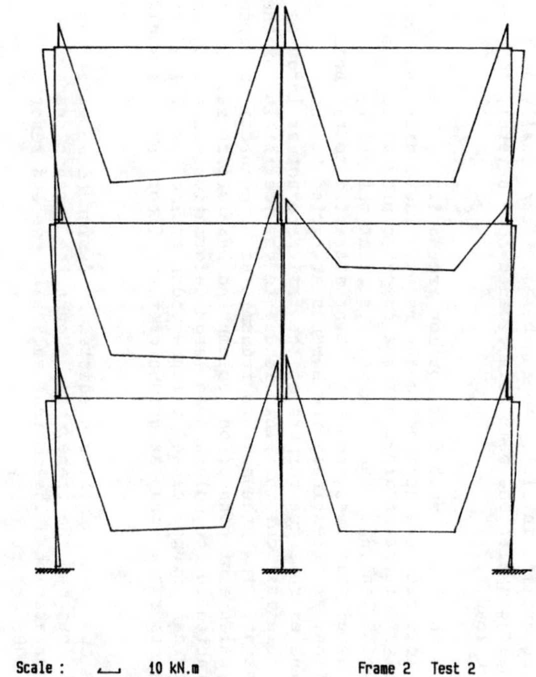

Figure 6. Measured moment distribution around frame 2

As with the subassemblage tests the loading was undertaken in two stages; load being applied initially to the beams, first equally up to service load conditions with one beam being loaded more severely to create a form of pattern loading.

Figs 5 and 6 show the moment distribution around frames 1 and 2 after the full design load had been applied to the beams. These distributions correspond closely to those obtained using a computer program developed at the Polytechnic of Milan [3]. The beam end restraint offered by the connections can be clearly seen from the moment attracted to the beam-column intersection. Interestingly the moments attracted to the connections at the extreme top corners of the frame are not as large as those at the intermediate column. This is due to the flexibility of the column and illustrates that the moments transmitted by connections depend on the relative balance of connection, column and beam stiffness. Due to the moments attracted to the column head large deformations in the upper lifts of the external columns caused failure under applied axial load to occur in these lifts rather than at ground or first floor level as would be expected.

Axial loads in excess of those predicted by design recommendations were sustained throughout the frame, even though the moments transmitted to the column were greater than the nominal values of the end reaction multiplied by a 100m eccentricity.

CONCLUSIONS

The in-plane behaviour of joints as determined from individual connection tests and as parts of subassemblages and complete frames is shown to be consistent. Experimentally determined two dimensional response of the subassemblages and frames are predicted to an acceptable degree of accuracy by computer programs developed at the University of Sheffield and the Polytechnic of Milan. The semi-rigid action of conventional beam to column connections has been shown experimentally to provide restraint against in-plane column failure, to reduce beam deflections and to increase beam load carrying capacity by reduction of the midspan moment.

ACKNOWLEDGEMENTS

The work reported herein forms part of a continuing investigation into the effects of semi-rigid joint action on the performance of steel frame structures supported by SERC, CIRIA and CONSTRADO (now subsumed into the SCI). In particular acknowledgement is made to the BRE who were major partners in the frame tests and in whose laboratories those tests were conducted. The authors are also grateful for assistance provided by staff of the department and particularly by Mr A M Rifai.

REFERENCES

1 Jones, S.W., Kirby, P.A. and Nethercot, D.A., "Effect of Semi Rigid Connections on Steel Column Strength" J. of Construct. Steel Research Vol 1, No 1, 1980, pp. 38-46.

2 Nethercot, D.A., Kirby, P.A. and Rifai, A.M., "Design of Columns in PR Construction Analytical Studies" to be published in Can. J. Civ. Engg., 1987. V13, N4 AUG 87 p485-497

3 Poggi, C. and Zandonni, R., "Behaviour and Strength of Steel Frames with Semi-Rigid Connections" Flexibility and Steel Frames, ed. W.F. Chen, ASCE, 1985, pp. 57-76.

4 Davison, J.B., Kirby, P.A. and Nethercot, D.A., "Semi-rigid Connections in Isolation and in Frames" State of the Art Workshop, Connections Strength and Design of Steel Structures, E.N.S., Cachan, May 1987. THIS VOLUME p195

5 Davison, J.B., Kirby, P.A. and Nethercot, D.A., "Column Behaviour in PR Construction - Experimental Behaviour", ASCE Convention, New Orleans, September 1986 (to be published in J. of Struct. Engg., ASCE, 1987). V113, SEP 87 p2032-2050

6 Rifai, A.M., "Behaviour of Columns in Sub-Frames with Semi-Rigid Connections" Ph.D. Thesis, University of Sheffield, 1987.

7 Bjorhovde, R., "Effect of End Restraint on Column Strength - Practical Applications" Engng. J. of AISC. First Quarter 1984, pp. 1-13.

8 British Standard Institution, "Structural Use of Steelwork in Building" BS 5950 Part 1 : 1985, BSI, London.

SIMPLICITY IN FLEXIBLY-CONNECTED FRAME ANALYSIS

Kurt H. Gerstle

and

Nevis E. Cook, Jr.

Department of Civil, Environmental, and Architectural Engineering
University of Colorado
Boulder, CO 80309
U.S.A.

ABSTRACT

Modeling of flexible connection behavior as elastic-perfectly plastic is proposed for analysis of building frames. Results based on this idealization are compared with those of fully non-linear analysis for three different examples. It is concluded that first-order effects of connection flexibility can be captured within engineering accuracy by linearly-elastic analysis under working loads, and by perfectly-plastic analysis at ultimate.

INTRODUCTION

Recent research [1-3] on flexible connection behavior, and on analysis of flexibly connected frames, has emphasized non-linear analysis. In general, non-linear analysis is unsuitable for office practice: Non-linear computer analysis is laborious and expensive, and the results are often so error-prone that their interpretation becomes very complex, more suited to an academic than a professional environment.

It is the aim of this paper to present a much simpler approach, and to document its suitability for design office practice and its relation to reality with several examples.

APPROACH

We will represent the moment-rotation relation of a range of flexible beam-column connections in steel building frames as elastic-perfectly plastic, and incorporate this formulation of the connection behavior in elastic-plastic frame analysis. With this premise, we will explore the feasibility of representing frame behavior under working loads by elastic, and the structure strength by plastic analysis. If it can be shown that the frame acts nearly linearly-elastic at working levels, and possesses sufficient ductility at ultimate, then simple, well-known methods will be available for the design of flexibly-connected steel frames.

CONNECTION BEHAVIOR

The behavior of typical beam-column connections in steel frames can be represented by moment-rotation curves such as those shown in Fig. 1 [4]. It should be noted that the range of rotations along the horizontal axis of Fig. 1 vastly exceeds any values to be expected under conditions prior to collapse. For instance, the end rotation of a simple-supported, prismatic elastic beam under uniform load causing a midspan deflection equal to 1/360 of its length will be θ = .0089 radians, and the rotation of a restrained beam will be much less than that. Similarly, the connection rotations of a frame subject to a lateral load causing sway equal to 1/300 of its height will be less than .0033 radians. Connection rotations under working conditions may thus be well within the near-linear range of the curves of Fig. 1.

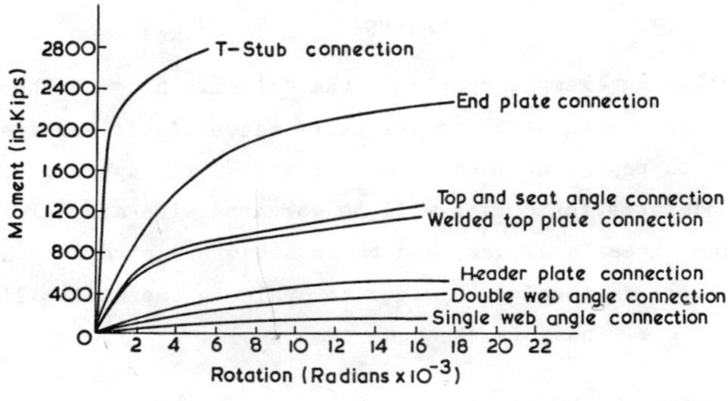

Figure 1. Moment-Rotation Relations

Steel frame connections commonly used exhibit great ductility, as shown, for instance, by the curves of Fig. 1. If it can be shown that the rotation capacity of the connections to be used exceeds the rotation demand necessary for full redistribution of moments, then it follows that plastic analysis is suitable for determination of the frame strength.

The connection response under load histories such as unloading and load reversal, as shown in Fig. 2, is needed to predict frame response to more general loadings such as lateral load reversal. We will make the classical elastic-perfectly-plastic assumptions shown dashed in our analysis.

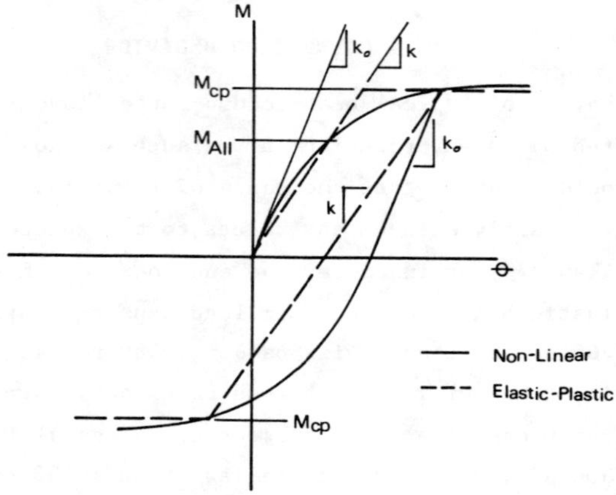

Figure 2. General Connection Behavior

ANALYSIS

In the following example problems, the "plastic hinge method" [5,6], a piecewise-linear analysis of the structure sequentially modified by internal hinges to represent local plastification, is applied. The resulting load-deformation curves will be compared with similar curves from "exact" non-linear analyses, and their behavior at working and ultimate loads will be studied in order to evaluate the suitability of elastic, and plastic, analysis at these two levels.

Example 1

A uniformly-loaded W18x40 steel beam, 30' (9.15 m) long, of yield strength 36 ksi (247.3 N/mm^2), supported by flexible connections of different types at each end, as shown in Fig. 3, is considered. This size beam was chosen because a number of connection test results of beams of this size are available.

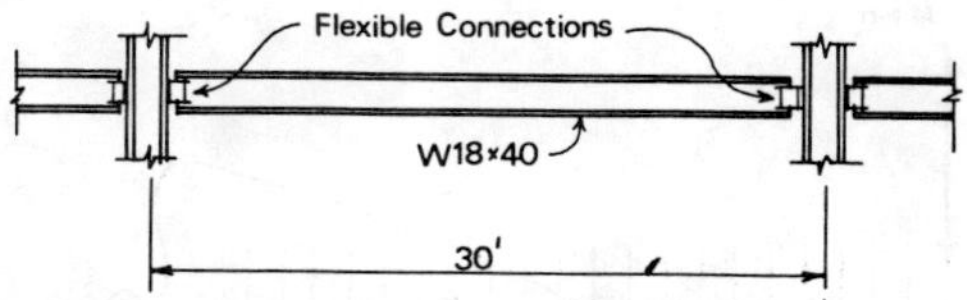

Figure 3. Structure of Examples 1 and 2

The allowable load of w_{all} = 1.0 k/ft (14.6 N/mm) on this beam is governed by the deflection criterion Δ/L = 1/360, assuming simple beam action.

Four different connections for which test data were available were considered in this analysis. Their stiffnesses, which are given in Table 1, cover the range of flexible connections $.05 \leq EI/kL \leq 2$, as suggested in Reference 7.

The moment-rotation curves for these four connections, taken from the references listed in Table 1, are plotted in Fig. 4. The intersection of the beam line with these curves indicates connection moments and rotations under a working load of 1 k/ft (14.6 N/mm). The elastic stiffness k of each of the connections was determined as the secant modulus to this working point; it, as well as the initial stiffness k_0, the assumed plastic connection moment M_{cp}, and the stiffness ratio EI/kL are given in Table 1 for the four different connections. They cover the entire previously defined range of flexible connections.

TABLE 1. Connection Properties

No.	Connection Type	k_0, k-in/rad	k, k-in/rad	M_{cp}, k-ft	EI/kL
1	3/8" double web angle, 4 bolts [8]	27 x 10^3	22 x 10^3	16	2.24
2	3/8" double web angle, 5 bolts [8]	96 x 10^3	70 x 10^3	40	.70
3	Top and seat angles, 150 x 90 x 12 mm [9]	144 x 10^3	115 x 10^3	64	.43
4	Extended end plate to W18x35 Beam, t=1" [9]	642 x 10^3	552 x 10^3	160	.09

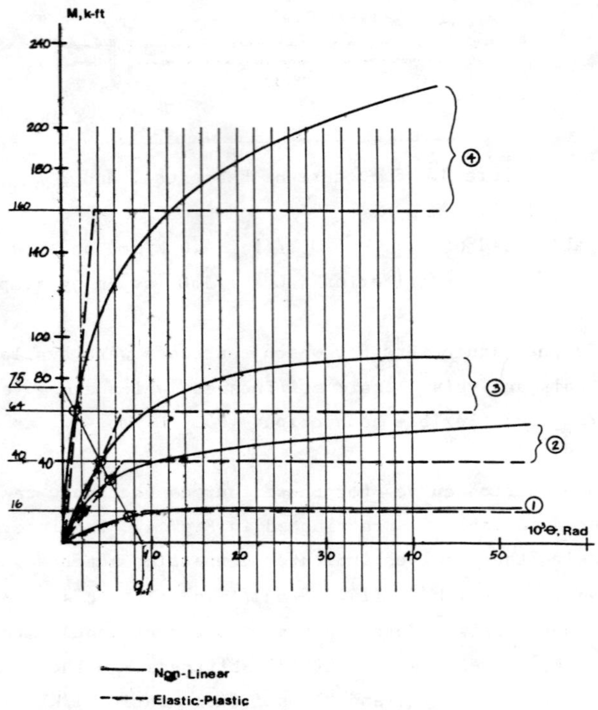

Figure 4. Exact and Elastic-Plastic Moment-Rotation Relations

Using relations of statics, geometry, elastic-plastic beam behavior, and moment-rotation relations shown in Fig. 4, the beam of Fig. 3 was analyzed for its midpoint deflection Δ under uniform load w, and the results plotted in Fig. 5. The solid curves are the results of a fully non-linear analysis based on the solid lines of Fig. 4, while the dashed curves represent the results of a plastic-hinge analysis based on the dashed elastic-plastic curves of Fig. 4. Additionally, the plot is bounded by dashed lines representing the limiting conditions of simply-supported and fully fixed-ended beam behavior.

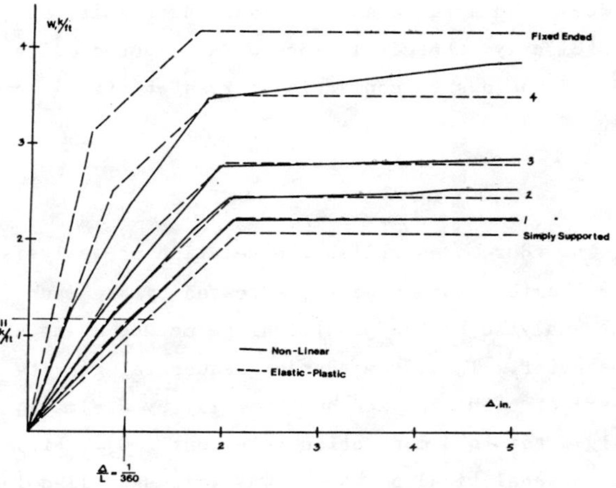

Figure 5. Load-Deflection Relations, Example 1

Example 2

The beam shown in Fig. 3, of length L = 30' and loaded with a uniform working load of 1 k/ft (14.6 N/mm), is joined to columns of story height H = 12' of an unbraced frame at both ends. Due to a lateral load V applied to each column, the columns rotate through equal angles θ_c, thereby permitting the frame to rack sideways, as shown in Fig. 6. During this stage, the leeward connection will load further, while the windward connection will unload according to the relations shown in Fig. 2.

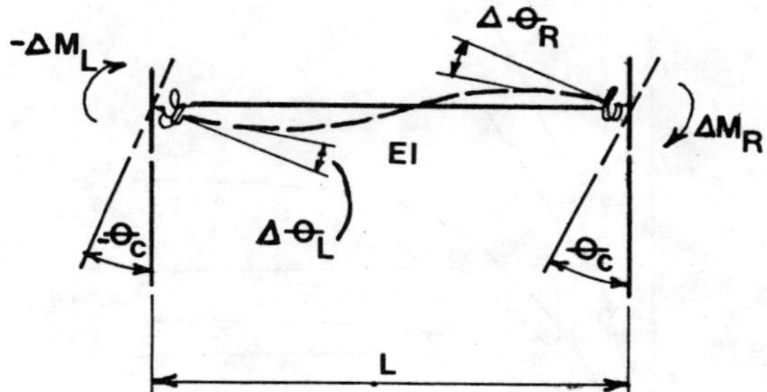

Figure 6. Beam of Frame in Sway

A non-linear analysis based on this connection behavior was devised to determine the connection moments and rotations, as well as the column rotation θ_c due to an applied shear V, and the results are shown solid in Fig. 7. Similarly, these relations were computed using the plastic-hinge method and the dashed connection curves of Fig. 4, and plotted dashed in Fig. 7

Example 3

Lastly, in order to establish the relation of analysis and experiment, a flexibly-connected frame tested by Stelmack [10] and shown in Fig. 8 was analyzed by the plastic-hinge method using elastic-plastic connection behavior. This frame, under sequential gravity and lateral loads, had earlier been analyzed non-linearly by Stelmack, using experimentally determined connection relations [10]. Fig. 8 shows experimental and analytical plots of sway versus applied lateral load H. We observe that stiffness and strength of the frame are predicted closely by the simple plastic-hinge method.

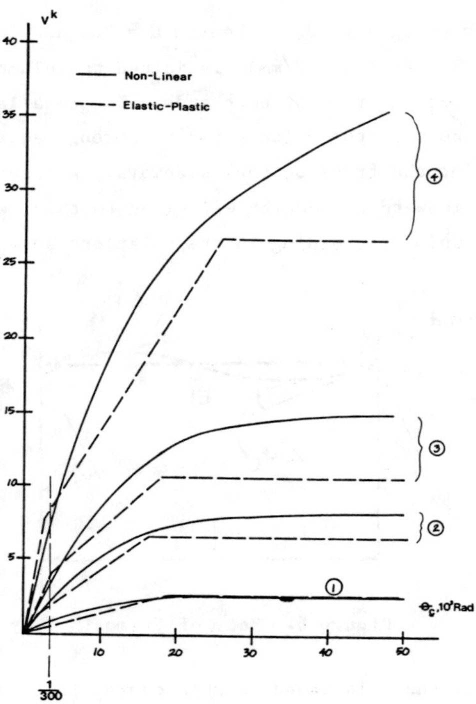

Figure 7. Shear-Rotation Relations, Example 2

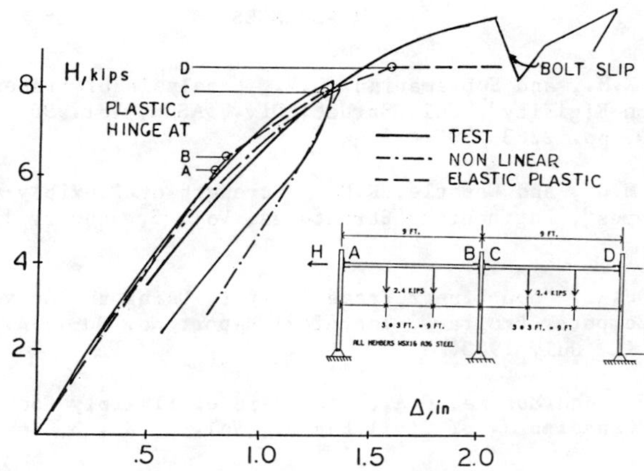

Figure 8. Lateral Load-Sway Relation, Example 3

CONCLUSIONS

The results of the three example problems indicate the following for flexibly-connected frames:

1. In general, results of elastic-plastic analyses reflect the behavior predicted by non-linear analysis to within reasonable engineering accuracy.
2. Connection rotations at working levels are within the near-linear range; elastic analysis can be used to predict structure behavior under working loads.
3. Beyond working levels, the plastic-hinge method will give sufficiently close results. This analysis can be used when excessive deformations or P-Δ effects need to be considered.
4. Structure strength can be predicted by classical plastic analysis.

Lastly, the need for further testing and analysis of connections to obtain a full range of moment-rotation curves for use in steel frame design cannot be overemphasized.

REFERENCES

1. Romstad, K.M., and Subramanian, C.V., "Analysis of Frames with Partial Connection Rigidity", Jnl. Struct. Div., ASCE, Vol. 96, No. ST11, Nov. 1970, pp. 2283-2300.

2. Ackroyd, M.H., and Gerstle, K.H., "Strength of Flexibly-Connected Steel Frames", Engineering Structures, Vol. 5, January 1983, pp. 31-37.

3. Edinger, J.A., "Non-Linear Frame Analysis Using the Ansys General Purpose Computer Program", Technical Report No. 83-1, A.I.S.C., Chicago, IL, July 1983.

4. Frye, M.J., and Morris, G.A., "Analysis of Flexibly Connected Steel Frames", Canadian J. of Civil Engrs., Vol. 2, No. 3, Sept. 1975, pp. 280-291.

5. Hodge, P.G., Jr., "Plastic Analysis of Structures", McGraw-Hill Book Co., New York, 1959.

6. A.S.C.E., "Plastic Design in Steel", Manual No. 41, 2nd Ed., 1971.

7. Ackroyd, M.H., and Gerstle, K.H., "Behavior of Type 2 Steel Frames", Jnl. Struct. Div., ASCE, Vol. 108, No. ST7, July 1982, pp. 1541-1556.

8. Richart, R.M., Private Communication, 1987.

9. Goverdhan, A.V. "A Collection of Experimental Moment-Rotation Curves ...", M.S. Thesis, Vanderbilt University, Nashville, Tenn. 1983.

10. Stelmack, T.W., Marley, M.J., and Gerstle, K.H., "Analysis and Tests of Flexibly Connected Steel Frames", Jnl. Struct. Eng., ASCE, Vol. 112, No. 7, July 1986, pp. 1573-1588.

SIMPLIFIED DESIGN OF FLEXIBLY-CONNECTED BUILDING FRAMES

M.H. Ackroyd
Techtrol, Ltd.
Montreal, Quebec

ABSTRACT

This paper develops a simple design procedure for unbraced building frames that takes advantage of beam-to-column connection stiffness in reducing girder sizes, and at the same time, avoids premature column plastification typified in most Type 2 frame designs.

MOTIVATION

Studies of the behavior of Type 2 Designs (1) have shown that exterior columns receive a portion of "fixed-end" moment from the gravity loaded girders that does not get accounted for in the Type 2 assumptions. Correspondingly, the girders tend to be overdesigned, because Type 2 assumptions do not take advantage of the rotational restraint actually provided by the "simple" connections. In some frames above nine stories tall, these analysis approximations can result in the onset of frame instability at load factors below 1.3, but for all frames, these approximations will result in oversized girders. Therefore, one would like to modify the prediction model to account for the beneficial effects of connection stiffness on girder forces and its detrimental effects on exterior columns. Also, one would like to avoid having to rely on sophisticated computational tools, as that would defeat the original spirit of simplicity of Type 2 design.

APPROACH

A thorough study of the analysis results for the frames studied above showed that the deformed shapes of the frames were similar to the shape one would obtain by superposing two shapes: (1) the portal shear model associated with wind loading, and (2) a gravity model having inflection points at midheights of exterior columns and essentially no flexure of the interior columns.

Based on these observations, it appears that wind moments are predicted well by the standard portal method of approximate analysis. However, a new model for gravity loads needs to account for the transfer of moment from girders into the exterior columns, as well as to account for beneficial "flexible-end" moments on girder ends. This paper proposes a model for gravity loading of Type 2 frames that can satisfy these criteria as well as provide the basis for simple design aids, suitable for manual computation.

Two subassemblages have boundary conditions that kinematically map onto the suggested gravity model, above. For interior bays of a frame, it seems appropriate to neglect the column rotations and consider only the gravity-loaded girder with flexible rotational springs connecting it to rigid supports. For exterior bays of a frame, it seems appropriate to assume inflection points at midheights of columns and hence to model only half the column above and half the column below.

The notation used herein for frame components is:

- Ig = moment of inertia of girder
- Ic,a = moment of inertia of column above
- Ic,b = moment of inertia of column below
- ha = story height of column above
- hb = story height of column below
- l = span of girder
- k = representative stiffness of flexible connections (both connections identical)

It should be noted that the exterior bay model subsumes the interior bay model in light of the fact that, as moments of inertia of the columns tend toward infinity, the right end of the model tends toward a rigid support as in the interior bay model. Therefore, only the exterior bay model needs to be analyzed as the general model of behavior.

The analysis used here is the widely used "direct stiffness" approach, where individual elemental models are combined to give the needed structural model. For this structure, only three elemental models are necessary: a standard prismatic beam element represents the girder, rotational springs represent the flexible connections, and prismatic beams with far-end-pinned represent the half-columns. The elements and associated stiffness matrices are shown below.

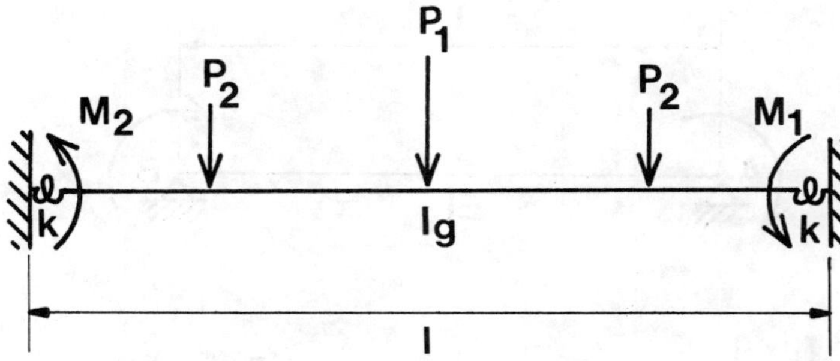

GRAVITY MODEL FOR INTERIOR BAYS

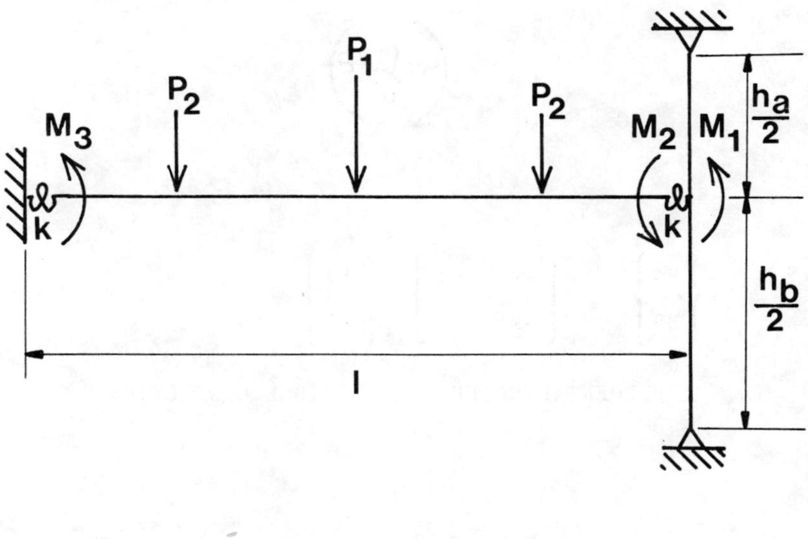

GRAVITY MODEL FOR EXTERIOR BAYS

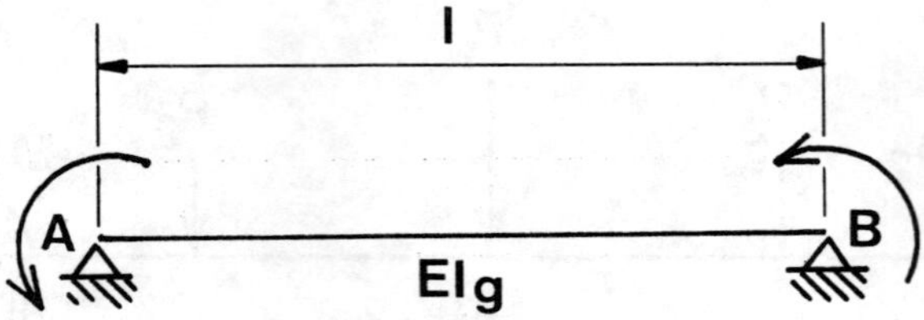

$$\begin{Bmatrix} MA \\ MB \end{Bmatrix} = \begin{bmatrix} \frac{4EIg}{l} & \frac{2EIg}{l} \\ \frac{2EIg}{l} & \frac{4EIg}{l} \end{bmatrix} \begin{Bmatrix} \theta A \\ \theta B \end{Bmatrix}$$

ELEMENTAL MODEL FOR GIRDER

$$\begin{Bmatrix} MA \\ MB \end{Bmatrix} = \begin{bmatrix} +k & -k \\ -k & +k \end{bmatrix} \begin{Bmatrix} \theta A \\ \theta B \end{Bmatrix}$$

ELEMENTAL MODEL FOR FLEXIBLE CONNECTION

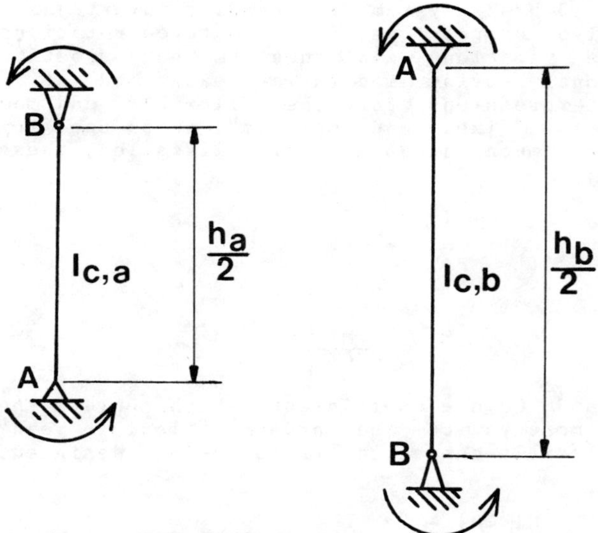

$$\begin{Bmatrix} MA \\ MB \end{Bmatrix} = \begin{bmatrix} \frac{3EIc}{1} & 0 \\ 0 & 0 \end{bmatrix} \begin{Bmatrix} \theta A \\ \theta B \end{Bmatrix}$$

ELEMENTAL MODEL FOR HALF COLUMNS

Using the direct stiffness method, the assembled structural stiffness equations become:

$$\begin{Bmatrix} M1 \\ M2 \\ M3 \end{Bmatrix} = \begin{bmatrix} k + \frac{3EIc}{\Sigma h/2} & -k & 0 \\ -k & k + \frac{4EIg}{1} & \frac{2EIg}{1} \\ 0 & \frac{2EIg}{1} & k + \frac{4EIg}{1} \end{bmatrix} \begin{Bmatrix} \theta 1 \\ \theta 2 \\ \theta 3 \end{Bmatrix}$$

Now, if one assumes that only symmetric gravity loads are applied to the girder, then the loading to be applied to the structural model is simply the conventional fixed-end beam moments, Mf, specified in degrees of freedom 2 and 3:

$$\begin{Bmatrix} M1 \\ M2 \\ M3 \end{Bmatrix} = \begin{Bmatrix} 0 \\ +Mf \\ -Mf \end{Bmatrix}$$

Next, this 3-by-3 system is readily inverted in algebraic form to give expressions for the three rotations, 01, 02, 03. These rotations can then be back-substituted into the rotational spring models at each end of the girder to obtain expressions for the "flexible end moments" (as contrasted to "fixed end moments") to be used in the frame design. After much algebraic simplification, these expressions become:

$$Mi = \frac{Mf}{Ci}$$

$$Me = \frac{Mf}{Ce}$$

where Ci and Ce are coefficients which convert the standard fixed end moments to the proper "flexible end" moments for the flexibly-connected frame being designed. They are defined as

$$Ci = 1 + 2a - \frac{g}{3(g + 1 + 6a)}$$

$$Ce = 1 + 2a + \frac{2g(1+3a)}{3(1+6a)}$$

and the following nondimensional terms are defined:

Connection Stiffness Parameter, $a = EIg/kl$

Relative Flexibility Factor , $g = \frac{Ig/l}{\Sigma Ic/h}$

Note that the "Relative Flexibility Factor" is simply the inverse of the well-known relative rigidity factor, G, used for estimating effective lengths of columns using the Jackson/Moreland nomographs.

Note that as connection stiffness, k, and column stiffness, Ic/h, become indefinitely large, then both "a" and "g" approach zero, the coefficients, Ci and Ce approach 1, and the "flexible end moments" become the usual fixed end moments, Mf.

A final point of corroboration is obtained by letting "g" become infinite (far-end pinned condition) and "a" become zero (fully continuous beam). It can be seen that the right moment, Me, goes to zero, and the left moment, Mi, goes to 1.5Mf. This would also be given by superposing an equal--and-opposite "Mf" at the right end of the girder to nullify the fixed-end "Mf": the left moment becomes 1.5Mf with the 0.5 carry-over factor.

The above expressions are easily automated on a programmable calculator. Alternatively, a plot of realistic ranges of the Connection Flexibility Parameter and Relative Flexibility Factor can provide a simple design aid for obtaining flexible end moments on girders, as illustrated below.

PROPOSED DESIGN PROCEDURE

The above development for predicting flexible end moments can be readily incorporated into the usual Type 2 design procedure, by appending refinement design calculations to the Type 2 results. This section summarizes the proposed Modified Type 2 Design procedure. Verification of its safety and economies that can be achieved in its use are described in reference 1.

MODIFIED TYPE 2 DESIGN PROCEDURE

STEP 1: FRAME DEFINITION (same as Type 2)

STEP 2: LOADING DEFINITION (same as Type 2)

STEP 3: ASSUMPTIONS
Assume column effective length factor for in-plane buckling and moment amplification, Kx = 1.5

STEP 4: ANALYSIS (same as Type 2)
Wind loading - use the Portal Method.
Gravity loading - assume girders are pinned.

STEP 5: DESIGN (same as Type 2)
Use allowable stress design provisions, Part 1 of AISC.
Use either extended flange plate connections or Type 2 wind connections (top-and-seat angles with web clips).
DESIGN CONNECTIONS FOR WIND MOMENT ONLY.

STEP 6: GRAVITY RE-ANALYSIS

a. Calculate initial elastic stiffness of connection, ki (e.g., inch-kips/radian).

b. Calculate representative connection stiffness:
Use k = ki for flange plate connections.
Also compute plastic moment capacity of flange plates, My (e.g., inch-kips).
Use k = 0.5*ki for other types of connections.

c. Calculate Connection Flexibility Parameters for girders
$a = EIg/kl$

d. Calculate Relative Flexibility Factors at exterior joints
$$g = \frac{Ig/l}{\Sigma Ic/h}$$

e. Determine flexible end moment coefficients for girders

$$Ci = 1 + 2a - \frac{g}{3(g + 1 + 6a)}$$

$$Ce = 1 + 2a + \frac{2g(1 + 3a)}{3(1 + 6a)}$$

f. Calculate flexible end moments of all girders.

Mi = Mf/Ci (not to exceed My, if flange plates)
Me = Mf/Ce (not to exceed My, if flange plates)

g. Adjust frame moment diagrams for new flexible end moments.

Girder midspan moments
Mss - 0.5*(Mi + Me)
Exterior column axials
Pg - (Mi - Me)/1
Interior column axials
Pg + (Mi - Me)/1
Column moments

$$Me * \frac{Ic/h}{\Sigma Ic/h}$$

REPEAT STEPS 5 AND 6 UNTIL NO CHANGE

REFERENCES

1. Ackroyd, M.H., "Design of Flexibly-Connected Steel Building Frames", Final Report to American Iron and Steel Institute, Project No. 333, Rensselaer Polytechnic Institute, Troy, N.Y., November, 1985.

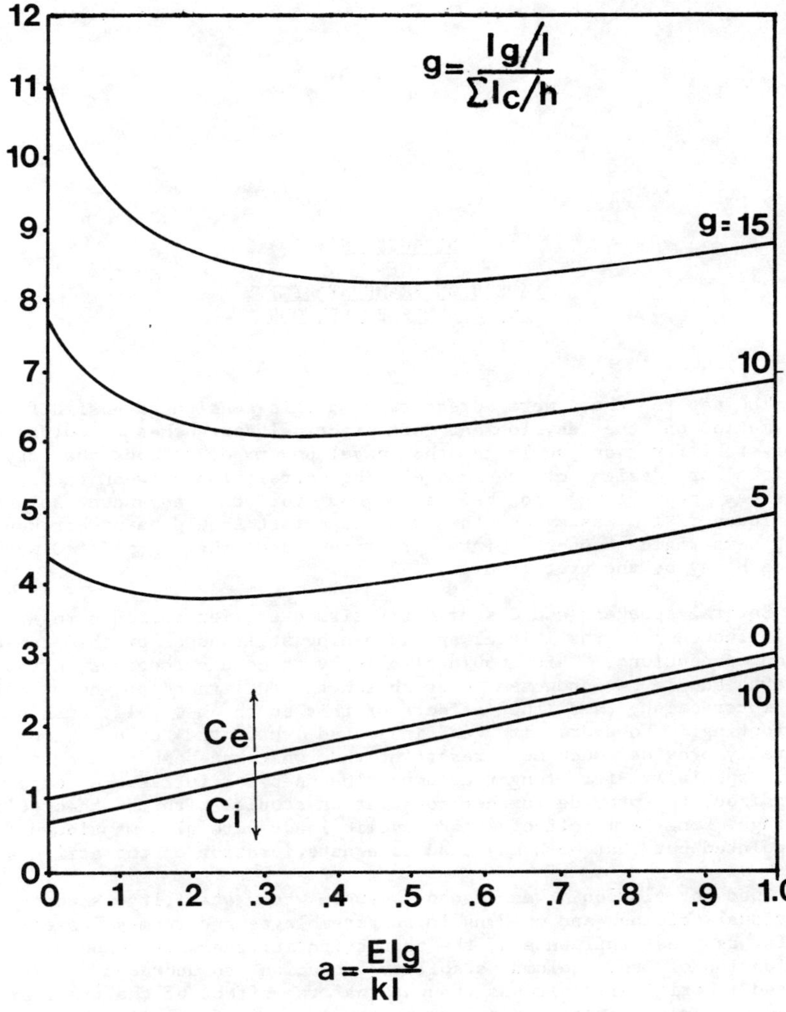

DESIGN AID FOR FLEXIBLE END MOMENT CALCULATIONS

DISCUSSION OF PAPERS ON FRAME STABILITY AND SIMPLIFIED METHODS

The papers that were presented in this session gave significant information on the development of practical approaches to column and frame stability, as well as the development of methods that may be feasible for design office usage. The correlation between theory and tests has been found to be very good for the advanced solution techniques ; it was agreed that a major effort should be undertaken to bring semi-rigid concepts into practice, and the simplified method approach may be the preferred way.

Several speakers and discussers affirmed earlier comments regarding the influence of the initial and unloading stiffnesses of the beam-to-column connections. This would also apply to column footings, although research should be conducted to establish the full range of moment-axial load interaction, and the effects of this on the M-Φ-relationship for the footing. However, it was indicated that the column footings generally provide much more restraint than what has been assumed in the past, especially for longer columns. The need for full-scale tests was emphasized, to provide further correlation studies with the theoretical findings. Long-term effects and cyclic loads were also mentioned ; it was pointed out that both may lead to a deterioration of the stiffness.

Good correlation was found between the stability studies of individual columns and columns in subassemblages and frames, especially insofar as the influence of the connection stiffness is concerned. The application of the column stability solution to unbraced frames was debated ; originally it was thought that the effect of the transfer of the beam moment into the column at the instant of buckling would make for an unconservative application. However, the frame subassemblage and full-scale tests had demonstrated that the buckling capacity and behaviour were much less sensitive to the moment transfer, which would appear to indicate that the column stability solution may have wider applicability. This is a subject in need of further research.

There was general agreement that the development of simplified frame analysis and design methods had the potential for offering economy in design as well as construction. For example, taking advantage of the end restraint would lead to lower girder stress levels. However, several discussers noted that serviceability checks would be critical. It was also pointed out that revised drift limits would have to be developed for frames with semi-rigid connections, recognizing that the commonly used (average) limit of 0.0025 for the ratio of building drift to building height had been arrived at for structures with rigid connections. In other words, if the computations of the drift values are more accurate, then the drift limits will have to be revised. This would appear to be a suitable topic for a major research effort.

Summarizing, there seems to be substantial agreement on the buckling characteristics and controlling parameters for members and frames. Ultimate limit states are well defined, the solutions are less sensitive to connection stiffness variations than previously thought, and tests and theory compare favorably.

SESSION Nº5

DESIGN REQUIREMENTS

LRFD CONNECTION DESIGN
IN USA

Robert O. Disque
Director of Building Design Technology
American Institute of Steel Construction

ABSTRACT

The American Institute of Steel Construction has been researching a limit state design procedure for structural steel since 1969. The principle investigator has been Dr. T.V. Galambos, then of Washington University and now at the University of Minnesota. The design specification and Manual of Steel Construction was completed in 1986. The USA is now in the process of converting from the traditional allowable stress design (ASD) to a limit state procedure, termed Load and Resistance Factor Design (LRFD). This paper presents the basics of how connections are designed using LRFD.

Basic Behavior

All steel connections have a unique Moment-Rotation Relationship. See Figure 1. The characteristics of the curve depend on the thickness of the material, size and number of bolts, configuration of detail material, etc. The state of the art does not permit a designer to predict the behavior of the connection, except in a few well defined cases. LRFD classifies frames according to the type of connection - partially restrained (PR) or fully restrained (FR). The designer must decide which of these he/she has before analizing the frame.

Fig. 2 shows examples of FR connections. Fig. 3 shows examples of PR connections.

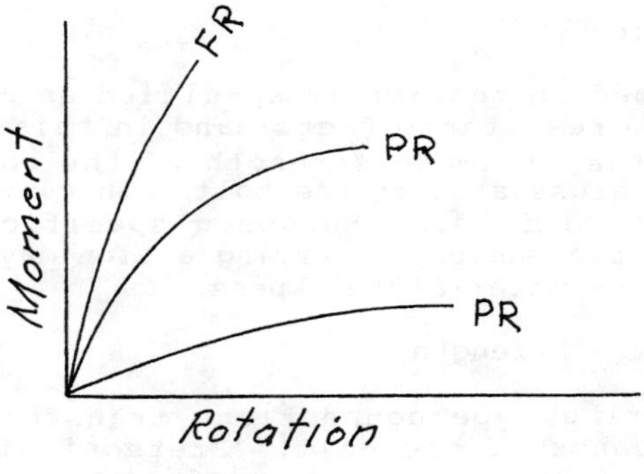

FIG. 1

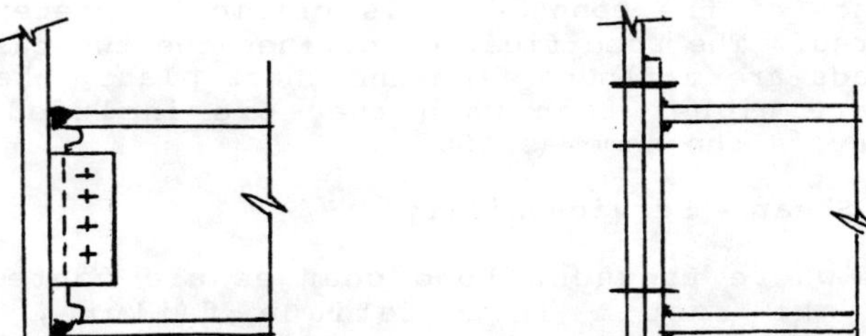

FIG. 2

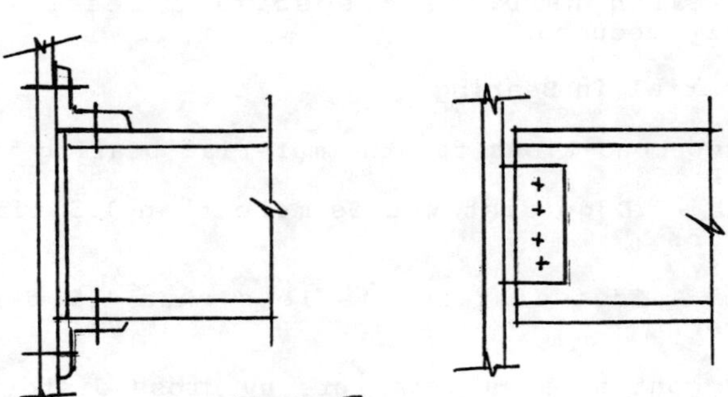

FIG. 3

Bolts in Tension

The design load in tension is specified as $\phi(.75)F_u(A_g)$, where ϕ is the resistance factor and in this case equals 0.75. F_u is the ultimate strength of the bolt material and A_g is the gross area of the bolt. In contrast to what may be permitted in some European specifications, the bolts must be pretensioned. Prying action may be a factor in bolt size and material thickness.

Bolts in Shear - Strength

Bolts in bearing type connections transfer shear load through the shank of the bolt. Pretensioning does not affect this capacity. In many applications, therefore, pretensioning is not required. In these cases, the connection is termed "snug-tight". In cases where the performance of the connection is critical, pretensioning is required. The specification defines the two cases. If the threads are excluded from the shear plane, the design values are higher than when they are included. This philosophy is the same as ASD.

Bolts in Shear - Serviceability

In cases where frequent load changes are contemplated which might result in a fatigue failure, bolted connections should be designed as slip-critical, formally called friction-type. Since this type of loading only occurs at working loads the connection is designed at service loads rather then at factored loads as is the usual case with LRFD. The specified resistance is appropriately reduced.

Bolts - Material in Bearing

There are two conditions for the material bearing value.

Condition 1 - Edge distance is more than 1.5 times the bolt diameter.

Condition 2 - Edge distance is less than 1.5 times the bolt diameter.

Condition 1 contemplates a failure by gross distortion in the material. Condition 2 protects failure by tearing out a "slug" of material between the bolt and the end of the material.

For condition 1 the design load is specified as 1.8 x F_u x bolt diameter x t (material thickness). For Condition 2, the design load is 0.75 x F_u x Edge Distance.

Because of the two conditions, connection design may require that the bolt loaded next to an edge may have a different bearing value than the other bolts.

For Condition 1, the specified value is ϕ x 2.4 x F_u x d x t, with ϕ = 0.75. If the designer is not concerned with hole distortion, 3.0 may be substituted for 2.4.

Bolts - Block Shear

Block shear is a failure mode first codified in the AISC Manual in 1980. The LRFD procedure is different but more logical. There is no longer an edge distance requirement normal to the load but block shear must be checked for either of two possible failure modes: tension yielding and shear fracture or tension fracture and shear yielding. The larger of the two governs because block shear is a tearing phenomenon and the fracture term is the determining one and it is not obvious which is that term. Tension yield is taken as 0.75 F_y; shear fracture = $0.45F_u$. Tension fracture = 0.75 x F_u; shear yield = $0.45F_y$.

Bolts - Interaction

Interaction equations are specified for both strength at factored loads and slip at service loads. For strength, the design tension is,

$$\phi F'_t = 85 - 1.8f_u < 68$$

The slip critical allowable shear is,

where t_b is the pretension $\phi V' = \phi V \times (1 - T/T_b)$

Bolts - Eccentric Loads

When a bolt group is loaded eccentrically (Fig. 4), rotation occurs about an "instantaneous center". The procedure for determining the design load is as follows.

1. Assume location of instantaneous center (i.c.).

2. Determine Ri for the heaviest load bolt (farthest from the center of gravity) from,

$$R_i = R_{ult} \left(1 - e^{-10\Delta} \right)^{0.55}$$

3. R_i for all the bolts is calculated based on determined from the linear ratio of the bolt distance from the i.c.

4. R and the moments for all bolts are summed and the statics for the group is checked.

5. The location of the i.c. is iterated until equilibrium is satisfied.

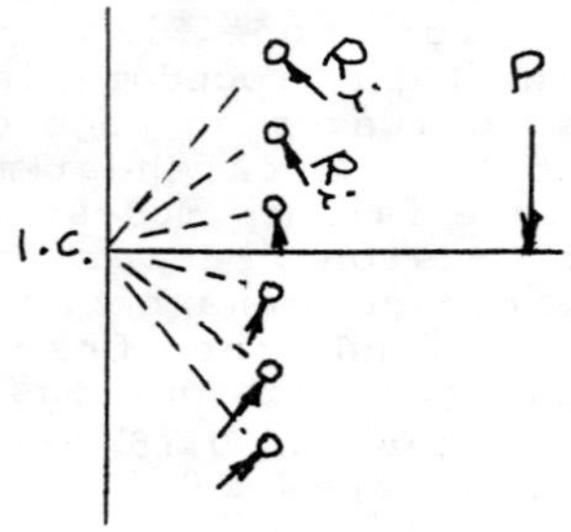

FIG.4

Bolts - Prying Action

For certain devices where a bolt is in tension (Fig. 5) the bolt is subjected to additional load. This, of course, adds to the bending in the plate. Although, the phenomenon is complex, AISC has simplified it by assuming that the bolt receives additional tension. In reality, the additional load is probably a combination of tension and bending. Fig. 6 diagrams the procedure as formulated by AISC.

The first extreme is when the plate is very thin (α = 1.0) and in double curvature. In this case the bolt is loaded the most. The second extreme is with a thick plate (α = 0.0) in single curvature. Here the bolt receives only the calculated applied load. It is necessary that α be assumed to be within 0.0 and 1.0.

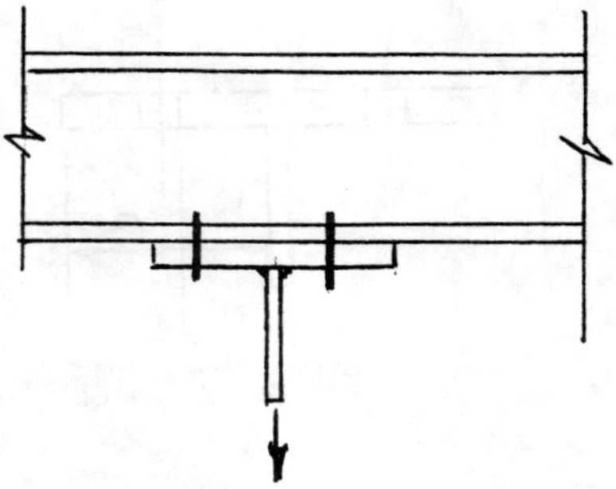

FIG. 5

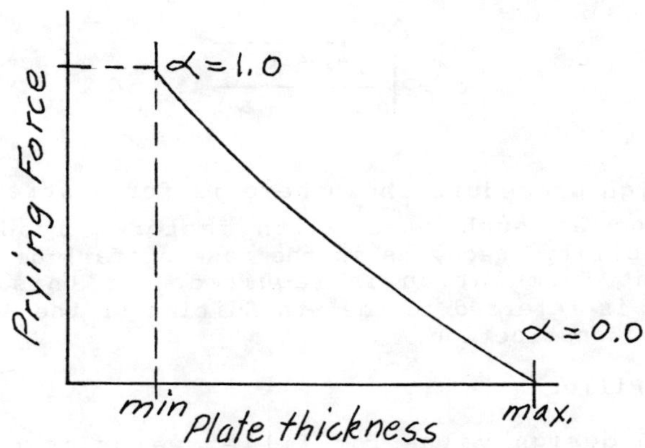

FIG. 6

The dimensions used in the formulas are shown in Fig. 7

The formulation follows:

$$\delta = 1 - d'/p$$

$$\rho = b'/a'$$

$$\beta = 1/\rho \times (B/T - 1)$$

$$\alpha = 1/\delta \times [\beta/(1-\beta)]$$

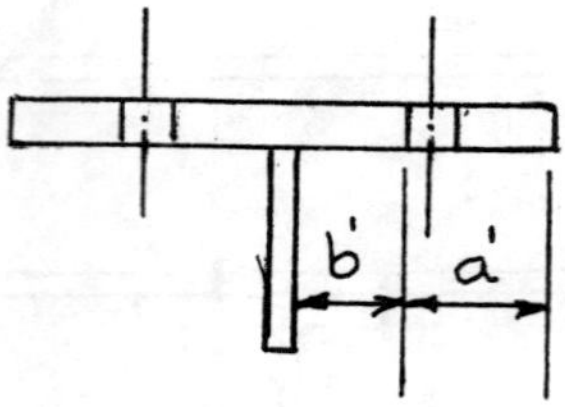

FIG. 7

The plate thickness is calculated to resist bending and is,

$$t_f = \sqrt{\frac{4,44\ T_b}{p\ F_y\ (1 + \delta\alpha)}}$$

The design procedure shown here is for a strength limit state and as such, used with factored loads. For a serviceability check, as in the case of fatigue loading, a different formulation is required. In this case, the designer is referred to the 8th Edition of the AISC Manual of Steel Construction.

Welds - Fillet

The LRFD design value for fillet welds is exactly 1.5 times that for allowable stress design. Identical designs result when the live load - dead load ratio is 3.0. This design value is,

$$\phi \times 0.45 \times F_{EXX}$$

where,

$$\phi = 0.75$$

F_{EXX} = Classified strength of weld metal

A study of tests on fillet welds indicates that the LRFD specified value is very conservative - especially when the load is transverse to the longitudinal axis of the weld.

Welds - Eccentric Loads

An instantaneous center procedure, similar to that used for bolts, has been developed for eccentrically loaded weld groups. However, the welded case is very much more complicated because the strength deformation curve for fillet welds depends on the angle to which the weld is loaded. As seen in Fig. 8 it was assumed that the ultimate strength value was a straight line and equaled the specified value of 0.6 x F_{EXX}. When the deformation on a particular weld segment was small the computer program followed the load - deformation curve down the real curve as indicated in Fig. 8. When the strength of the weld group was determined, a ø of 0.75 was applied.

Groove Welds

The design strength of groove welds is essentially the same in LRFD as in ASD, that is, it matches the strength of the material it joins. The effective throat dimension depends on the welding process, welding position and the included angle at the root of the groove.

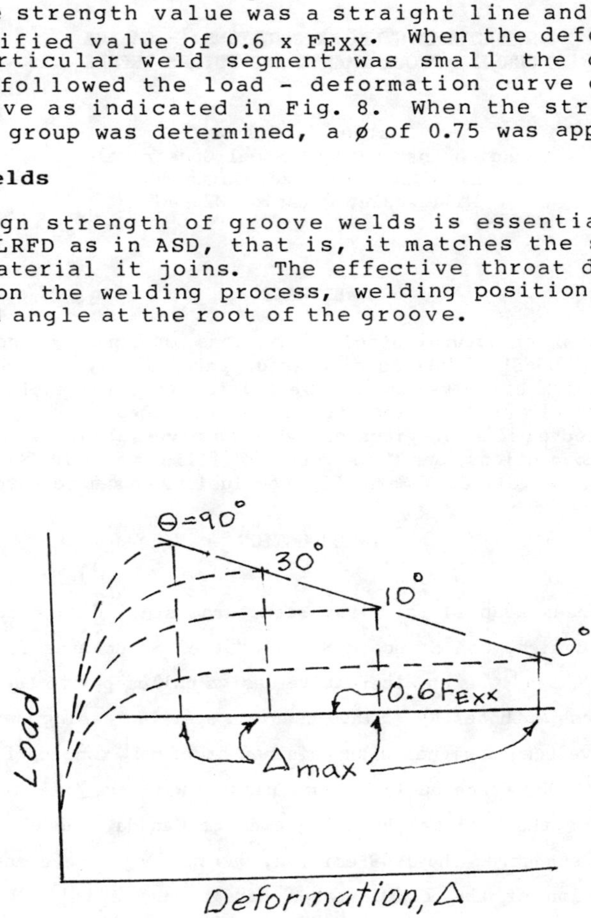

FIG.8

STRUCTURAL STEEL CONNECTIONS IN CANADA
CURRENT REQUIREMENTS AND FUTURE TRENDS

Michael I. Gilmor
Canadian Institute of Steel Construction
201 Consumers Road, Suite 300
Willowdale, Ontario, M2J 4G8
CANADA

ABSTRACT

The design of structural steel connections in Canada as covered under CSA Standard CAN3-S16.1-M84 is discussed, particularly as related to the ultimate strength of bolts and welds, and to the serviceability of slip-resistant connections. The results of recent research on partial joint penetration groove welds, longitudinally and transversely loaded fillet welds, gusset plate connections, and the strength of fillet welds in HSS truss joints are presented, as well as some of the conclusions drawn to date.

INTRODUCTION

In 1974 Canada adopted its first structural steel design code based on limit states design, CSA Standard S16.1, "Steel Structures for Buildings-Limit States Design", as an alternative design philosophy to the traditional allowable stress methodology. Since then, an ever increasing number of steel structures have been designed with this new standard, ranging from shopping centres to a 40 000 tonne nuclear power plant. With the introduction of the 1985 edition of the National Building Code of Canada, the older allowable stress design standard, the 1969 edition, was no longer referenced, leaving the 1984 edition of the limit states design standard [1] as the only structural steel design standard in Canada.

Although the formats differ from their allowable stress counterparts, the clauses related to connections covered similar areas but have been expanded where new knowledge was available. In S16.1, strength requirements for bolts in tension, in shear, and in combined shear and tension are given. The strength of the plate elements within a bolted joint are written as bearing requirements. Serviceability requirements of bolted connections, where slip at service level loads is critical to the usefulness of the structure, are written in terms of the total slip resistance of the joint.

For welds, resistances are given for complete and partial joint penetration groove welds, fillet welds, as well as plug and slot welds, in terms of the strength of the weld metal and the base metal, at the joint.

S16.1 permits simple construction, special simple construction, and continuous construction. Special simple construction permits the use of connections to resist gravity loads on the basis of simple construction and lateral loads by distributing the moments due to lateral loads among selected joints of the frame. However the Standard qualifies this method and its use seems limited to smaller structures, as there are no rules set down in S16.1 for the method of analysis nor are the behavioural assumptions given for various connection geometries.

For continuous construction, the Standard does provide the requirements for rigid joints similar to those in AISC's Specification.

CURRENT REQUIREMENTS

Bolted Connections

S16.1 is concerned primarily with high strength bolted connections and describes their tensile and shear resistances in terms of the ultimate tensile strength of the bolt, F_u, the cross-sectional area of the bolt based on the nominal diameter, and the following ratios, as appropriate, the tensile stress area to nominal area, the root area to nominal area, shear stress to tensile stress. The resistance factor, the number of bolts, and the number of shear planes complete the specification of the joint's resistance.

For serviceability, the slip resistance is stated in terms of the ultimate tensile strength of the bolt, the nominal area of the bolt, and a factor relating slip probability for the surface type and condition, tensile stress area to nominal area, and preload to ultimate tensile load.

The bearing resistance of the plate material within a bolted joint is stated in terms of the ultimate tensile strength of the plate, its thickness and the end distance of the bolt.

Bolted joints in shear must meet the shear and bearing requirements and may also meet the slip requirements. However, slip-resistant joints are considered to be the exception rather than the rule.

Installation of high strength bolts is restricted to turn-of-nut method or the use of direct tension indicators.

For inspection, the emphasis is on the observation of the actual bolt installation. For bearing-type connections subject to shear, visual inspection, that the bolt has been tightened, is all that is necessary.

Welds

Under limit states design the technical requirements for welds are virtually unchanged from allowable stress standards. Depending upon the type of load and its orientation, the weld resistance is taken as

i) the same as the base metal,

ii) the lesser of the shear of the base metal on the fusion face,

or

iii) shear on the effective throat of the weld as a function of the area of weld, the ultimate tensile strength of the weld metal, based on its electrode classification, and the ratio of shear to tensile strength of the weld metal.

The Standard does permit the designer, for the weld group in a connection, to use an ultimate strength analysis method to determine the resistance, instead of the traditional elastic analysis method, but no method is given.

Restrained Members

Members retrained either due to full or partial end fixity must be designed for shear and moment. S16.1 provides requirements for column web stiffeners when beams frame into H-type column members based on the 1959 work of Graham, Sherbourne, Khabbaz and Jensen.

For corner and "T" type joints, the designer is cautioned to detail the joint to minimize the risk of lamellar tearing.

Bolts and Welds in Combination

While bolts and welds are permitted in the same shear plane, the number of bolts are to be determined based on slip resistance at service loads, and the ultimate resistance of the connection is limited to the greater of that of the weld or bolt group. This recognizes that the load-deformation response of bolts differs from that of welds. Other than performing a load-deformation analysis of the connection as a whole, this conservative approach, to take the larger of the two individual resistances, is used.

RECENT RESEARCH RESULTS

As the cost of fabricated, erected structural steel is influenced heavily by the man-hours required for fabrication, it is essential to ensure that connections are both economical and reliable. In the past few years a number of research projects have been undertaken in Canada which deepen our knowledge on connection behaviour and reliability. This portion of the paper will touch on some of their results.

Partial Joint Penetration Groove (PJPG) Welds

According to any standard in North America for welding, AISC [2], AWS [3], or CSA, a PJPG weld is permitted to carry, in tension perpendicular to the weld axis, only two-thirds of the load of a complete joint penetration groove (CJPG) weld of similar weld size and electrode strength. This produces considerable fabrication difficulties when backing bars cannot be complete or when the root pass cannot be backgouged and rewelded, in order to qualify as a CJPG weld. Thus a joint with as much as 99.9% penetration can, by specification, carry only 66% of the load in the parent plates.

In 1977, Popov and Stephen [4], in a limited series of tests, investigated the strength of pairs of PJPG welds, in static and cyclic behaviour, in splices of W14 columns. Welds from complete to only 23% penetration were tested.

Recently, Kennedy and Gagnon [5] tested 75 specimens with groove welds of 5 different penetrations, ranging from 20% to 100% of the plate thickness, in plates of two steel grades, some singly and some in pairs. The PJPG welds tested singly were intended to demonstrate any influence of eccentricity of load with respect to the weld axis. All welds were single V 45° bevel SMAW welds made with 3.2 mm diameter E48018 electrodes.

For the specimens tested singly, the mean failure stress of the weld was 1.15 times the ultimate stress of full penetration welds with a CoV of 10.2%. For welds tested in pairs the ratio was 1.17 with a CoV of 11.2%.

Kennedy and Gagnon noted that their results were consistent with those of Popov and Stephen, as well as Lawrence and Cox, and drew the following conclusions:

(a) "...such welds on a unit area basis are as strong as full penetration welds...";
(b) "Welds tested singly (and therefore eccentrically) were as strong as double welds tested concentrically. The specimens with eccentrically loaded welds deflected sideways reducing the eccentricity on the welds as well as the ductile straining produced a mean uniform stress distribution across the weld throat."; and,
(c) "As the percent penetration decreases the restraint to lateral deformation increases with the result that the weld strength increases over that proportionate to the weld area."

For design, Kennedy and Gagnon recommended that "PJPG welds loaded in tension be designed on the basis of the percent penetration multiplied by the ultimate tensile resistance of the plate. Root passes must be properly made to achieve uniformity of depth and the desired penetration."

Fillet Welds

Current weld design standards base the design for fillet welds on the strength of longitudinal fillet welds regardless of the direction of loading [1-3]. Although CSA Standards S16.1 and W59 [6] permit an ultimate strength analysis none is given. In 1971, Butler and Kulak [7] proposed that ultimate strength be calculated from the load-deformation response of the weld as a function of the angle of load using the instantaneous centre of rotation model.

Although strengths of fillet welds have been studied since the early 1930's, few tests have been done on fillet welds loaded at intermediate angles. Butler and Kulak , Clark [8], Holtz and Harre, Swannell and Skewes [9], and Biggs et al [10] have contributed to the experimental data while, Neis [11], and Marsh [12] have attempted to develop mathematical models.

In a carefully designed and instrumented test programme involving 42 specimens (seven different loading angles, two weld sizes with three specimens for each combination), Miazga and Kennedy [13] showed that the strength of fillet weld, determined by the vector sum of the longitudinal and transverse stresses, is overly conservative.

Miazga and Kennedy demonstrated that an analysis method based on the maximum shear stress failure criteria is in good agreement with the test data, with a test-to-predicted ratio of 1.01. Further, the maximum shear stress failure criteria is better than methods based on either maximum normal stress or von Mises strain energy of distortion failure criteria.

In their tests, all specimens failed in the weld while the plates were not strained beyond the yield strain. Miazga and Kennedy also examined the role, if any, of frictional forces, as proposed by the International Institute of Welding(IIW), in the analysis of the results and concluded that they "do not appear to be a significant factor in the analysis".

In an examination of the measured weld deformations, they stated: "By normalizing the deformations of the welds, measured in the direction of loading, to the gauge lengths used to obtain the deformation, it is concluded that the fillet weld ductility is essentially independent of the loading angle, and that the deformations are proportional to the gauge length".

Gusset Plate Connections

While gusset plates have been used for decades, a rational design method has not been fully developed. Since Whitmore's [14] effective width concept not much has been reported until work by Bjorhovde [15], and Richard [16].

These recent programmes have combined physical tests with finite element models to better define tensile behaviour. However, in certain circumstances concerns have been raised when a single gusset plate lapped to a connection plate were subject to compressive loads. Cheng [17] has shown, based on limited tests, that the effective width approach can grossly over-estimate the collapse load of the connection while ASSHTO's limit of $930/\sqrt{F_y}$ for an unsupported edge is a conservative limit in such situations.

An obvious way to avoid the problem is to detail the connection to ensure that a stiffener, even to the connection plate, is provided. Certain shapes can be coped or blocked to ensure a flange or leg is retained to act as a stiffener on one side of the connection. Alternatively, an angle can be attached to the connection plate to provide the necessary stiffness.

HSS Truss Connections

While much research has been done on the joint strength of HSS members connected in a truss, less attention has been paid to the actual performance of the connectors, usually the welds. When fillet welds are used, a weld leg size of 1.2 times the wall thickness has been recommended. However, in many situations, chords and webs are sized for more than strength. Often the joint strength rules ensure that the web members are larger than required solely to carry calculated loads. Thus a design methodology relating the size of the fillet weld to the calculated force to be carried by the truss member would be more satisfactory and more economical.

Work is currently underway at the University of Toronto under Jeff Packer to investigate the strength and behaviour of fillet welded HSS truss connections. Tests on isolated joints have been completed [18] and full scale truss tests are now being planned.

It is hoped that this research will develop rational design rules for fillet welded HSS truss joints to complement the IIW joint strength rules.

CONCLUSIONS

While adopting a limit states design standard has helped focus both the code writers and the designers attention on the actual failure mechanisms of connections, research continues to point to areas where the margins of safety are inconsistent and improvements can be made with significant cost savings. Controlling weld volume in any structural steel joint results, not only in reduced fabrication costs and improved ability to compete with concrete structures, but also in reduced shrinkage and distortions in the work piece, and more favourable residual stresses.

REFERENCES

1. Canadian Standards Association, CAN3-S16.1-M84, Steel Structures For Buildings (Limit States Design), Canadian Standards Association, Rexdale, 1984

2. American Institute of Steel Construction, Load and Resistance Factor Design Specification for Structural Steel Buildings, American Institute of Steel Construction, Chicago, 1986

3. American Welding Society, Structural Welding Code - Steel, American Welding Society Inc., Miami, 1986

4. Popov, E.P. and Stephen, R.M., Tensile capacity of partial penetration groove welds, Jour. Struc. Div., American Society of Civil Engineers(ASCE), New York, 1977, 103, ST9, pp.1721-1729

5. Gagnon, D.P. and Kennedy, D.J.L., The ultimate strength of partial joint penetration groove welds, paper to be delivered at the Canadian Engineering Centennial Convention, Canadian Society of Civil Engineering, Montreal, 1987

6. Canadian Standards Association, Welded Steel Construction (Metal-Arc Welding), W59, Canadian Standards Association, Rexdale, 1984

7. Butler, L.J. and Kulak, G.L., Strength of fillet welds as a function of direction of load, Welding Research Supplement, Welding Reasearch Council, May, 1971, pp. 231s-234s.

8. Clark, P.J., Basis of design for fillet-welded joints under static loading, Proc. Conf. on Improving Welded Product Design, The Welding Institute, Cambridge, England, Vol 1, pp. 88-96

9. Swannell, P. and Skewes, I.C., The design of welded brackets loaded in plane: Elastic and ultimate load techniques, Civil Engineering Report, University of Queensland, Brisbane, Australia, Feb., 1977

10. Biggs, M.S., Crofts, M.R., Higgs, J.D., Martin, L.H., and Tzogius, A, Failure of fillet weld connections subject to static loading, Proc. Conf. Joints in Structural Steelwork, Pentech Press, London, England, 1981,pp. 1.92-1.109

11. Neis, V.V, New constitutive law for equal leg fillet welds, Jour. Struc. Div., ASCE, Vol 111, No. 8, pp. 1747-1759

12. Marsh. C., Strength of aluminum fillet welds, Welding Journal, Welding Research Council, Vol. 64, No. 12, pp. 335s-338s

13. Miazga, G.S. and Kennedy, D.J.L., Behaviour of fillet welds as a function of the loading angle, Structural Engineering Report, University of Alberta, No. 133, Mar., 1986

14. Whitmore, R.E., Experimental investigation of stresses in gusset plates, Engineering Experimental Station Bulletin, University of Tennessee, No. 16, Knoxville, 1952

15. Bjorhovde, R. and Charkrabarti, S.K., Tests on full-sized gusset plate connections, Jour. Struc. Div., ASCE, Vol, 111, No.3, Mar., 1985, pp. 667-684

16. Richard, R.M., Rabern, D.A., Hormby, D.E., and Williams, G.C., Analytical models for steel connections, Proc. Behaviour of Metal Structures W.H. Munse Symposium, ASCE, May 17, 1983, pp. 128-155

17. Cheng, J.J. R., and Hu, S.Z., Compressive tests of gusset plate connections, paper presented to Annual Meeting, Structural Stability Research Council, Mar. 25, 1987

18. Frater, G.S., Weldment design for hollow structural sections, M.A.Sc. Thesis, Dept. Civil Engineering, University of Toronto, Dec., 1986

PRACTICAL CONSIDERATIONS IN THE DESIGN OF FRAMES WITH PR CONNECTIONS

Stanley D. Lindsey, Ph.D., S.E.
Stanley D. Lindsey & Associates, Ltd.
1906 West End Avenue
Nashville, TN 37203 USA

ABSTRACT

The codification process that will allow structural engineers to design buildings using PR connections has been completed in the United States with the publication of the LRFD Specification. The design of such frames requires a different design approach due to the non-linear response of the frames. The writer's firm has had experience in the practical application of the frames using PR connections. The purpose of this paper is to emphasize some of the most important factors for the practicing engineer to evaluate when designing buildings with PR connections.

One of the most promising structural concepts now receiving a great amount of attention in both research and in practical applications is the use of PR connections in frames. In the United States with the recent publication of the current limit design specifications[1] - LRFD - use of actual connection restraint has been recognized. Through the LRFD specifications, the designer has the guidelines to produce safe designs that employ PR connections. The field is now open. While the design rules are there through LRFD, the actual experience and recorded performance of frames in real buildings with PR connections are far behind the technical design aspects. It is true that frames using other assumptions of connection behavior, such as the Type 3 semi-rigid of the ASIC specifications, have been used for many years. However, frames using the actual moment rotation curves for the connections have not yet been designed in great numbers in the United States. Thus, a limited performance base is available to the profession. This paper will describe some of the important design aspects that have been learned through the writer's actual design of buildings with PR connections.

Very importantly, a word of caution is given. The design of buildings with PR connections requires a special level of design consideration to produce a safe and economical end result. The design problem is non-linear and that one fact alone sets the design apart from the normal design process used by the majority of today's engineers. The entire design process for a system that responds in a non-linear fashion as compared to one that has a linear response is different. Unless done properly, a non-linear system will not always have a sufficient degree of safety and may well be unsafe. Earlier papers[2,3] by the writer have given some specifics concerning inelastic vs. elastic designs.

One starts any design with PR connections with moment-rotation curves for the connections. An immediate observation is made. The prediction of the behavior of the system will only be as accurate as the connection moment rotation curves that are used. The designer must know the limits of his curves. He must not extrapolate beyond known limits and he must not change the connection from the tested connection that generated the moment rotation curve. While these points seem obvious when mentioned, in the real world of the practitioner, it is very easy to allow a slight modification to a connection detail, say, for the convenience of the fabricator. Even the simplest change can substantially affect behavior. So practically speaking, the designer must watch very, very carefully to see that his design in the field is the same as his design in his computer. Whereas in buildings with other types, conformance of final design to intended design may not be as important, but in buildings with PR connections, it is of paramount importance. If what is used is not the exact same as what was designed, the result could mean disaster.

The analysis routines used for frames with PR connections are extremely important. Assuming valid moment-rotation curves, the predicted behavior used for design must accurately portray actual behavior in a fashion sufficiently accurate for design. The writer has found that an elastic second order analysis using non-linear moment rotation curves and factored loads will produce sufficient results for design. The second order analysis should include, as a minimum, frame P-delta, loss of column stiffness due to axial load on columns, individual member P-delta, and inelastic moment rotation curve capabilities. In addition, it should have the computation routines in the program for column K-factors for each loading. Programs of this type have worked well in the writer's design practice. More elaborate analysis routines, for instance, to include partial plastifications of column cross sections has not been found to be necessary for design. Sufficient accuracy for design of buildings does seem to require such elaborate programs. For research applications;

however, it is a different matter and for these instances it is wise to use the most accurate tool available to verify research results.

Another important design aspect is to match the proper type PR connection to the frame that is to be designed. As an example, a very low building several bays deep in the direction of light lateral wind loading does not require a connection that will develop a large moment in the connection. Therefore, a practical design will use the connection with just the right amount of restraint, such as a pair of web angles or a very thin end plate. Top and bottom seat angles or thick end plates are not needed. The writer has found the best economies come from using the connection that furnishes the least restraint but yet yields a safe structure. The more the restraint, the more costly the connection so, therefore, it is wise to use the least possible restraint in the connection to achieve the least expensive connection cost. As lateral loads increase in relation to gravity loads, as the building becomes less and less deep in the direction of loading and as the building becomes taller, more and more connection restraint is demanded. The art of matching available connection restraints to required demand is presently being learned by designers. As the experience is gained, it will be possible to produce a reference of past successes and failures for the profession to use. The PR connection selection will become an integral part of the design process.

A very important design consideration in frames with PR connections is the drift of these frames. The less the connection restraint, the more the drift. Much research has been focused on the ultimate loads needed on frames with PR connections for collapse and shake down. From a practical designer's view, allowable drift limits for damage or occupant comfort seem to always come before a frame is even close to collapse. The fact is especially true if one adopts the philosophy of using the minimum connection restraint required for the actual demands. The writer's experience has been that almost without fail, drift requires the most connection restraint, not gravity loads. In the writer's designs, drift is controlled by selectively adjusting the available connection restraint to the demand. If one follows that procedure, a safe building design controlled by allowable drift will result.

One final consideration must be mentioned and that is individual and overall stability. For overall stability, any analysis routine for frames with PR connections should indicate that unstability exists when the convergence criteria of the routine is not met. If a valid solution does not exist, then the design is not valid. Individual member restraint; i.e., effective length, is needed to design compression members. One must be able to accurately compute the effective restraint offered to a column by the beams connecting with PR connections to that column. Column K factors are directly influenced by the available connection restraint. The smaller the available restraint, the larger the K factor. Since available connection restraint is variable under different loads, K factors will vary with loading. This variance is due generally to the non-linear shape of most moment rotation curves. In practical terms, the designer must be very careful to use the proper procedures to determine his K factors, lest he can have a very unconservative design. The less stiff the connection selected by the designer, the greater this problem becomes. Therefore, practical designs must always be precise in this area of computation.

It is possible to design frames with PR connections and produce safe economical structures. Many factors must be considered and must be carefully designed to avoid having a structure that does not perform as intended. Hopefully, this paper has presented some practical aspects for consideration by designers when confronted with the task of designing frames with PR connections.

REFERENCES

1. Manual of Steel Construction, Load & Resistance Factor Design, American Institute of Steel Construction, Chicago, Illinois, 1986.

2. Lindsey, S. D., Ioannides, S.A., and Goverdhan, A.V., "An Analysis and Design of Beams Partially Restrained Connections," American Institute of Steel Construction, Engineering Journal, Volume 22, No. 4., 1985.

3. Lindsey, S.D., "Design of Frames with PR Connections," to be published in the Constructional Steel Research (Special Issue).

OBSERVATIONS ON THE SEISMIC BEHAVIOUR OF STEEL CONNECTIONS AFTER THE MEXICO EARTHQUAKES OF 1985.

Enrique Martinez-Romero
Consulting Structural Engineer
Av. Nuevo Leon 54-2nd. floor
Mexico, D.F. 06140 MEXICO

ABSTRACT

The Mexico City Earthquakes of September 19 and 20, 1985 affected about 3000 buildings, 330 of them suffered either collapsed or severe damage. Since there are about 1'200,000 buildings in Mexico City, while the number of affected number is large, its percentage is very small. Only 12 Steel buildings received important damage. This paper describes the principal types of steel construction existing in Mexico City at the time of the Earthquakes, particularly as regards to its connections, and relates them with its construction time. A recopilation of the Seismic Codes development is given and the connection's seismic behaviour is discussed. Proposed implementations of the new Code to incorporate the use of Flexible Connections are discussed, based on the evidences left by the earthquakes.

INTRODUCTION

Seismic provisions for the design of buildings in Mexico City, have been incorporated into the Design Codes since 1942. Several revisions of them have taken place since then, based principally on the observations of building damages and frame behaviour under extreme earthquake loading. The major earthquakes occurred in 1943, 1957, 1962, 1979 and 1985, served to calibrate previous provisions and to incorporate changes in the newest versions (1).

The construction of steel-framed buildings in Mexico City started back in 1910 with the execution of Museums, Theatres, Railroad Stations and Monuments. The first multistory-steel-framed office and hotel buildings appeared in the late 20's, all of them built with ASTM A-7 type of structural steel, using built-up-columns and I-beam sections in simple construction. Lateral load resistance was accomplished either thru shear action of the multiple masonry filler walls or by elementary double-angled X-bracing systems.

By the mid 30's some steel framed buildings incorporated knee-braces to the simple beam-to-column connections, in order to eliminate the need of some of the interior filler walls, without losing its lateral load resistance.

By the late 40's and early 50's most of steel framed multistory buildings were incorporating moment-frame action by the use of moment beam-to-column connections in riveted construction. Welded-moment connection appeared in the 60's and high-strength bolded connections in the 70's.

No specific provisions were made in the Building Codes to account for the flexibility of semi-rigid connections in particular. However, begining with the Emergency Regulations issued on 1957, the recognition of the inelastic frame action in the seismic behaviour of building frames, originated a series of different approaches to building seismic design.

The observations of the damage produced by the September 1985 earthquake on several steel-framed buildings, most of them built before 1957, contributed in a very significant manner to understand the behaviour of different types of beam-to-column connections. Even though most of them were not specifically designed bearing in mind its amount of flexibility and rotation capacity, the assessment of their seismic behaviour displayed interesting parameters, some of which are discussed in the present paper.

TYPES OF BEAM TO COLUMN CONNECTIONS.

Figure 1 shows some of the most common beam-to-column connections used in riveted construction in the 30's decade. In some cases, encasement in concrete of both beams and columns for fire protection purposes, served -- also as a joining material with the filler masonry walls. This fire proo-- fing added significant strength to both the lateral and vertical loading - capacity of the steel frame. Standard diagional angle-X-bracing. normally with high slenderness ratio, substituted the wall in some instances, and fire proteciton was not always supplied.

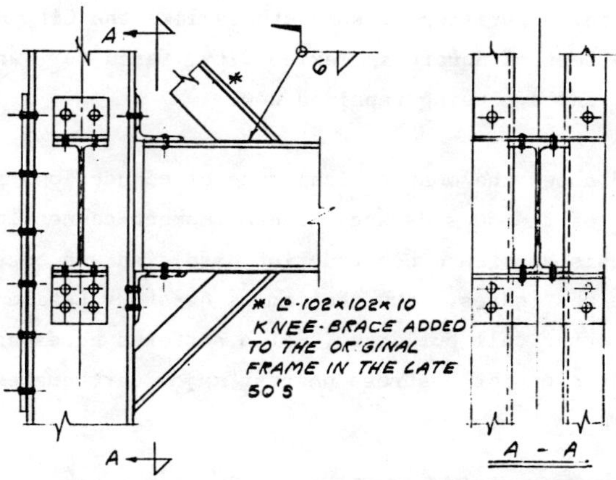

Figure 1. Riveted beam-to-column connection in Simple-Construction.

Some buildings like the Hotel Del Prado, in Mexico City, had been re-- trofitted in the late 50's, after being moderately damaged by the 1957 -- earthquake. The steel frame received welded knee braces under and above the beams, to improve the lateral load capacity of the bare frame. Such a rein forcement proved to be satisfactory inspite of being executed in a very -- crude way. (see elements marked with * in Fig. 1)

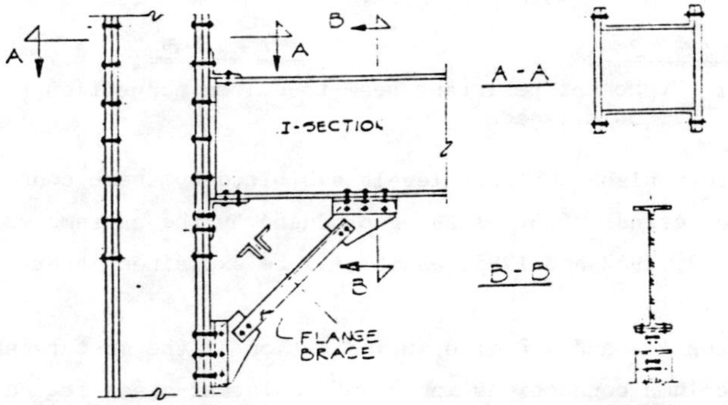

Figure 2. Typical beam-to-column connection with flange brace

Figure 2 represents the first serious intent to attain lateral load resistance of the steel buildings in the 40's without the need of filler masonry walls. They also used riveted construction and A-7 Steel. This type of construction displayed large drifts during the macro-earthquake of September 1985, as a result of large joint rotations. It was often found the knee brace either buckled or failed in tension, and some of the buildings which survived, suffered permanent lateral deflections which condemned them for demolition. Surprisingly, some others like the City Hall building complex, at the Central Square in Mexico City, fared very well showing up some damage only, and are being repaired now.

Figure 3 shows the most typical type of connections found in the steel construction of the 50's decade. Riveted moment-connections of this types in moment-resistant-frame construction displayed and outstanding seismic behaviour in most cases, like the famous 44-story Latino Americana Tower and several other tall buildings, which suffered no damage at all, nor - - given evident signs of distress on past major earthquakes.

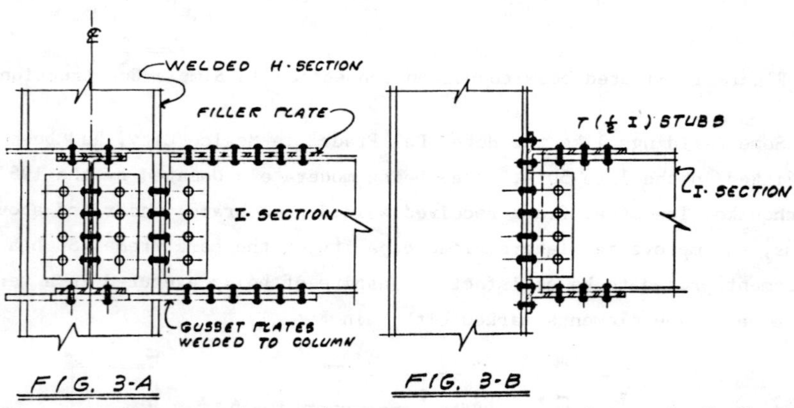

Figure 3. Moment-resistant beam-to-column connection typical of the 50's decade

The very high ductility levels exhibited by these connections overcame the large demands of overloading produced by the extreme earthquake suffered in 1957, 1962 and 1985, as it will be explained later.

Figures 4 A and 4 B give an indication of the most recent types of -- beam-to-column connections in moment resistant-steel framed buildings.

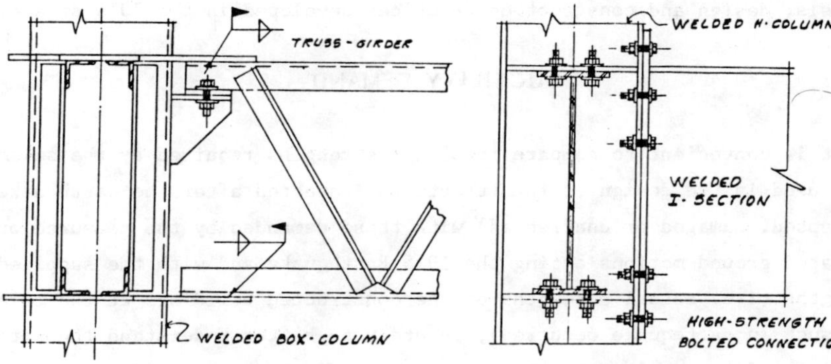

FIG. 4-A FIG. 4-B

Figure 4. Welded beam-to-column connection (2) and H.S. bolted connection used in moment-resistant.

With the exception of the five-building "Pino Suarez" complex, built with moment resistant space frame in which one of the three 23-story towers collapsed upon one of the two 14-story buildings (remaining the other three in the range of severely damaged to moderately damaged), after the September 1985 earthquake, all other buildings having beam-to-column connections similar to the one shown in Fig. 4-A, displayed a very satisfactory ductile behaviour (2,3).

In the case of this particular building complex, an overloading of some floors, the proximity of the building's natural period with the ground motion, and the very long duration of the almost constant pulses of the ground, produced successive deterioration of the joints in the inelastic range, causing either buckling or fracture of some beam-truss diagonal which propicitated a collapse mechanism through large interstory drifts, enhanced P-Δ effect of columns, local buckling of the box-column flange plates, and rupture of the joining weldements.

Practically all the rest of buildings with these types of connections, had a very satisfactory ductile behaviour, although in some instances, large lateral deflections and insuficcient clear distances between adjacent buildings, propiciated further damage by pounding. Buildings with connections of the type shown in Fig. 4-B, displayed a remarkably good seismic behaviour with no damage suffered. This was also the result of a better --

analysis, design and construction practices developed in the 70's decade.

DUCTILITY DEMAND

It is convenient to compare first the strengths required by the seismic codes used in the design of the structures inspected after the earthquake (collapsed, damaged or unaffected) with those demanded by the measured or estimated ground motions during the 1985 Earthquake and with the supplied strengths (i.e. actual strengths of the constructed buildings at the time that such an earthquake occurred), in order to better understand the actual behaviour of the various types of connections included in this document.

The damage assessment carried out on the steel buildings located in the most severely affected area of Mexico City, identified 102 engineered steel buildings (4). The most common type of framing found in this survey was the MRFO (Moment-Resistant-Frame-Only), being the second most common the MRFD (Moment-Resistant-Frame with Diagional Bracing). Therefore, the following discussion concerns mostly to these types of structures.

Based on the seismic coefficients specified by the various codes and on some simplifying but consistent assumptions, the demanded yielding seismic coefficient Cy, corresponding to the demanded nominal strengths (Vy/W) were estimated on its upper and lower bounds (5). Figures 5-A and 5-B show these values of Cy compared with the linear elastic demand strength by the earthquake motion, with 5% damping.

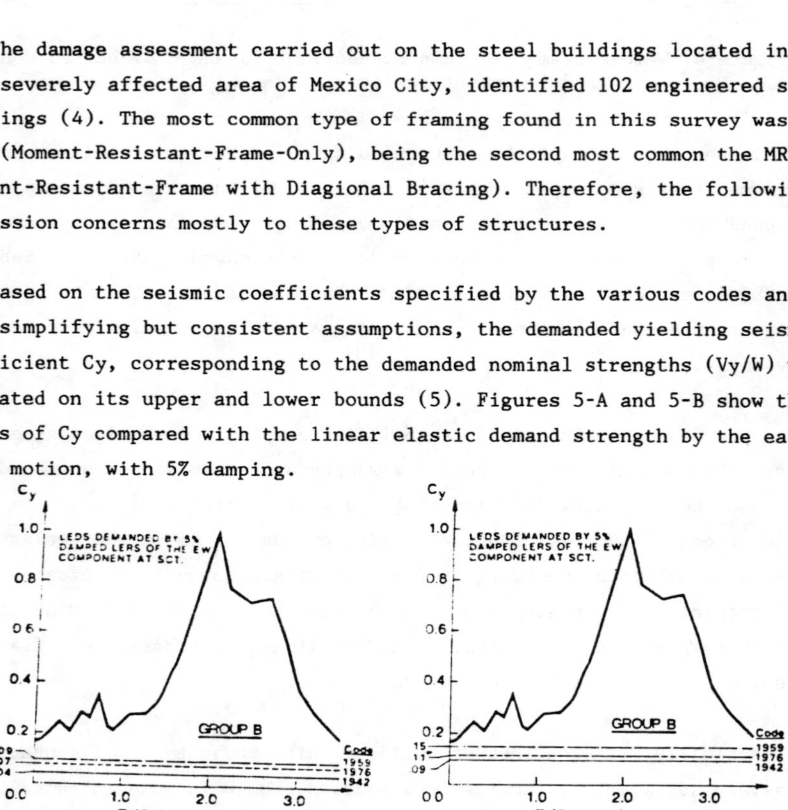

(a) LOWER BOUND OF ESTIMATED C_y REQUIRED BY CODES (GRAVITY STRESSES ARE NEGLECTED)

(b) MOST PROBABLE UPPER BOUND OF C_y (FULL EFFECTIVENESS OF GRAVITY STRESSES)

Figure 5. Comparison of Linear elastic demanded strength (LEDS) demanded by the 5% damped Linear elastic response-spectra (LERS) of the recor--ded E-W component at SCT station (5).

It can be noted that the strength demanded by the earthquake overex--- ceeded the upper bounds of the minimum nominal yielding strengths required by the seismic codes since 1942. However inspite of this clear evidence, a lot of buildings fared quite well. The most satisfactory explamation of it lies in two main reasons: a) The strength supplied by the filler walls, fa cades and non-structural elements, normally not considered in designs, and b) a large amount of inelastic (plastic) redistribution of the internal -- forces and moments due to the ductile behaviour of the connections.

The first point is a well recognized effect in which the actual buil-- dings will always supply strengths 30 to 180% larger than the theoretical design strength of the bare framing only (6). The second one is of interest to us now since it recognizes the ductile behaviour of the structure. This mere fact moved the Mexican Government to incorporate the inelastic design spectra into the Building Codes of 1976. The observation of the damage to buildings in September 1985 produced a substantial increase of the seismic coefficients and a more stringent requirements for the ductile behaviour - of frames. Thus the inelastic response spectra (IRS) obtained by dividing the linear elastic response spectra (LERS) by the so-called ductility coef- ficient Q, given by the newest codes, presents a more realistic situation which explains better the survival of so many buildings in Mexico City. (Fig. 6). According to the 1987 Code, (7) the recomended ductility values for steel moment-resistant-frames were reduced from the 6 to 2 range spe- cified in 1976 to the range of 4 to 1. Similarly, the seismic coefficient C corresponding to the soft-grounded area was increased from 0.24g given by the 1976 Code to 0.40g.

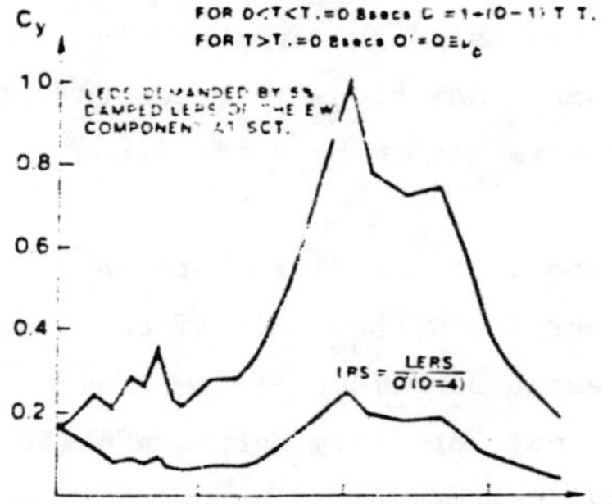

Figure 6. Inelastic response spectra (IRS) derived from the 5% damped LERS of the E-W component at SCT, divided by Q=4.

We could well summarize that the New 1987 Seismic Code, the demanded strength for ductile- moment-resistant frames, manda- te a 167% increase in the li-- near elastic demand response - spectra, but also a decrease - in the value of the ductility ratio Q from the 6-4 range for the same type of frames, to the 4-3 range, for the moment-resis tant-frames. All of this represent an increase of 2.5 and 2.2 times the demanded strength required by the 1976 Code, for the DMRF and the MRF respectively. In addition a further increase of the importance factor from 1.3 to 1.5 essential buildings, allows an additional margin to the upgraded demanded strength.

FUTURE DESIGN TRENDS AND RESEARCH NEEDED

At this point, the seismic behaviour of steel framed buildings has been superficially identified with the type of construction. Its supplied strength has been approximately compared with the evolution of the Code's strength demands and the 1985 earthquake's demand, suggesting that the enormous difference between the supplied elastic strength and the demanded elastic strength was furnished in great part by the many incursions of the steel - frame into its inelastic behaviour. It was also pointed out that in many - instances, the beam-to-column behaviour displayed large ductility values - (4 or more) during the long duration of the seismic event, by means of - - several stable hysteresis cycles.

However, it is difficult to assess the elastic rotations of the connections and consequently the deformational behaviour of such connections. -- This of course, deals both with the connection flexibility and the panel-zone-deformations, both of which were indeed responsible in a good part of the large drifts observed in some frames, but scarce energy dissipation.

A revision of the 1987 Building Code is now under consideration, to incorporate sound basis for the limit-state-design philosophy which accounts for the flexibility of the beam-to-column connection, in order to rely better in the actual reserve of inelastic strength of the steel frame.

The elastic strength of the frame should be sufficient to resist the design earthquake (10-year recurrence period) without significant non-structural damage, whereas the ductile inelastic behaviour of the frame, should act as a second-defense line, to resist extraordinary earthquake (50-year recurrence period), with some structural damage, but without collapse.

Here it will be helpful to evaluate the highly redundant type of construction represented by the Moment-Resistant-Space-Frames, with boxed columns and trussed girders...How much ductility does this type of construction provides when subjected to several cycles of extreme earthquake loading?... How does the collapse mechanism of this type of construction develops in -- sucssessive pulses of lateral loading?... What should the required strenght of the side fillet welds be in a boxed columns to avoid local buckling of the flange plates, as in the case of the Pino Suarez building complex?... -- and finally, what b/t ratios should be recommended for the boxed-columns plates, to allow a satisfactory hysteretical behaviour during long-duration earthquakes?...

The questions above stated represent only a few of the many unknows -- presently intriquing our understanding of the exceptionally good performance of many engineered steel buildings which fared quite well during the -- September 1985 earthquake in Mexico City, like the new Loteria Nacional- - building which had every argument to have been damaged (irregular shape, -- corner building, first story soft, about two seconds natural period, etc.), and had not a single window broken. And by the same talken, the required englightment of our knowledge to ascertain the causes of the damage and collapse suffered by similar types of buildings like the Pino Suarez Complex, in the same event.

Even though this topic has been partially focussed by some researchers (8), and numerous literature is already available on connection flexibility in Steel Frames (9), future research is urged on this matter.

REFERENCES

1. Departamento del Distrito Federal. "Normas de Emergencia para las Construcciones en el Distrito Federal". Octubre 1985.

2. Hanson R.D., Martin H.W., and Martinez-Romero, E., "Performance of Steel Structures in the September 19, and 20, 1985 Mexico City - - Earthquakes". Proceedings fo the AISC's National Engineering Conference, Nashville, Tenn. June 1986.

3. Martinez-Romero E. "The behaviour, overview of damage and Retrofit of Steel Buildings after the Earthquakes of September, 1985 in Mexico City". Proceedings of the Pacific Structural Steel Conference Auckland, N.Z. August 1986.

4. Martinez-Romero, E. "Damage assessment of Steel Buildings in Mexico City, after the earthquakes of 1985". Proceedings of the ASCE's International Conference: The September 1985 Earthquakes of Mexico City Lessons Learned and Factors involved". Mexico City, September 1986.

5. Bertero V.V. "The Mexico Earthquake of September 19, 1985 Performance of Building Structures". 2nd. International Conference of Engineering and Technology. Asociación Mexicana de Ingenieros Mecánicos y Electricistas and Society of Hispanic Professional Engineers. Mexico City, August 1986.

6. Brokken, S.T. and Bertero V.V., "Studies on the effects of Infills in Seismic Resistant Construction"., Report No. UCB/EERC-83/11. -- Earthquake Engineering Research Center, Univ. of California, Berkeley, CA. October 1981

7. Departamento del Distrito FEderal. "Reglamento para las Construcciones en el Distrito Federal, Edición 1987". (to be released in june 1987).

8. Rosenbaum, D.B. and Richard, R.M., "Flange Stress and Strains in W-sections welded to Box Columns". The Departament of Civil Engineering and Engineering Mechanics. University of Arizona. Tucson, -- Arizona. April 1986.

9. Chen, W.F., editor "Connection Flexibility and Steel Frames". - - Proceedings of a session sponsored by the Structural Div. of the ASCE. Detroit Michigan, October 1985.

MODIFIED EUROCODE 3 DESIGN RECOMMENDATIONS FOR HOLLOW SECTION LATTICE GIRDER JOINTS

J. Wardenier
Department of Civil Engineering
Delft University of Technology
Delft, The Netherlands

ABSTRACT

The Eurocode 3 recommendations have been extensively discussed in all international committees dealing with the design of hollow section joints. Based on these discussions and additional research, the IIW Subcie XV-E, the Cidect Joint Working Croup and ECCS-TC-10 have proposed a common modification. The modified recommendations are presented in this paper.

INTRODUCTION

The design recommendations for hollow section joints in Eurocode 3 have been based on the IIW-XV-E recommendations published in 1981. Based on the results of new research and re-analyses carried out in the last years, the IIW-XV-E recommendations for hollow section joints have been revised.

The various international committees (IIW-XV-E, Cidect Joint Working Group, ECCS-TC-10) working in this field have produced a common document with proposed modifications for Eurocode 3. These modified recommendations are given in this paper. Although the IIW/ Cidect recommendations (to be published) cover a larger range of joints for simplicity the modified Eurocode 3 recommendations cover the same types as previously.

1. Hollow section lattice girder joints

1.1 Scope

In this section design rules are given to determine the static strength of single plane joints in lattice structures composed of hollow sections (being round, square or rectangular in shape) or combinations of these with open sections.

The general provisions for the scope given in clause ..... apply, except that the wall thickness is limited to a minimum of 2.5 mm.

1.2 Definition

Single plane joints in lattice structures signify, in this context, connections between members situated in a single plane which transmit primarily axial loads.

The axial load distribution in a lattice girder may be determined on the assumption that the members are connected by pin-joints. Secondary moments in the joints, caused by the actual bending stiffness of the joints, and eccentricities within the limits given in subclause 1.4 can be neglected.

The gap g is defined as the distance measured along the length of the connecting face of the chord between the toes of the adjacent brace members. The overlap o_v is expressed as q/p x 100% as shown in figure 1.

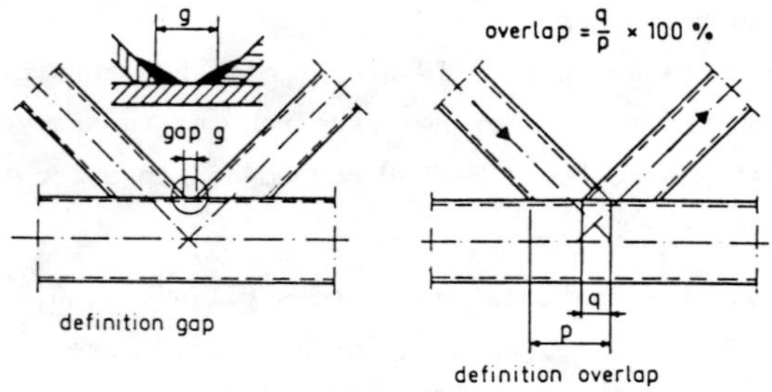

Figure 1. Gap and overlap of joints

1.3 Welds

In welded joints, the connection should be established over the entire perimeter of the hollow section by means of a butt weld, a fillet weld, or a combination of the two.

Fillet welds should satisfy the following conditions :

$a \geq t$ for $f_y \leq 280$ N/mm^2

$a \geq 1.2\ t$ for $f_y > 280$ N/mm^2

where a = throat thickness of the fillet weld to be determined in accordance with subclause .....

This requirement may be waived where departure from them can be justified in regard to strength and to deformation and/or rotational capacity.

1.4 Field of application

The calculation rules given in this section can be applied provided the following conditions are observed :

i. The angle between the chord and the brace member or between brace members should be at least 30°.

ii. Moments resulting from eccentricities may be neglected in calculating joint strength if the eccentricities are within the following limits :

$$-\ 0.55\ d_o \leq e \leq 0.25\ d_o$$

$$-\ 0.55\ h_o \leq e \leq 0.25\ h_o$$

where e = eccentricity
d_o = diameter of the chord
h_o = depth of the chord

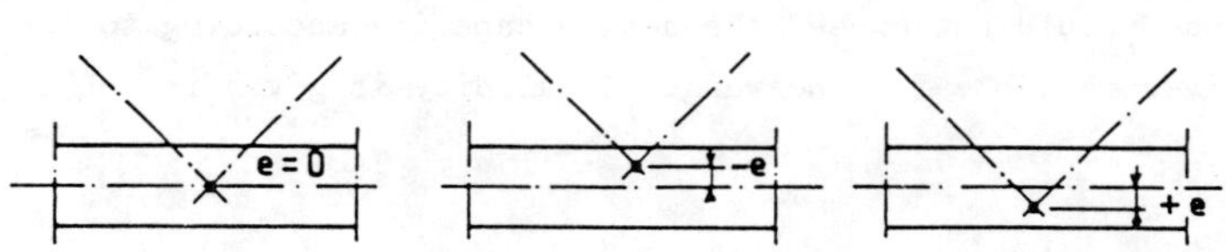

Figure 2. Eccentricity of joints

The members that meet at a joint must have their ends prepared in such a way that their cross-sectional shape is not modified.

1.5 Welded joints between circular hollow sections

The axial force(s) in the brace member(s) due to the design ultimate load should not exceed the design capacity according to the formulae given in Table 1.

1.6 Welded joints with square or circular hollow section brace members and a square hollow section chord.

The axial force(s) in the brace member(s) due to the design ultimate load should not exceed the design capacity according to the formulae given in table 2. Table 2a defines the range of validity of these formulae.

1.7 Welded joints between rectangular hollow sections

For joints between rectangular hollow sections and joints between square hollow sections outside the range of validity for 1.6 the design strength should be based on the following criteria where applicable :

i. chord face failure
ii. chord wall failure (yielding or instability)
iii. chord shear failure
iv. chord punching shear failure
v. brace failure with reduced effective width

The design strength may be determined according to: "Design recommendations for hollow section joints", IIW-document XV-...-87.

1.8 Welded joints with hollow section brace members and an I-section chord

The axial force(s) in the member(s) due to the design ultimate load should not exceed the design capacity according to the formulae given in table 3. The range of validity is given in table 3a.

TABLE 1

Design strength capacity of axially loaded welded joints between circular hollow sections

<table>
<tr><th>Type of joint</th><th colspan="2">Design strength (i = 1 or 2)</th></tr>
<tr><td>T, Y
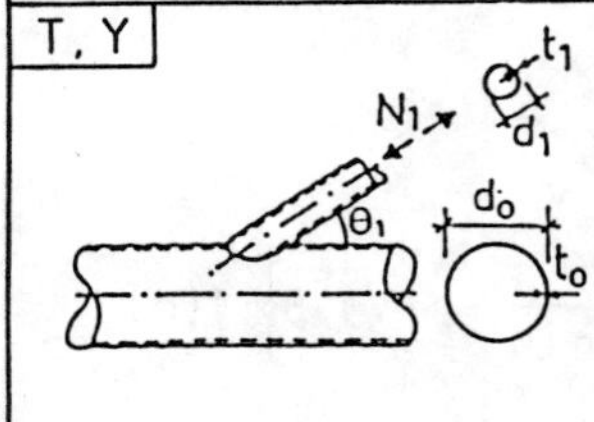
</td><td colspan="2">$\hat{N}_1 = \frac{f_{yo} \cdot t_o^2}{\sin \theta_1} \cdot (2.8 + 14.2\beta^2) \cdot \gamma^{0.2} \cdot (f(n^1)$</td></tr>
<tr><td>X
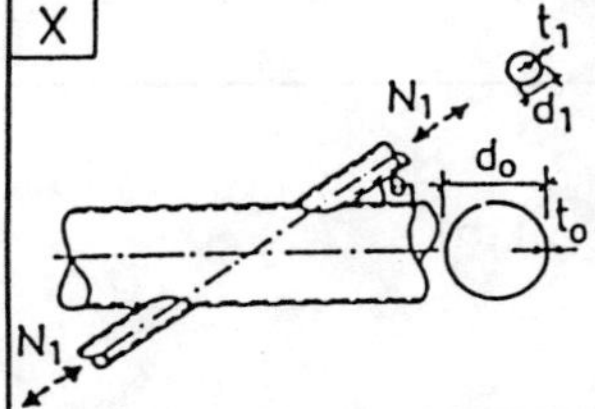
</td><td colspan="2">$\hat{N}_1 = \frac{f_{yo} \cdot t_o^2}{\sin \theta_1} \cdot \left(\frac{5.2}{1 - 0.81\beta}\right) \cdot f(n^1)$</td></tr>
<tr><td>K, N gap or overlap
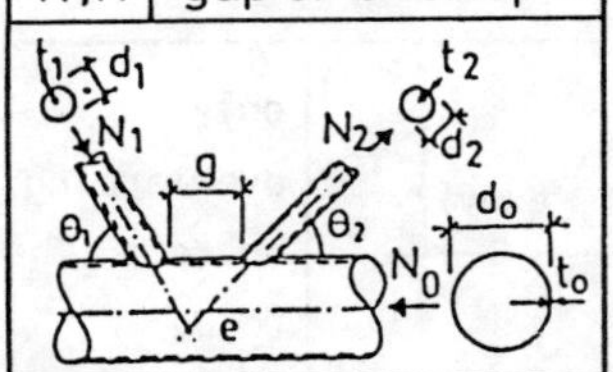
</td><td colspan="2">$\hat{N}_i = \frac{f_{yo} \cdot t_o^2}{\sin \theta_1} \cdot \left(1.8 + 10.2 \frac{d_1}{d_0}\right) \cdot f(\gamma \cdot g^1) \cdot f(n^1)$</td></tr>
<tr><td>General
Punching shear check for T, Y, X and K, N, KT joints with gap</td><td colspan="2">$\hat{N}_i = \frac{f_{yo}}{\sqrt{3}} \cdot t_o \cdot \pi d_i \cdot \frac{1 + \sin \theta_i}{2 \sin^2 \theta_i}$
i = 1, 2 or 3</td></tr>
<tr><td rowspan="2">Functions</td><td>$f(n') = 1.0$ for $n' \geq 0$
$f(n') = 1 + 0.3\,n' - 0.3\,n'^2$ for $n' < 0$
≤ 1.0</td><td>Note:
n' is negative for compression</td></tr>
<tr><td colspan="2">$f(\gamma, g') = (\gamma)^{0.2}\left[1 + \frac{0.024\gamma^{1.2}}{\exp.(0.5\,g' - 1.33) + 1}\right]$</td></tr>
<tr><td rowspan="2">Validity range</td><td colspan="2">$0.2 < \frac{d_i}{d_o} \leq 1.0$ | $\frac{d_i}{2t_i} \leq 25$ | $\gamma \leq 25$ | $-0.55 \leq \frac{e}{d_o} \leq +0.25$</td></tr>
<tr><td>For X-joints : $\gamma \leq 20$</td><td>$g \geq 2t_o$ or $O_v \geq 25\%$</td></tr>
</table>

TABLE 2

Design strength capacity of axially loaded welded joints with square or circular brace members and a square hollow section chord

Type of joint	Design strength (i = 1 or 2)	
T,Y,X 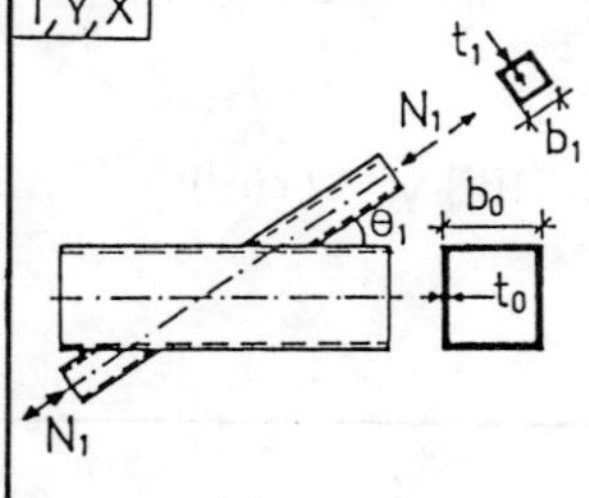	$\hat{N}_1 = \frac{f_{yo} \cdot t_o^2}{(1-\beta) \cdot \sin\theta_1} \left\{ \frac{2\beta}{\sin\theta_1} + 4\sqrt{1-\beta} \right\} \cdot f(n)$	
K,N, gap 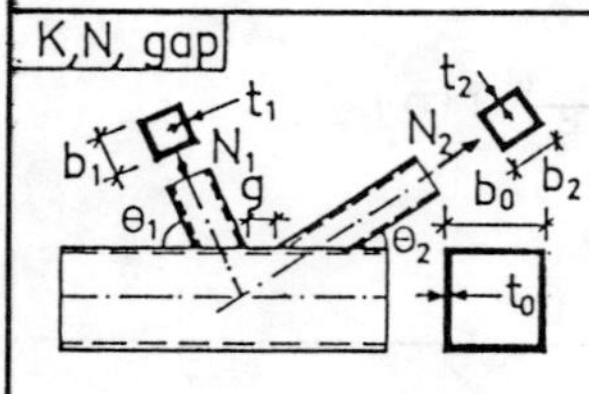	$\hat{N}_i = \frac{8.9 f_{yo} \cdot t_o^2}{\sin\theta_i} \cdot \beta \cdot \gamma^{0.5} \cdot f(n)$	
K,N, overlap 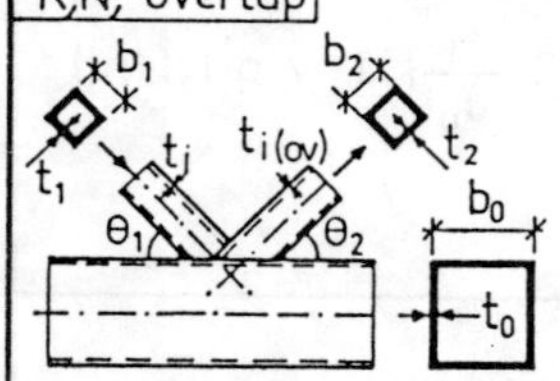	100% overlap $\hat{N}_i = f_{yi} \cdot t_i \cdot \left\{ 3b_i - 4t_i + b_{e(ov)} \right\}$ $50\% \leq Ov < 100\%$ $\hat{N}_i = f_{yi} \cdot t_i \left\{ 2b_i - 4t_i + b_e + b_{e(ov)} \right\}$	only overlapping brace to be checked
Joints with circular brace members	multiply the formulae with $\frac{\pi}{4}$ and replace b_i by d_i (i = 1 or 2)	
Functions	$f(n) = 1.0$ for a tension force in the chord $f(n) = 1.3 - \frac{0.4}{\beta}\lvert n \rvert$ for a compression force in the chord ≤ 1.0 $b_e = \frac{10}{b_o/t_o} \cdot \frac{f_{yo} \cdot t_o}{f_{yi} \cdot t_i} \cdot b_i$ $b_{e(ov)} = \frac{10}{(b_i/t_i)_{ov}} \cdot \frac{(f_{yi} \cdot t_i)_{ov}}{f_{yi} \cdot t_i} \cdot b_i$	
Validity range	See Table 2a.	

TABLE 3
Design strength capacity of axially loaded welded joints between hollow section braces and an I or H section chord

Type of joint	Design strength (i=1or 2)	
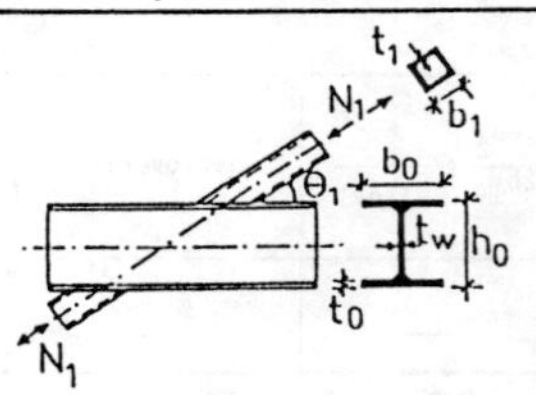	$\hat{N}_1 = \frac{f_{yo} . t_w . b_m}{\sin \Theta_1}$ ― ― ― ― ― ― ― ― $\hat{N}_1 = 2 f_{y1} . t_1 . b_e$	
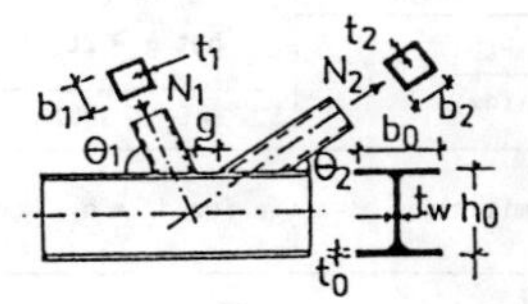	$\hat{N}_i = \frac{f_{yo} . t_w . b_m}{\sin \Theta_i}$ $\hat{N}_i = 2 f_{yi} . t_i . b_e$ ― ― ― Note * ― ― $\hat{N}_i = \frac{0.58 . f_{yo} . A_Q}{\sin \Theta_i}$ $\hat{N}_{o(gap)} \leq (A_o - A_Q) . f_{yo} + A_Q . f_{yo} \sqrt{1 - (\frac{Q}{Q_p})^2}$	
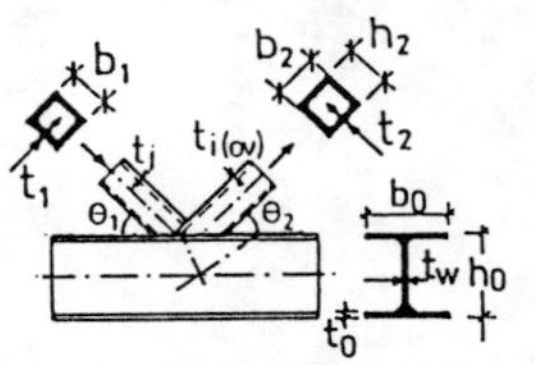	100 % overlap $N_i = f_{yi} . t_i . \{2h_i - 4t_i + b_i + b_{e(ov)}\}$	only overlapping brace to be checked
	$50\% \leq Ov < 100\%$ $N_i = f_{yi} . t_i . \{2h_i - 4t_i + b_e + b_{e(ov)}\}$	
Functions	RI-joints: $b_m = \frac{h_i}{\sin \theta_i} + 5 (t_o + r_o)$ $b_m \leq 2t_i + 10 (t_o + r_o)$ CI-joints: $b_m = \frac{d_i}{\sin \theta_i} + 5 (t_o + r_o)$	
	$b_e = t_w + 2r_o + 7 \frac{f_{yo}}{f_{yi}} \cdot t_o$	$b_{e(ov)} = \frac{10}{(b_i/t_i)_{ov}} \cdot \frac{(f_{yi} \cdot t_i)_{ov}}{f_{yi} \cdot t_i} \cdot b_i$
	$A_Q = A_o - (2 - \alpha) b_o \cdot t_o + (t_w + 2r_o) t_o$ RI-joints : $\alpha = \sqrt{\frac{1}{1 + \frac{4g^2}{3t_o^2}}}$ CI-joints : $\alpha = 0$ $Q = (N_i \cdot \sin \theta_i)_{max}$ $Q_p = 0.58 f_{yo} \cdot A_Q$	
Note *	Not to be checked if : $0.75 \leq (\frac{d_1}{d_2} \text{ or } \frac{b_1}{b_2}) \leq 1.33$; $\beta \leq 1 - 0.03 \gamma$ and $g' \leq 20 - 28 \beta$	
Validity range	See Table 3a.	

TABLE 2a
Range of validity of Table 2

Type of joint	Joint parameters (i = 1 or 2)					
	$\frac{b_i}{b_o}$	$\frac{b_i}{t_i}$; $\frac{d_i}{t_i}$ compression	$\frac{b_i}{t_i}$; $\frac{d_i}{t_i}$ tension		$\frac{b_1+b_2}{2b_1}$; $\frac{b_i}{b_{1(ov)}}$	gap/overlap
T,Y,X	$0.25 \le \frac{b_i}{b_o} \le 0.85$*	$\frac{b_i}{t_i} \le 1.25\sqrt{\frac{E}{f_{yi}}}$ ≤ 35	$\frac{b_i}{t_i} \le 35$	* $10 \le \frac{b_o}{t_o} \le 35$	—	—
K and N with gap	$\ge 0.1 + 0.01\,\frac{b_o}{t_o}$			* $15 \le \frac{b_o}{t_o} \le 35$	* $0.6 \le \frac{b_1+b_2}{2b_1} \le 1.6$ *	$0.5(1-\beta) \le \frac{g}{b_o} \le 1.5(1-\beta)$ but $g \ge 2t_o$
K and N with overlap and $t_{i(ov)} \ge t_i$	≥ 0.25	$\frac{b_i}{t_i} \le 1.1\sqrt{\frac{E}{f_{yi}}}$		$\frac{b_o}{t_o} \le 35$	$\frac{b_i}{b_{i(ov)}} \ge 0.75$	$50\% \le O_v \le 100\%$
Joints with circular braces	$0.4 \le \frac{d_i}{b_o} \le 0.8$	$\frac{d_i}{t_i} \le 1.5\sqrt{\frac{E}{f_{yi}}}$	$\frac{d_i}{t_i} \le 50$	further limitations as above for $d_i = b_i$		

* Outside this range of validity other failure criteria may be governing, i.e. punching shear, effective width, side wall failure, chord shear or local buckling.

TABLE 3a
Range of validity of Table 3

Type of joint	Joint parameters (i = 1 or 2)				
	$\frac{b_i}{b_{i(ov)}}$	$\frac{h_i}{b_i}$	b_i/t_i ; d_i/t_i compression	b_i/t_i ; d_i/t_i tension	$\frac{h_w}{t_w}$
T,Y,X	--	--	$\frac{b_i}{t_i} \le 1.1\sqrt{\frac{E}{f_{yi}}}$	$\frac{b_i}{t_i} \le 35$	$\frac{h_w}{t_w} \le c\sqrt{\frac{E}{f_{yo}}}$
K and N with gap	--	1.0	$\frac{d_i}{t_i} \le 1.5\sqrt{\frac{E}{f_{yi}}}$	$\frac{d_i}{t_i} \le 50$	with: c = 1.2 for X joints c = 1.5 for T, Y, K and N joints
K and N with overlap	≥ 0.75				

SYMBOLS

A_i	:	cross sectional area of member i (i = 0, 1, 2)
A_Q	:	effective cross sectional area of the chord for shear
E	:	elastic modulus of steel
$N_{o_{gap}}$	:	axial load in the cross section of the chord at the gap
$\hat{N}_i$	:	joint design strength capacity based on load in member i (i = 1 or 2)
O_v	:	overlap
Q	:	shear force ($Q = N_i \cdot \sin \theta_i$)
Q_p	:	shear yield capacity of a section
a	:	throat thickness of a fillet weld
b_i	:	external width of a square or rectangular hollow section (RHS) for member i (i = 0, 1, 2)
b_e	:	effective width for brace to chord connection
b_{ep}	:	effective punching shear width
$b_{e(ov)}$	:	effective width for overlapping brace to overlapped brace connection
b_m	:	effective width for the web of the chord
c	:	coefficient
d_i	:	external diameter of a circular hollow section (CHS) for member i (i = 0, 1, 2)
e	:	eccentricity of a joint
f_o	:	maximum compressive stress in the chord due to axial force and bending moment
f_{op}	:	maximum compressive stress in the chord excluding the stress due to the horizontal brace load components
f_{yi}	:	design value for the yield stress in member i (i = 0,1,2)
g	:	gap between the braces of a K or N joint
g'	:	gap g divided by the wall thickness of the chord
h_i	:	external depth of a section, for member i (i = 0, 1, 2)
h_w	:	depth of the web of an I or H section chord
i	:	integer used to denote member of joint, i = 0 designates chord and i = 1-2 the brace members. Normally i = 1 refers to the strut and i = 2 to the tie.
n, n'	:	$n = \frac{f_o}{f_{yo}}$; $n' = \frac{f_{op}}{f_{yo}}$
(ov)	:	index used for overlapped brace
r_o	:	radius of an I- or H-section
t_i	:	wall thickness of member i (i = 0, 1, 2)
t_w	:	web thickness of an I- or H-section
α	:	factor giving the effectiveness of the chord flange for shear
β	:	(average) brace to chord diameter ratio or width ratio $(\frac{d_1}{d_o}, \frac{d_1 + d_2}{2d_o}, \frac{b_1}{b_o}, \frac{b_1 + b_2}{2b_o})$
γ	:	half chord width, or half chord diameter to wall thickness ratio $(\frac{d_o}{2t_o}$ or $\frac{b_o}{2t_o})$
γ_m	:	joint coefficient
θ_i	:	included angle between a brace member (i = 1 or 2) and the chord

THE RESPONSE OF STEEL STRUCTURES WITH SEMI-RIGID CONNECTIONS TO SEISMIC ACTIONS

Prof. Dr.-Ing. G. Sedlacek, Dipl.-Ing. M-S Koo
Lehrstuhl für Stahlbau, RWTH-Aachen
W. -Germany

and

Prof. Dr. G. Ballio
Department of Structural Engineering,
Politecnico of Milan
Italy

1. Method for the aseismic design in Eurocode 8

The Eurocode 8 provides unified rules for the aseismic design of buildings, which base on the response spectrum method. Steel structures according to these rules are designed as follows :

A structure is modelled as a linear elastic dynamic system, along which the masses due to dead weight and cooperating loads are distributed.

The natural vibration periods Ti are calculated and the response values R(Ti) are taken from the response spectrum for the relevant soil condition.

For a system as in fig. 1 a static horizontal force H, which acts parallel to the assumed seismic direction and is applied at the centres of the masses, may be determined with equation (1)

$$H = 2\ m.\ R(Ti).\ A\ /\ q \qquad (1)$$

and a safety assessment in view of the limit of the assumed elastic model, i.e. the first attainment of yield stress in the extreme fibres of the cross-sections, may be conducted. In equation (1) A is the maximum ground acceleration applicable to the relevant seismic zone and q is the behaviour factor.

This behaviour factor takes account of the energy dissipation due to the real nonlinear structural behaviour, which depends on the topology and loading of the structural system, the second order effects and the plastic rotations of the members and their connections in case of local ductility.

The behaviour factor also allows for the real dynamic limit states in relation to the fictitious limit state of initial yielding assumed in the safety assessment. Hence the q -factor performs a correction factor for the linear dynamic model and its limit states.

In this paper the influence of semi-rigid connections to the q -factor will be treated.

2. Method for the calculative determination of the q -factors

A rational procedure for the determination of q -factors has been proposed by Ballio, Perotti, Rampazzo, Setti /2/.

According to this method the q -factor is determined by its definition in Eurocode 8, which is represented by the equations (2-a, 2-b, 2-c) :

$$a = q_a \cdot A \qquad (2\text{-a})$$

$$\delta = q_\delta \cdot \delta_{el} \qquad (2\text{-b})$$

$$q_a = q_\delta = q \qquad (2\text{-c})$$

In equ.(2-a) the q_a -factor describes the relation between the maximum ground acceleration a at the real limit state and A as applied in the safety assessment with the linear response method ; in equ. (2-b) the q_δ -factor is used to derive the deflections δ at the real limit state from the deflections δ_{el} calculated for A with the linear response method. By the definition in Eurocode 8 both the q_a and the q_δ -factors are equal, equ(2-c).

This definition results in a linear dependence between the applied maximum ground acceleration and the maximum deformation of the structure, or between q_a and q_δ as demonstrated in fig. 2.

This linear dependence can be confirmed by calculating the q_a - q_δ graph for several q_a - values with the time history method the accelerogram being defined such that it fills the assumed response spectrum, see fig. 3, and an unlimited elastic constitutive material law and no second order effects being taken into account.

In applying a more realistic elasto-plastic material model and a second order theory the time-history method produces a nonlinear relationship between q_a and q_δ with values $q_\delta < q_a$ for small and moderate q_a -values and $q_\delta > q_a$ for greater q_a-values.

The cross point of the nonlinear graph obtained by the time-history method with the straight line gives maximum q -values that fulfill the definition of equ. (2-a), (2-b), (2-c) on the safe side.

Though this definition of the q -values is rather artificial, the q -values so obtained normaly are close to the physical limits caused by sway down effects and represent safe side values in view of other limit state as lowcycle-fatigue or fracture of connections, fig. 4.

These limits of course have to be checked separately, unless their attainment is prevented by special detailing rules, e.g. fracture strength of the net section > yield strength of the gross section and maximum b/t - ratios.

In the table 1 some numerical values of the q -values for single and double storey-frames, calculated for different natural periods T(sec) of the structures and different strong motion periods t(sec) of the applied accelerograms are indicated and compared with the q -values as proposed in the first draft of Eurocode 8.

The probleme yet to be solved is how these q -values are controlled by the behaviour of semi-rigid connections.

3. Modelling of semi-rigid connections in view of their behaviour in structual systems excited by earthquakes.

The hysteretical behaviour of bolted end-plate connections in terms of moment-rotation dependences has been experimentally determined /4/. In assuming stable loops the hysteretical behaviour can be idealized according to fig. 6 and be modelled by the truss model of Stutzki, see fig. 7.

In applying this truss model to frames with semi-rigid connections the q -values can be determined in the way described obove.

In fig. 8 for a single storey frame the δ-a - dependence is calculated for rigid connections and semi-rigid connections.

This example demonstrates that semi-rigid connections cause a greater deflections and hence higher natural vibration periods Ti which may result in a smaller response R(Ti).

In addition the q -values are greater (by about 20 %) than those for rigid connections.

A third advantage may be realised by the fact that the ultimate moment capacities of rigid connections are smaller than those of the connected beams and columns, which makes them passiv controlling devices that protect the columns and the beams from being overstressed.

This may be the philosophy for developing specific semirigid connections for the aseismic design, e.g. friction controlled connections with slotted holes etc, that behave rigidly due to friction grip for earthquake events in the serviceablity range and assure energy dissipation by slip-effects without yielding in the ultimate range, with background safety due to "strain-hardning effects" by the bearing forces at the ends of the slotted holes, fig. 9 /6/, /7/.

References

/1/ EuroCode 8 : common unified rules for structures in seismic regions, Commission of the European Communities, Eur. 8850, 1983

/2/ Ballio, G. / Pertti, F. / Rampazzo, L. / Setti, P.: Determinazione del coefficiente di struttura per costruzioni metalliche sogette a carichi assiali. 2. convegno nazionale l' ingegneria sismica in italia, rapallo 1984

/3/ Koo, M-S : Untersuchung zum Einfluss der Bebendauer,Strukturausbildung und des Verhaltens von Verbindungen auf die Sicherheit von Stahlbauten bei Erdbebenbeanspruchungen, Diss. RWTH-Aachen 1987

/4/ Indagina sperimentale sulla resistenza e duttilita di collegamenti strutturadi, institute of construction theory of the university of Pisa, Monografia 5

/5/ Stutzki, Ch. : Traglastberechnung räumlicher Stabwerke unter Beräcksichtigung verformbarer Anschlüsse, Diss. RWTH-Aachen, 1982

/6/ Roik, K. / Dorka, U. : Vibration Control of structures under earthquake loading by friction-grip elements, Part 1: How to put brakes on a building, Earthquake eng. and struc. dyn. (demnächst)

/7/ Roik, K. / Dorka, U. : Optimales Systemverhalten unter Erdbebenbelastung durch Reibelemente mit "dimensionierbarer Dissipation", Ruhr-Uni Bochum SFD 151-Berichte Nr.6, März 1987

Draft of Eurocode 8/1984		
Condition	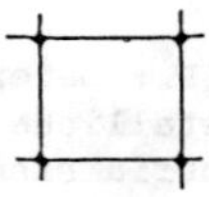	q
Ductility level	plastic reserves	
III	$\alpha > 1{,}2$	6
II	$1{,}1 \leqslant \alpha \leqslant 1{,}2$	4
I	$1{,}0 \leqslant \alpha < 1{,}1$	2
Elastic		1

Results of the numerical determination of q-factors

strong motion period / nat. vibration period	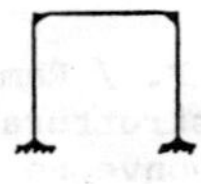			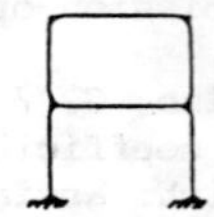		
t(sec) / T(sec)	4	7	14	4	7	14
1,5	8	4	3,5	7	6	5
1,0	5,5	3,5	3,5	4,5	3,5	3,5
0,6		2,5			3,5	
0,4		2,0			2,5	
0,2		1,5			2,0	
Elastic		1			1	

Draft of Eurocode 8/1984		
Condition	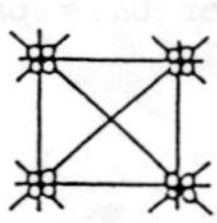	q
Ductility level		
III	$N > 0{,}8 N_y$ for at least 50% of all diagonals	5
II	relat. slenderness $\bar{\lambda} \leqslant \sqrt{2}$	3
I	fracture ≫ yielding $R_u > R_y$	2
Elastic		1

Results of the numerical determination of q-factors

strong motion period / nat. vibration period	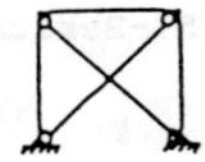			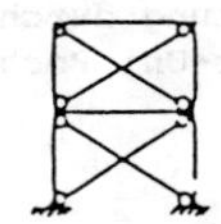		
t(sec) / T(sec)	4	7	14	4	7	14
1,5						
1,0						
0,6						
0,4						
0,2		1,5			2,0	
Elastic		1			1	

Table 1: Comparison of q-values according to the draft EC 8 /1984 and as numerically determinated

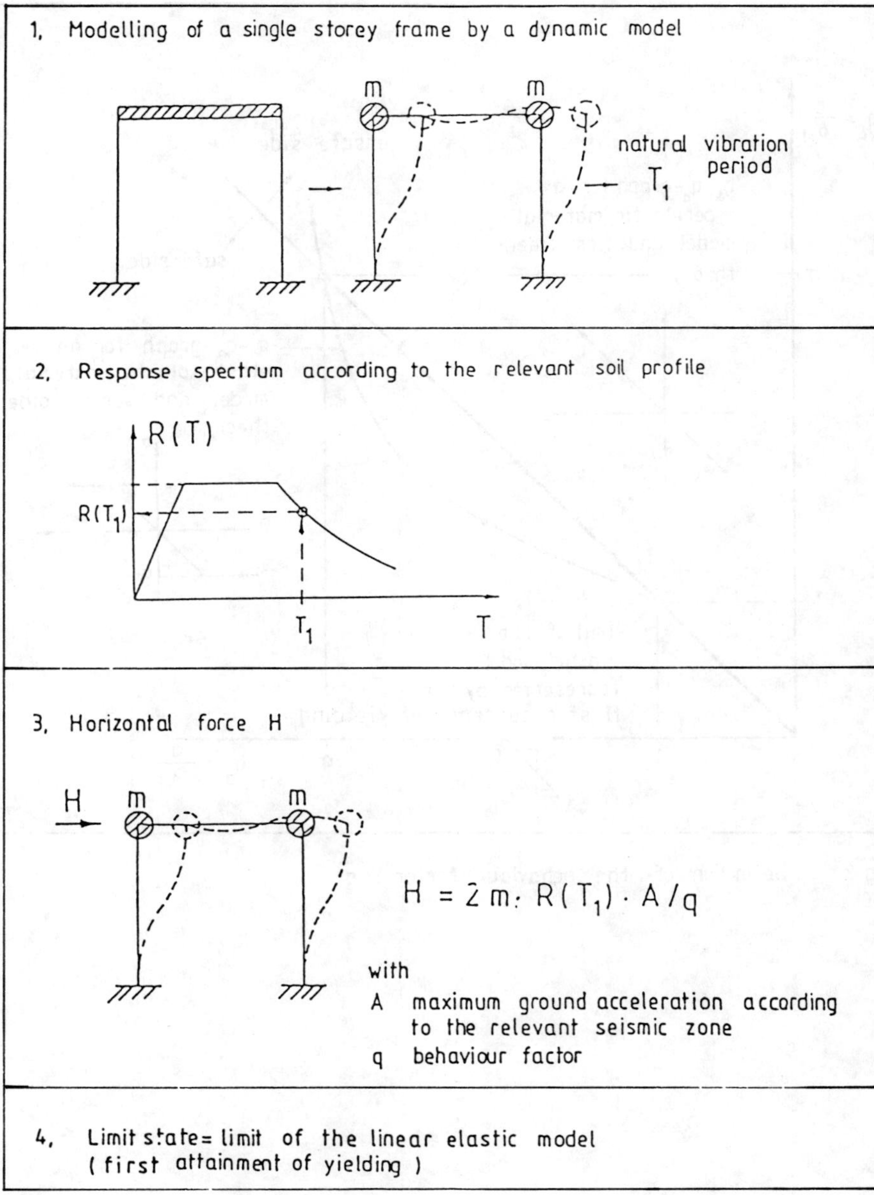

fig 1: The response spectrum method and its elements

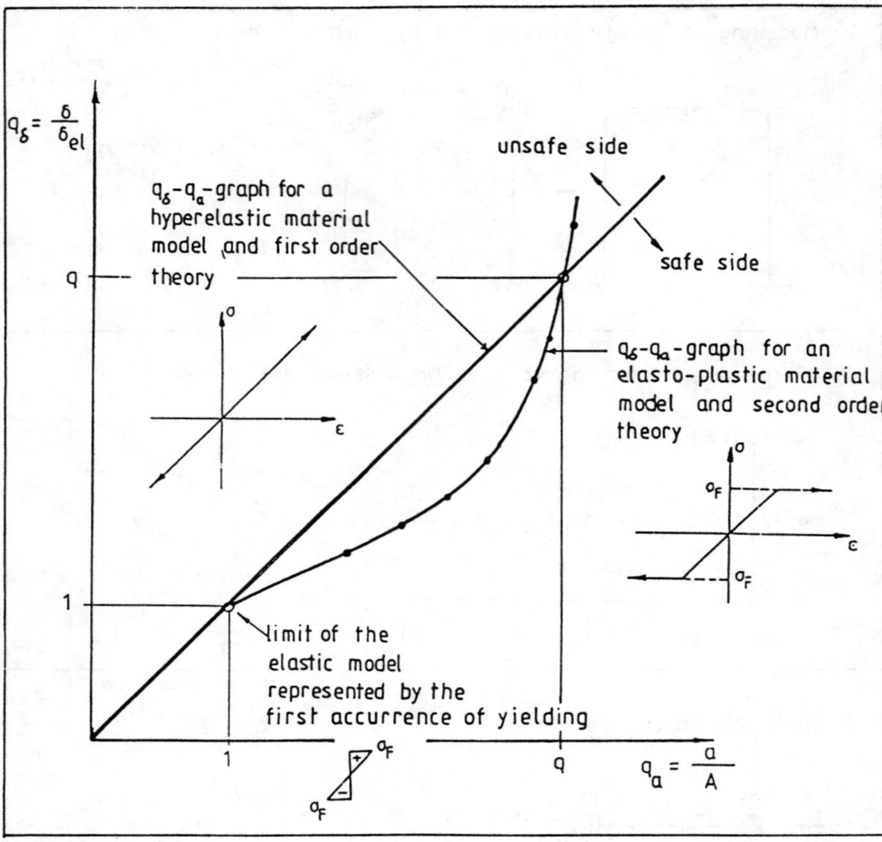

fig 2: Definition of the behaviour factor q

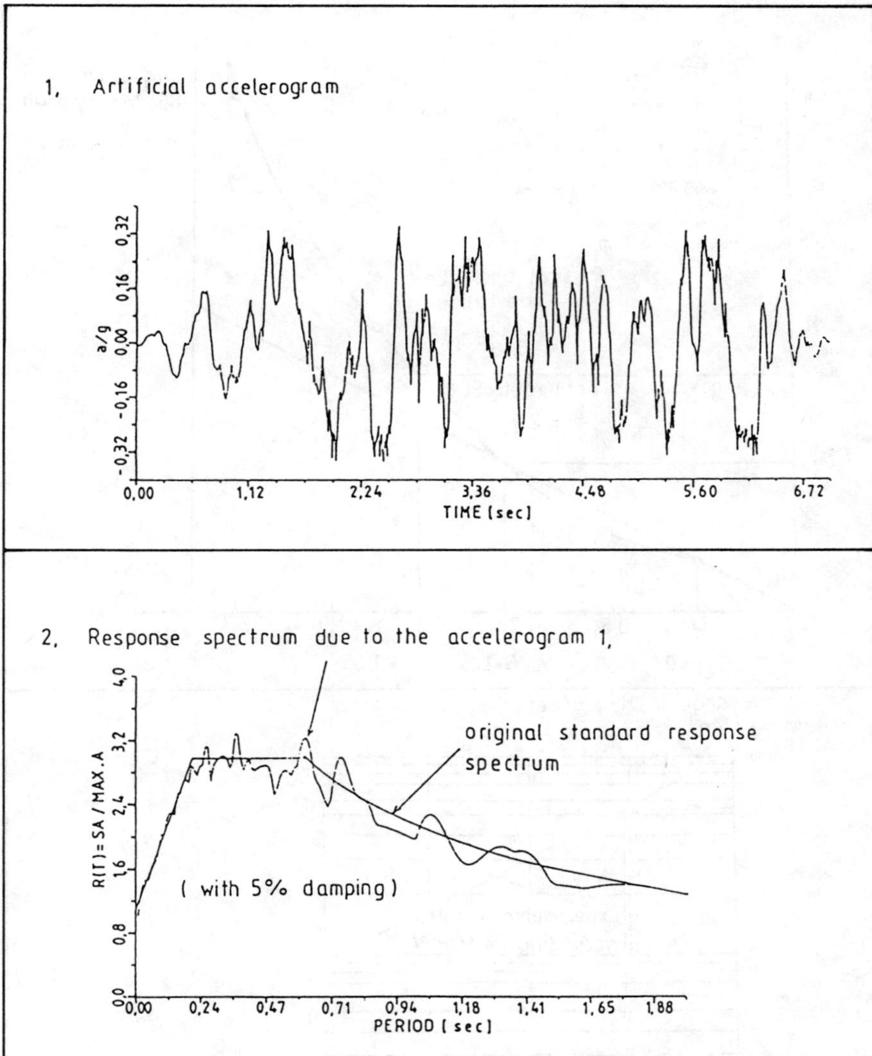

fig 3: Example for an artificial accelerogram with a strong motion period of 7sec that fills the response spectrum

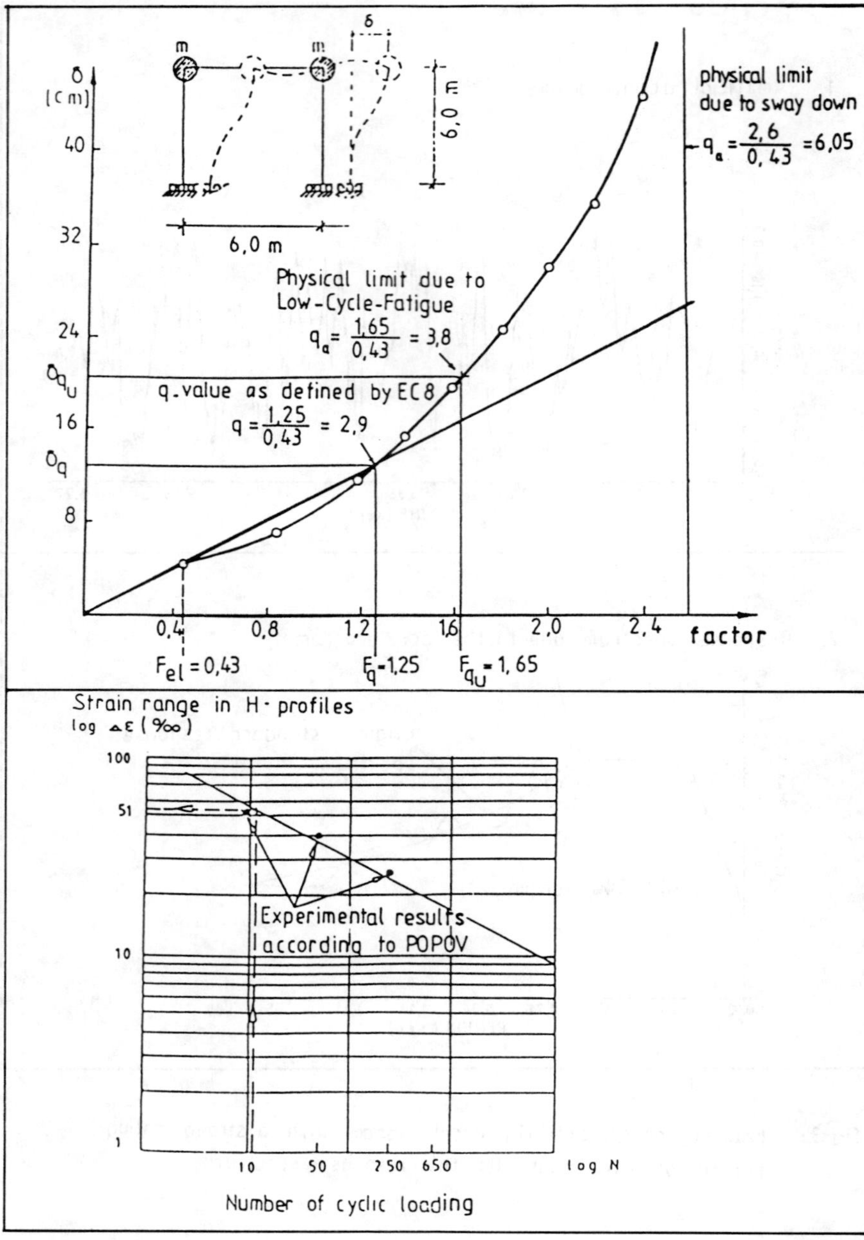

fig 4 Comparison between the q-value as defined in Eurocode 8 and the real physical limits

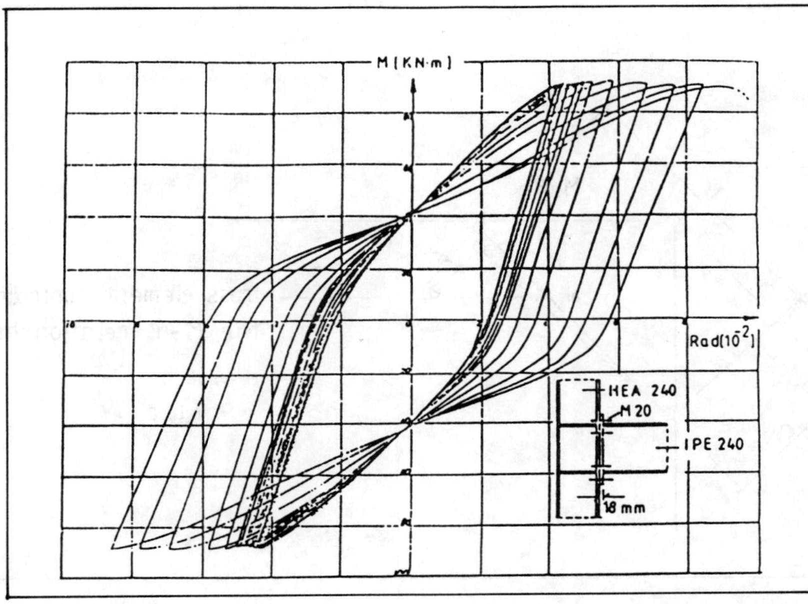

fig 5 : Moment-rotation hysteresis for bolted end-plate connections, from /4/

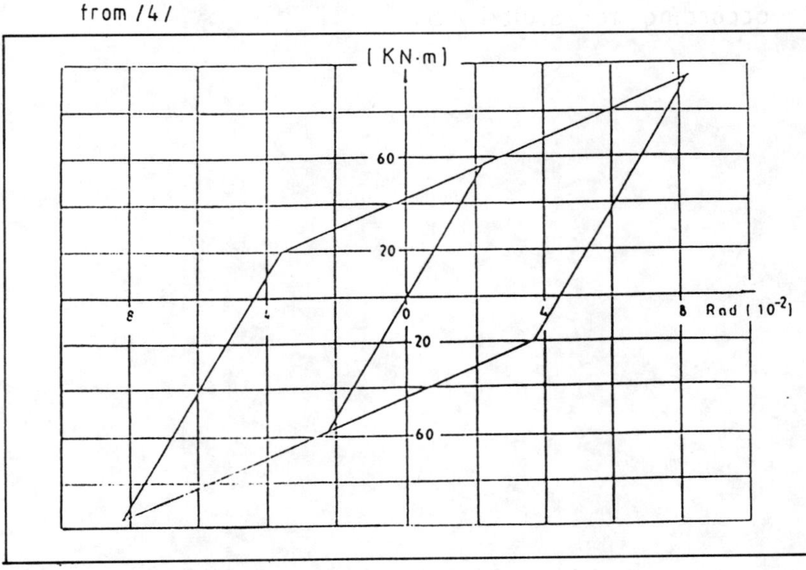

fig 6 : Idealisation of the hysteresis in fig. 5

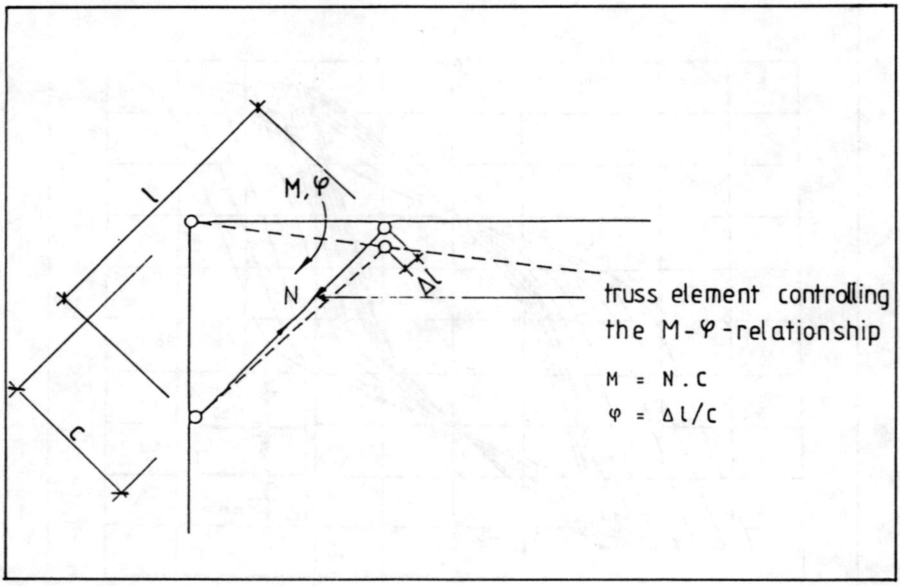

fig7: Modelling of semi-rigid connections by substitute trusses according to Stutzki /5/

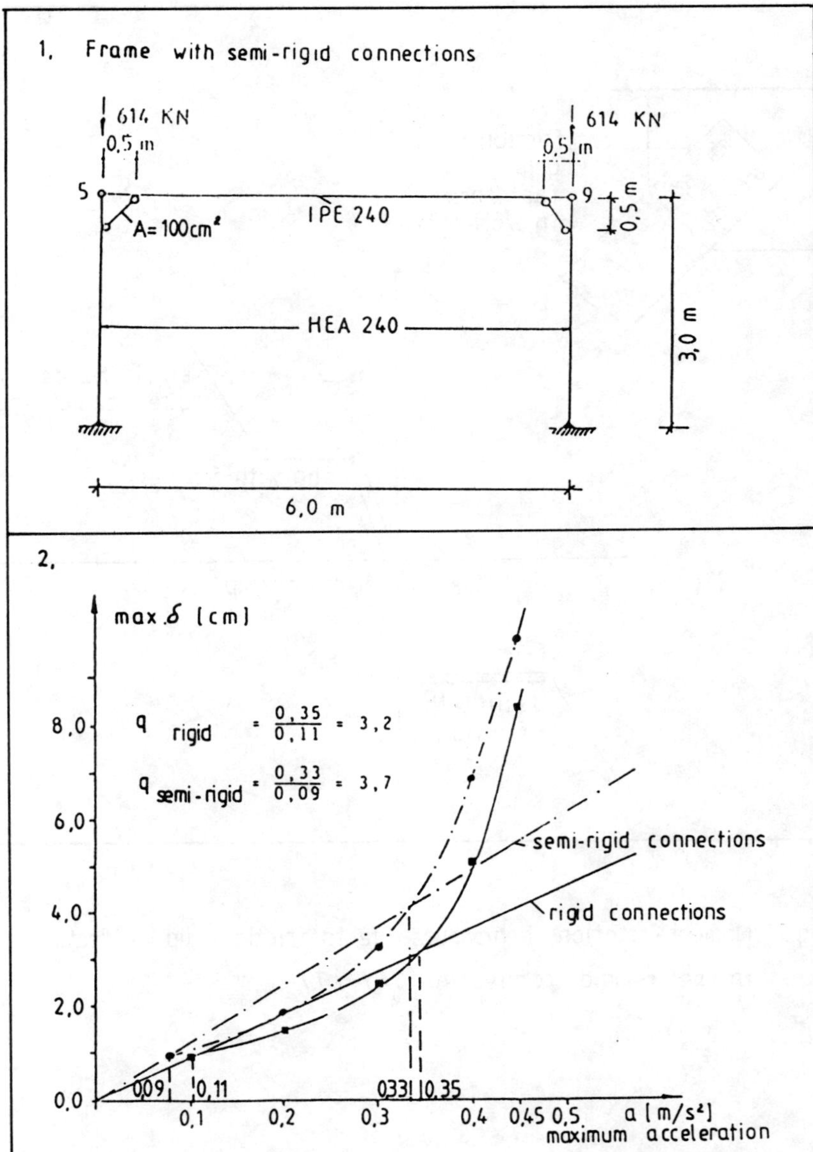

fig 8 : Comparison of the δ - a graphs for a frame with rigid and semi-rigid connections

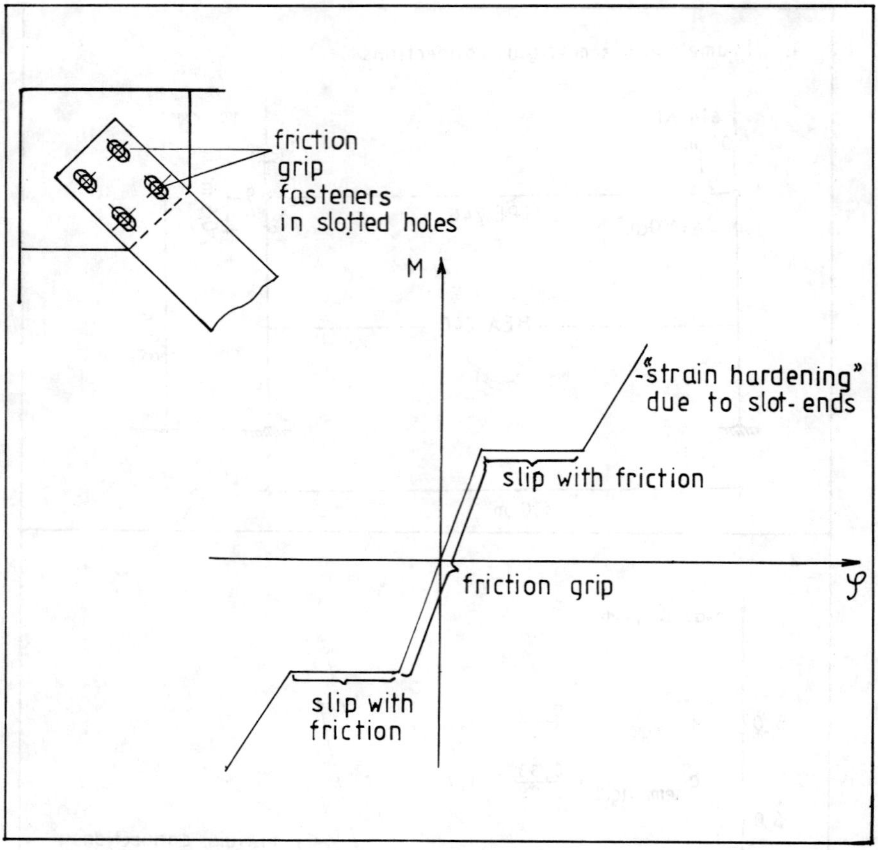

fig 9: Moment-rotation behaviour due to friction slip effects in semi-rigid connection" /6/, /7/

DISCUSSION OF DESIGN REQUIREMENTS PAPERS

Numerous points were raised regarding individual connector capacities and the corresponding deformations that can be anticipated. For example, the American design criteria for bolts in bearing imply an ultimate deformation of approximately 12 mm (0,5 inches). This is not a serviceability limit, since it simply happens to be the deformation that will have taken place by the time the bearing failure occurs. However, it was agreed that various ultimate as well as serviceability limit states should be examined.

Special note was made of the block shear failure mode ; this is apparently only given in North American standards, and not in Europe. This concept has now been extended to gusset plate and other tension-type connections, both for bolted and welded joints.

Many discussers felt that the only realistic way in which semi-rigid design concepts could be entered into the codes would be with the use of limit states principles. The detailed structural design example that was presented illustrated this very clearly, not the least because serviceability criteria are currently very poorly defined for flexibly-connected frames. Nevertheless, serviceability in some form must be addressed, although a number of discussers questioned the wisdom of doing so through design code requirements.

Many comments were made regarding the introduction of new steel types and grades on the market, and how current design standards are utilized in such cases. Some concern was expressed that past research, which was mostly based on mild structural steels, had been used to prepare the design criteria that now are being applied for materials with significantly different properties. Although there was general consensus that this type of recognition is warranted, it was indicated that as long as the ratio of yield stress to ultimate stress is adequate (to facilitate stress redistribution), and the material ductility and weldability are satisfactory, it should be acceptable to use the current codes with new steel grades. However, some form of occasional verification may be warranted. By the same token, it was also suggested that design standards should establish scopes and extent of coverage, to add to their background information. This is particularly important for codes that address seismic design requirements essential, whose attendant material and structural ductility needs are essential to the adequate performance of the structure.

SESSION N°6

DATA BASE ORGANIZATION

NEEDS, CONCEPTS AND ORGANIZATION : A DISCUSSION

As outlined in the Introductory Notes of these Proceedings, the increasing amount of information on the response of various types of beam-to-column and column footing connections makes it imperative to provide for some organization and channeling of the findings, to maximize usage and minimize overlap. The need for harmonizing the data gathering was stressed by all participants, to the effect that the manner of presentation, the interpretation of the results and the views towards their eventual application should be kept in mind. To that end it was felt that accessing the data base through a computer network such as E.A.R.N., for example, would be of significant potential.

It was observed that one of the problems of using current data is the highly variable quality of the manner of reporting. Thus, too many reports do not offer sufficient details for the results that are presented. For example, connection fabrication details are often omitted. Similarly, although a large amount of data is available, important gaps do exist, especially when it comes to three-dimensional behaviour and strength. With a measure of data collection coordination, especially in the form of standardized data sheets (as discussed in the following), the efforts of disseminating the available and developing information can be vastly enhanced. Several discussers noted the data collection efforts that had already been started, but indicated that these were strictly just that, with relatively few attempts at analyzing or unifying the findings.

It was agreed that the eventual aim of any data collection organization should be to assist in the development of design requirements. Without such, it cannot be expected that the advanced concepts of frames with semi-rigid connections can get much beyond being the province of the very sophisticated designer.

Several points were made regarding the need to avoid any form of duplication of research efforts, and how an effective and easily accessible data base can serve this purpose. This should also be extended to inform potential users about on-going research work.

Summarizing, it was agreed that without a well-organized and-maintained data base, including clearly established limitations/extent of the collection effort, significant advances could not be expected to occur at a reasonable pace.

The data sheet format and possible connection classification approach that are presented in the following address these needs in detail. It is noted that it was agreed to restrict the data collection effort to beam-to-column connections and column footing connections under monotonic loading conditions. Data on fasteners would not be a part of the collection, nor would fatigue characteristics. Although both of the latter are needed, the limitations are necessary to make the task manageable.

DATA SHEET FORMAT PRESENTATION

Reidar Bjorhovde
Department of Civil Engineering
University of Pittsburgh
Pittsburgh, Pennsylvania, U.S.A.

and

Jacques Brozzetti
C.T.I.C.M.
Domaine de Saint Paul
Saint Rémy lès Chevreuse, France

ABSTRACT

This brief paper provides a discussion of the information that should be provided by a beam-to-column connection data sheet. It is emphasized that such sheets may be put to use by other researchers in similar or related studies, but also by designers for actual construction projects. Details of the connection that has been tested are given, along with material properties (test values) of members and connectors, drawing(s) of the connection test setup and loading procedure, and clear indications of how loads, moments, deflections and rotations were obtained. The data sheet should also give the complete moment-rotation (or load-deflection) curve, indicating service and ultimate loads and deformation magnitudes. Finally, it is suggested that the initial stiffness of the connection be provided, preferably on the basis of a partial unloading curve. A sample filled-out data sheet is given at the end, using the format that has been adopted by C.T.I.C.M.

INTRODUCTION

On the basis of the papers that have been presented at this conference, as well as the research data collection efforts that are underway in several locales around the world, it is clear that much information is currently available regarding the strength and behaviour of a number of types of beam-to-column connections. It is important to gather these data into a coherent and unified basic format, in order that several primary objectives can be attained, as follows :

(1) On a world-wide basis, to obtain a data base of experimental results for beam-to-column and column bases connections.

(2) With a view to the development of strength and design criteria and equations, to promote the mutual recognition of test data and the techniques for evaluation and modeling, in order that code writers and designers around the world can have access to the largest possible assembly of findings.

(3) In order to establish a reliable data base as well as procedures for the exchange of experimental results, to arrive at a format of data presentation that will facilitate their interpretation and use.

(4) On an ongoing basis, to evaluate research and development needs and to update the requirements as additional data become available.

(5) By noting the data that are required for subsequent usage, to promote improved methods of testing and reporting.

DATA SHEET PRESENTATION

The attached Data Sheet is described in the following. It is emphasized that it represents one possible format for such sheets ; however, the clarity of the presentation of the data makes it easy to use.

The first page contains three primary types of information :

(1) The source of the test data that are presented, in particular, the reference document that gives additional information about the test. It is critical that sufficient retrieval data are given, in order that the researcher/designer can obtain further information. Thus, authors, report or paper title, location, date, and where a copy may be obtained should be indicated.

(2) The steel grades of the different components of the connection (i.e. beam, column, connection plates and/or angles, bolts and welds). If at all possible, the material properties that are shown should be based on tests, and the applicable material standard (e.g., ASTM, CSA, NF, DIN, BS, etc.) and number should be given.

(3) Detailed measurements and geometric characteristics of the

connection and its elements. If at all possible, these data should be based on measurements, rather than referring to nominal sizes. Shape designations (e.g. HEA 200) should be noted wherever applicable. If measured sizes are not available, nominal figures must be given, and it should be noted that such is the case.

The second Data Sheet page summarizes the details of the test setup and the connection itself. It is important to indicate the following :

(1) Exact load application and measurement points, including relative distances.

(2) Form of load application (static or dynamic ; monotonic or cyclic) and measurement (load cell or otherwise).

(3) Forms and instruments of strain and displacement measurement and corresponding directions.

(4) If rotations are given, indicate how they were obtained, including any computational details.

(5) Manner of test control (i.e. load, stroke or displacement control).

(6) Detailed testing procedure.

The third Data Sheet page gives the load-deformation curve, including pertinent coordinates for points such as the end of the linear range, the ultimate load-carrying capacity, and the initial stiffness of the connection, and the failure mode. The latter should preferably be based on an unloading curve, for example, by unloading from one half of the projected ultimate strength of the connection to one tenth of this value before resuming the original loading procedure. The use of unloading curve data will likely eliminate the types of erractic behaviour that are sometimes encountered during the first application of load in connection tests.

The load-deformation curve may be given in the form of applied load vs. measured deflection, as is shown in the example. It may also be given as a moment-rotation curve, but it is critical that in this case the Data Sheet must also provide information on how the moment and rotation magnitudes were arrived at. To assure sufficient detail for subsequent uses, the load-deformation curve should be based on a minimum of 35 test points. If available, digitized test curve data should also be provided.

The last Data Sheet page provides additional observations or remarks which may be of value to the user. For example, it may be valuable to have some comments regarding connection ductility, second order (local deformation) effects, etc.

Data on the response characteristics of column footing connections should follow the general outline of the preceding. In this case it is also necessary to include information on the concrete and anchor bolt materials, the length and type of anchoring of the bolts, the use of

DATA SHEET no:	END PLATE BEAM TO COLUMN CONNECTION

specimen reference: A6	test no: 6	origin: CTICM

authors: *CEBTP-CTICM : Assemblages poutre-poteau par platine d'extrémité et par boulons à haute résistance. Dossier 511-503. Déc. 1980.*

STEEL GRADES: *E235*

- END PLATE : *(not measured)* f = N/mm²
- ATTACHED BEAM SECTION : *flange* f_y = *298* N/mm²
  COLUMN SECTION : *flange* f_y = *213* *N/mm2*
- H.S. BOLT STEEL GRADE : *Grade 10.9* f = N/mm²

f = N/mm²

DIMENSIONAL CHARACTERISTICS:

H.S. BOLT

- DIAMETER : *24* mm
- SECTION : *356* mm²
- PRELOAD : *254000* N

END PLATE DIMENSIONS

- THICKNESS e : *40* mm
- WIDTH B : *240* mm
- HEIGHT H : *1029* mm

- c_e : *33.03* mm
- t_e : *50* mm
- δ_e : *120* mm
- c_s : *31.03* mm
- t_s : *48* mm
- c_a : *42.69* mm
- t_a : *54* mm
- δ_a : *58* mm

ATTACHED BEAM PROFILE

- ROLLED SECTION REFERENCE : *IPE 600*

or

- WELDED SECTION : *(measured)*
- t_w : *12* mm
- h_w : *560* mm
- b_f : *220* mm
- t_f : *20* mm

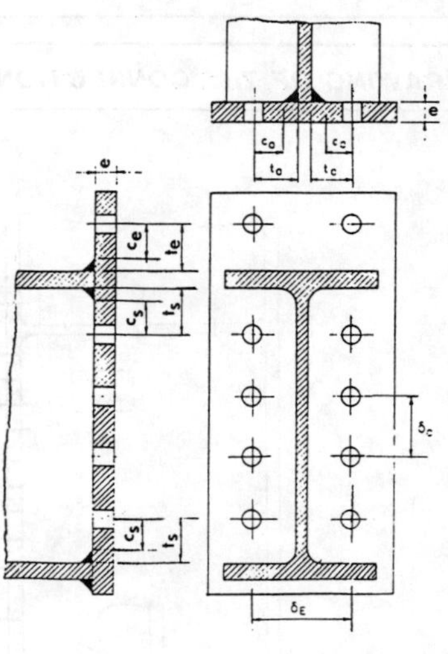

first page

EXPERIMENTAL TESTING ARRANGEMENT:

STATIC LOAD TEST

- MONOTONIC LOADING ☒
- REVERSED LOADING ☐

A6

$N = 1300$ kN

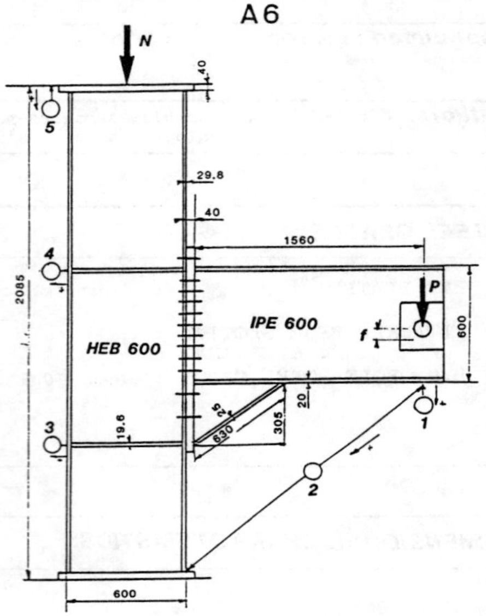

NOTE: Column ends are fixed

DRAWING OF THE CONNECTION :

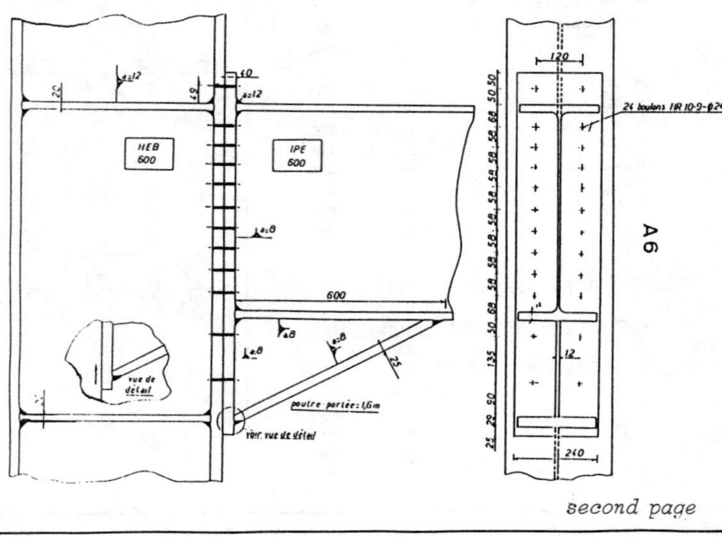

second page

LOAD BEHAVIOUR CHARACTERISTIC:

- MOMENT-ROTATION ☐
- LOAD-DISPLACEMENT ☒
- LOAD-ROTATION ☐

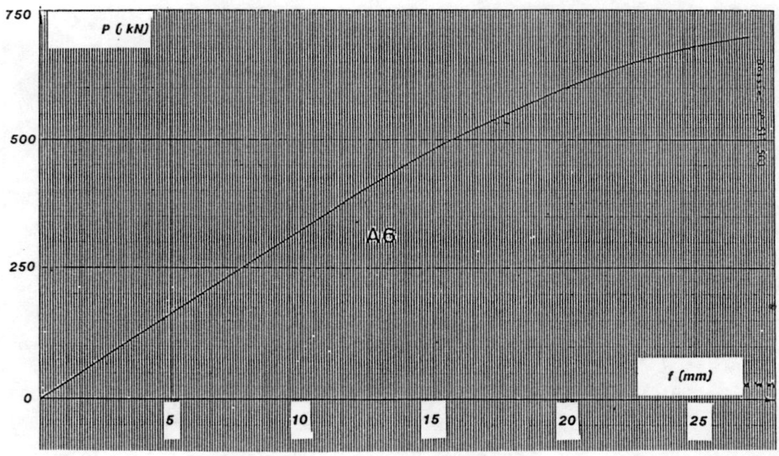

ULTIMATE STRENGTH CHARACTERISTIC:

- ULTIMATE CARRYING CAPACITY : *P = 700 kN (f_{max} = 27.02 mm)*
- FAILURE MODE : *Important plastification zone at the footing of the column. No plastification within the connection. No weld failure.*

third page

OBSERVATIONS:

The horizontal stiffener in the column web was welded at shop at the wrong position. See the figure.

fourth page

grout, and so on. These additional data can be included on Data Sheet pages 1 and 2.

SUMMARY OF WORKSHOP PROPOSAL

It is emphasized that the Connection Data Sheet should not give any form of analytical details. It is intended only as a vehicle for standardization of connection test data presentation and to facilitate the eventual work of the data user.

Contact individuals for the Data Sheet collection effort were identified during the workshop, as follows :

A. Beam-to-Column Connections :

For Europe : J. Brozzetti
C.T.I.C.M.
Domaine de Saint Paul
F-78470 Saint Rémy lès Chevreuse. France

For North America and Australia : R. Bjorhovde
Department of Civil Engineering
University of Pittsburgh
949 Benedum Hall
Pittsburgh, PA 15261. U.S.A.

B. Column Footing Connections :

For Europe : F.S.K. Bijlaard
TNO Institute for Building Materials and Structures
P.O. Box 49
NL - 2600 AA Delft. The Netherlands

For North America and Australia : D. Beaulieu
Département Génie Civil
Université Laval
Ste. Foy, Quebec G1K 7P4. Canada

Contact individuals for Japan will be determined at a later date. The assignment of the contact individuals is to collect the Data Sheets with the aim of eventually publishing a Data Sheet Collection Report.

Information regarding the collection effort at any one time can be obtained by contacting the individuals noted in the preceding.

CLASSIFICATION OF CONNECTIONS

Reidar Bjorhovde
Department of Civil Engineering
University of Pittsburgh
Pittsburgh, Pennsylvania, U.S.A.

Jacques Brozzetti
C.T.I.C.M.
Saint Rémy lès Chevreuse, France

and

André Colson
Laboratoire de Mécanique et Technologie
Ecole Normale Supérieure de Cachan
Université Paris VI
Cachan, France

INTRODUCTION

The following represents a presentation of connection classification concepts only, and is not intended for direct use at this time. However, it is felt that the principles of categorization are rational and relatively easy to use for practical purposes. Moreover, it will facilitate the classification of new connections or for connections where M-Φ-curves are not currently available.

BASIC PRINCIPLES

A non-dimensional format is preferred, where the vertical axis gives the ratio M/M_p and the horizontal axis gives Φ/Φ_p. These terms designate the following :

M = connection moment

M_p = fully plastic moment of beam

Φ = connection rotation

Φ_p = beam rotation at beam moment = M_p (see Figure 1)

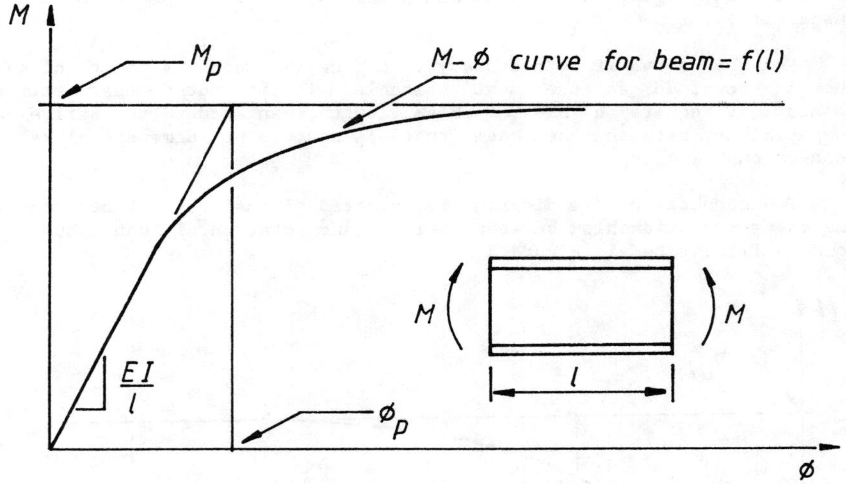

Figure 1 Definition of Beam M-Φ-curve

It is noted that the initial slope of the beam M-Φ-curve, as shown in Figure 1, is a function of the length of the beam element. Figure 2 demonstrates this further, for a range of beam lengths as a function of the beam depth.

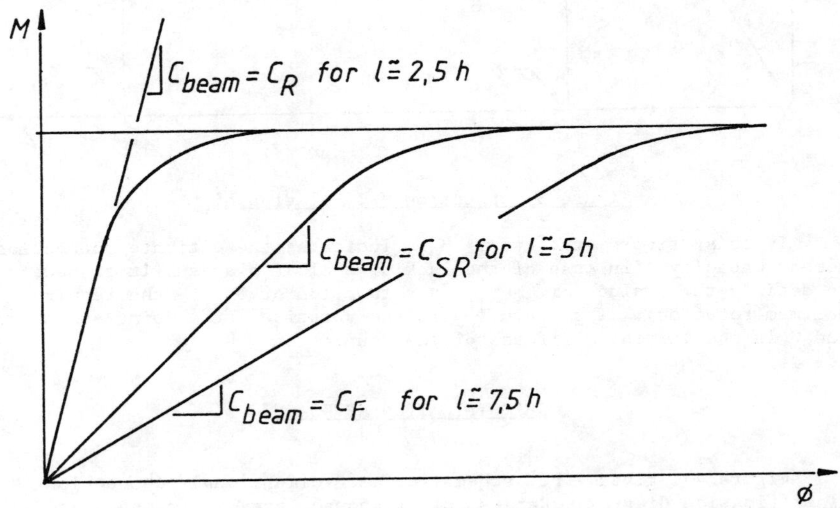

Figure 2 Beam M-Φ-curve as a Function of Beam Length

In this figure h indicates the depth of the beam ; C_{beam} is the initial slope of the beam M-Φ-curve ; C_R , C_{SR} and C_F indicate the stiffness of rigid, semi-rigid and flexible beam-to-column connections, respectively.

For the above it is necessary to define and choose a length of beam that is regarded as the most suitable. It is noted that stiffer connections require a shorter beam length, in order to arrive at compatibility between the beam that is used to non-dimensionalize the connection M-Φ-curve.

The connection M-Φ-diagram is plotted dimensionally first, to get the clear relationship between moment and rotation. It can assume the form as illustrated in Figure 3.

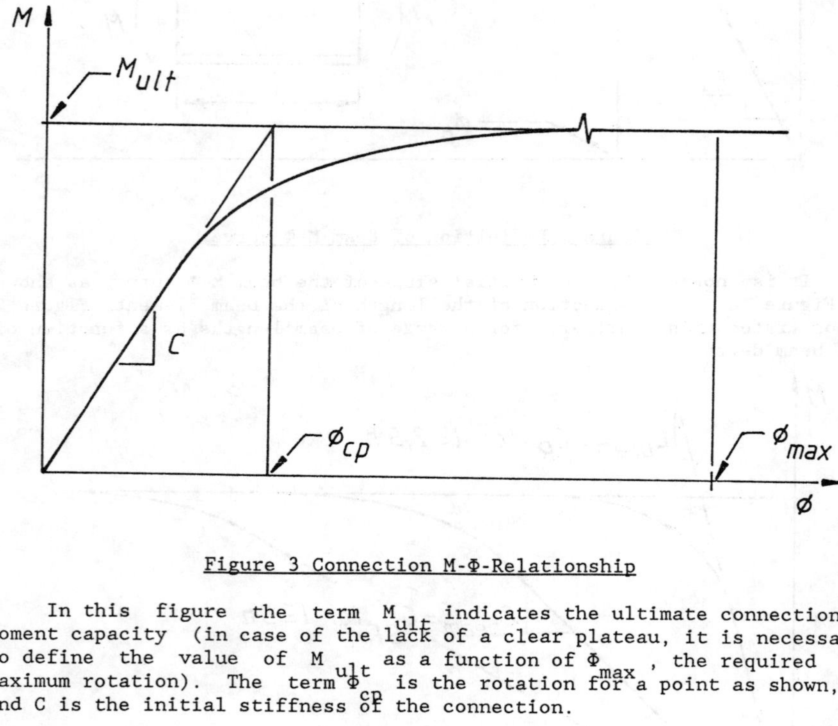

Figure 3 Connection M-Φ-Relationship

In this figure the term M_{ult} indicates the ultimate connection moment capacity (in case of the lack of a clear plateau, it is necessary to define the value of M_{ult} as a function of Φ_{max} , the required maximum rotation). The term Φ_{cp} is the rotation for a point as shown, and C is the initial stiffness of the connection.

CLASSIFICATION APPROACH

Figure 4 gives the schematic, non-dimensional connection M-Φ-classification diagram. Ranges of response are indicated, and it is evident that it will be easy to categorize new connections according to this scheme.

It is emphasized that this classification system is conceptual only at this stage. Further studies are expected to lead to quantified data for the classification diagram.

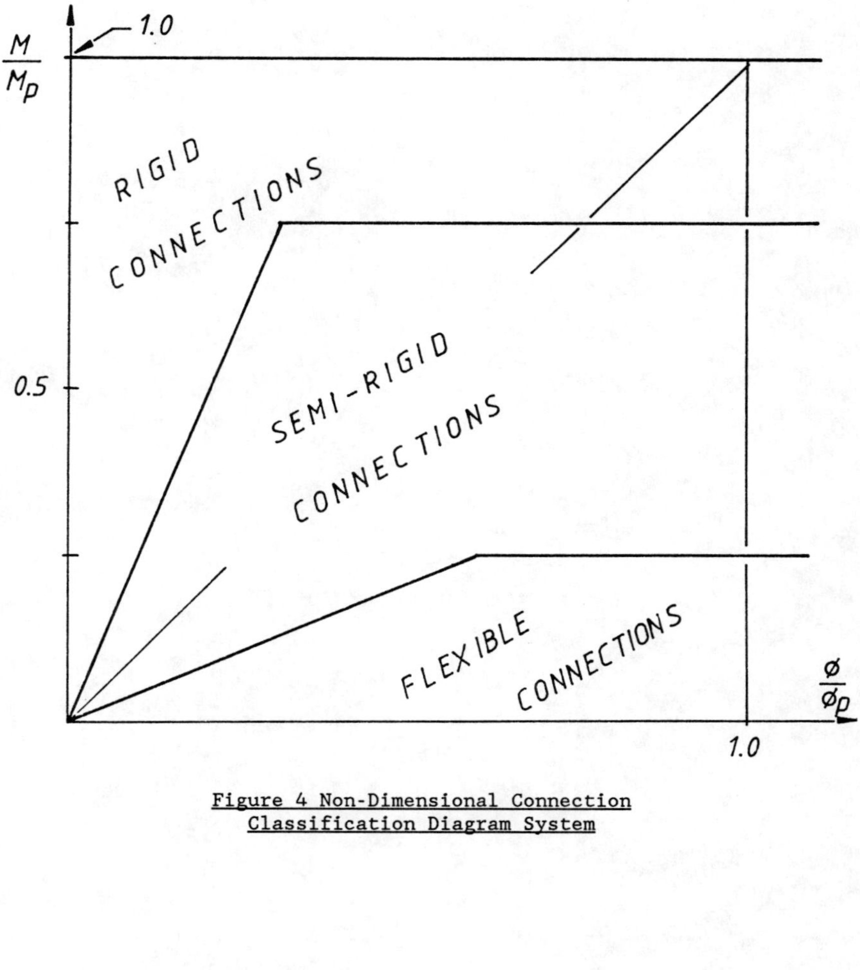

Figure 4 Non-Dimensional Connection Classification Diagram System

RESEARCH AND DEVELOPMENT NEEDS

INTRODUCTORY COMMENTS

The research and development needs that are given in the following were extracted from the observations of the speakers and the discussions that took place during the workshop. They represent the essence of the work of the research reporters, and although no explanatory notes are provided along with each subject, it is felt the topics give a realistic impression of the work that needs to be undertaken.

The research topics are listed in the order of the technical sessions of the workshop, and have not been prioritized in any form. Some of the subjects are somewhat broad in scope, and may in fact entail multiple research projects.

RESEARCH AND DEVELOPMENT NEEDS

1. Local Analysis of Joints

    (a) Ultimate limit states for connections with monotonically increasing moment-rotation relationship.

    (b) Three-dimensional connection analysis and tests.

    (c) Loading and unloading behaviour of typical building beam-to-column connections.

    (d) Development of standardized methods of connection testing and reporting.

    (e) Clarification of the prying phenomenon for bolted connections.

2. Mathematical Modeling

    (a) Two-and three-dimensional modeling of connections through the full range of behaviour.

    (b) Incorporation of strain-hardening and fracture considerations into connection modeling.

    (c) Finite element modeling of bolts in tension, including bolt pretension effects and contact considerations.

(d) Development of pre - and post - processing tools for computer analyses of connections.

3. Classification

(a) Data base organization, including fasteners, beam-to-column connections, tubular connections, composite connections, and fatigue data.

(b) Development of computerized data bank for ease of access and transfer.

(c) Integration of current data banks into one cohesive assembly.

(d) Preparation of connection data sheet

4. Frame Analysis

(a) Verification of theoretical results through full-scale subassemblage and frame tests.

(b) Serviceability and ultimate limit states analysis criteria for frames and their connections.

(c) Cyclic load behaviour of semi-rigid connections.

(d) Definition of bolt pretension to produce specific forms of connection slip behaviour.

5. Frame Stability and Simplified Methods

(a) Stiffness characteristics of column footing connections, including axial load and column effects.

(b) Verification of column footing analytical models.

(c) Initial and unloading stiffness properties for beam-to-column connections.

(d) Application of end-restrained column buckling theories to columns in unbraced frames.

(e) Moment transfer from beam to column during buckling of end-restrained subassemblages.

(f) Influence of axial load on subassemblage and frame stability.

(g) Drift characteristics of flexibly connected frames.

(h) Development of serviceability criteria for flexibly connected frames.

(i) Interaction of connection and bracing stiffness in braced frames.

6. Design Requirements and Codes

(a) Investigate and settle differences between the standards of various countries in their treatment of fastener strength and behaviour.

(b) Ductility criteria for high-strength bolts.

(c) Tension-shear interaction relationship for bolts.

(d) Clarify design criteria and approaches for prying.

(e) Classification system for connections according to M-Φ-relationship or similar.

(f) Effective width criteria for connections between tubes and other types of cross sections.

(g) Design criteria for use with high strength steels : Ductility, weldability, applicability of current requirements.

(h) Ductility requirements for connections in the presence of axial loads.

(i) Rotation capacity of connections under cyclic loading.

(j) Behaviour and strength characteristics for composite connections.